HIGH SCHOOL MATH MADE UNDERSTANDABLE

BOOK 2: MATH 11 & 12

WRITTEN BY: JEREMY MARTIN

Copyright 2021 by: Jeremy Martin

Published by: Jeremy Martin

ISBN: 9798671244236

Revision: 3

All rights reserved. This book may not be reproduced in whole or in part without written permission from the publisher, except by a reviewer who may quote brief passages in a review; nor may any part of this book be reproduced, stored in a retrieval system, or transmitted in any form or by any means, electronic, mechanical, photocopying, recording, or other, without written permission from the publisher.

All of the graphs in this book were created by using *DESMOS* web graphing application. Written permission was acquired from DESMOS prior to using it in this book. Each graph has an attribution to *DESMOS* at the bottom-left corner.

ACKNOWLEDGEMENTS

I would like to thank my mother (Carole Martin), my father (Daniel Martin), and my sister (Angelique Martin), for their support throughout High School and throughout the creation of this book. I would also like to thank Cordell Wiebe, the Math teacher at GROW Centre(the school where I graduated), in Williams Lake, BC, who helped me with Math whenever I got stuck or stumped with a new concept.

Formula Sheet

Trigonometry

$$\sin\theta = \frac{opp}{hyp} \qquad \cos\theta = \frac{adj}{hyp} \qquad \tan\theta = \frac{opp}{adj}$$

$$\csc\theta = \frac{1}{\sin\theta} \qquad \sec\theta = \frac{1}{\cos\theta} \qquad \cot\theta = \frac{1}{\tan\theta}$$

Pythagorean Theorem: $c^2 = a^2 + b^2$

Sine Law: $\dfrac{a}{\sin A} = \dfrac{b}{\sin B} = \dfrac{c}{\sin C}$

Cosine Law: $c^2 = a^2 + b^2 - 2ab\cos(C)$

$$b^2 = a^2 + c^2 - 2ac\cos(B)$$

$$a^2 = b^2 + c^2 - 2bc\cos(A)$$

Arc length: $a = r\theta$

Sum & Difference Identities:

$\sin(a+b) = \sin(a)\cos(b) + \cos(a)\sin(b)$

$\sin(a - b) = \sin(a)\cos(b) - \cos(a)\sin(b)$

$\cos(a + b) = \cos(a)\cos(b) - \sin(a)\sin(b)$

$\cos(a - b) = \cos(a)\cos(b) + \sin(a)\sin(b)$

$\tan(a + b) = \dfrac{\tan(a) + \tan(b)}{1 - \tan(a)\tan(b)}$

$\tan(a - b) = \dfrac{\tan(a) - \tan(b)}{1 + \tan(a)\tan(b)}$

Double Angle Identities:

$\sin 2\theta = 2\sin\theta\cos\theta$

$\cos 2\theta = \cos^2\theta - \sin^2\theta$
$= 2\cos^2\theta - 1$
$= 1 - 2\sin^2\theta$

$\tan 2\theta = \dfrac{2\tan\theta}{1 - \tan^2\theta}$

Log Laws

POWER RULE: $\log_a x^n$ is the same as $n\log_a x$

CHANGE THE BASE RULE: $\log_b a = \dfrac{\log_c a}{\log_c b}$

LAW OF MULTIPLICATION: $\log_a xy = \log_a x + \log_a y$

LAW OF DIVISION: $\log_a \left(\dfrac{x}{y}\right) = \log_a x - \log_a y$

RECIPROCAL RULE: $\log_a b = \dfrac{1}{\log_b a}$

EXTENDED POWER RULE: $\log_{a^n} b^n = \dfrac{\log b^n}{\log a^n}$

OTHER: $a^{\log_a x} = x$, $\log_a a^x = x$

Table of Contents

INTRODUCTION 9

Trigonometry (11)	13
Quadratic Functions	31
Radical Expressions & Equations	52
Rational Expressions & Equations	59
Absolute Value & Reciprocal Functions	66
Linear & Quadratic Inequalities	86
Transformations Of Functions	100
Radical & Rational Functions	126
Polynomials	156
Exponents & Logarithms	187
Circular Functions	257
Trigonometric Equations & Identities	345
Intervals	403
Conclusion	409
About the Author	411
Answers & Solutions	414

INTRODUCTION

Thank you for purchasing *High School Math Made Understandable Book 2*, I hope you enjoy reading this book just as much as I enjoyed writing it. I wrote this book with the aspiration of making High School Math understandable for everyone. This book covers material that is covered in Maths 11 & 12. This book is divided in 2 distinct parts (Math 11 & Math 12). Each chapter is divided into several sections. At the end of each chapter there are practice questions that you should attempt before moving on to the next chapter. Once you have completed the questions then you can check the answers in the *Answers & Solutions to Practice Questions* section that is found near the end of this book. The solutions section of this book gives you the step-by-step solution to the practice questions, this way if you get the wrong answer you can see where you went wrong. I strongly suggest you attempt the problem before looking at the answer and solution, by attempting the problem by yourself you will learn much more than just by copying answers from the back of the book. Now before we start, there is one more thing I want to tell you, just keep trying your best and keep on practicing, don't give up!

PART 1: MATH 11

Trigonometry (11)

Chapter 1

In grade 10, you learned a little bit about the Sine, Cosine and Tangent ratios. In this Chapter, you will be extending this knowledge even further. This is a very important chapter; therefore, the rate of progression in this chapter will be very gradual. We will be learning about angles in Standard Position (when the initial arm lies on the positive x-axis). First, we will look at how the four quadrants on the cartesian plane are divided.

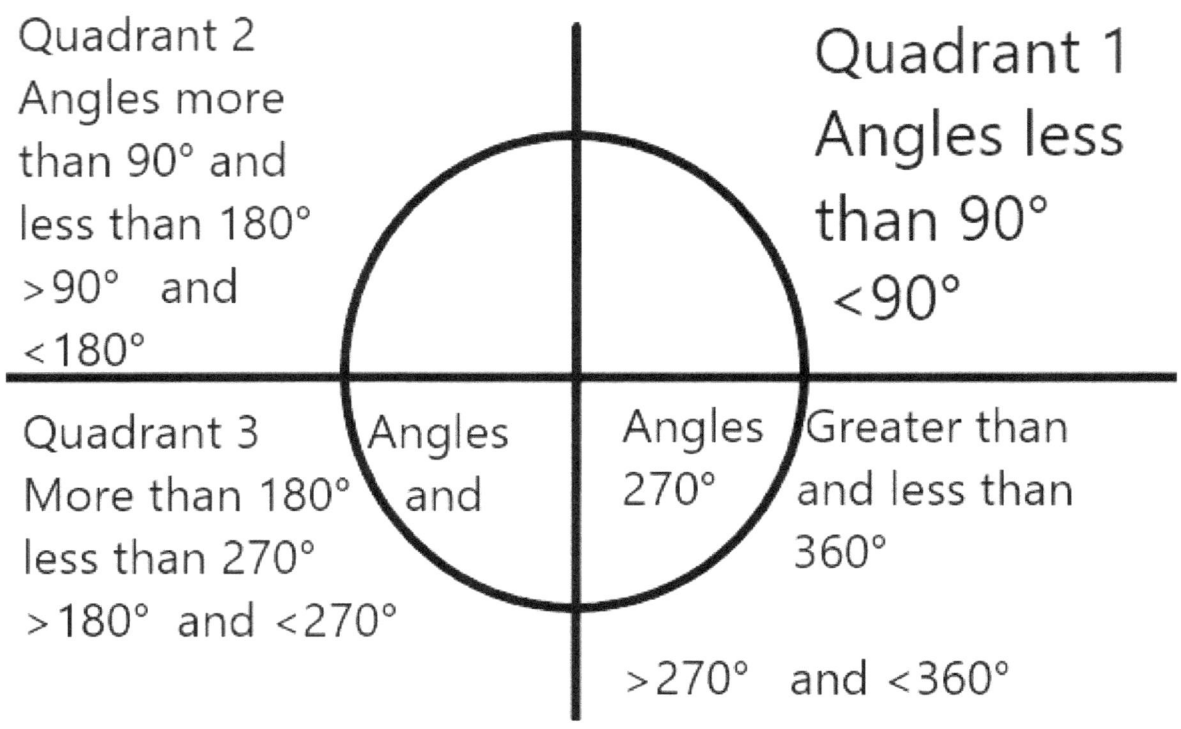

As you can see, there is a circle that is divided into four sections (the four quadrants), and each quadrant has a range of 90°. 90° lies on the positive y-axis, 180° lies on the negative x-axis, 270° lies on the negative y-axis, and 0° and 360° lies on the positive x-axis. When we sketch an angle in Standard position, something important to note is that when your angle is positive the rotation is counter clockwise, and when the angle is negative the rotation is clockwise.

θ (This symbol is called "Theta"), this fancy symbol simply means "Angle."

Here are two examples of drawing angles in standard position.

Example 1: $\theta = 45°$

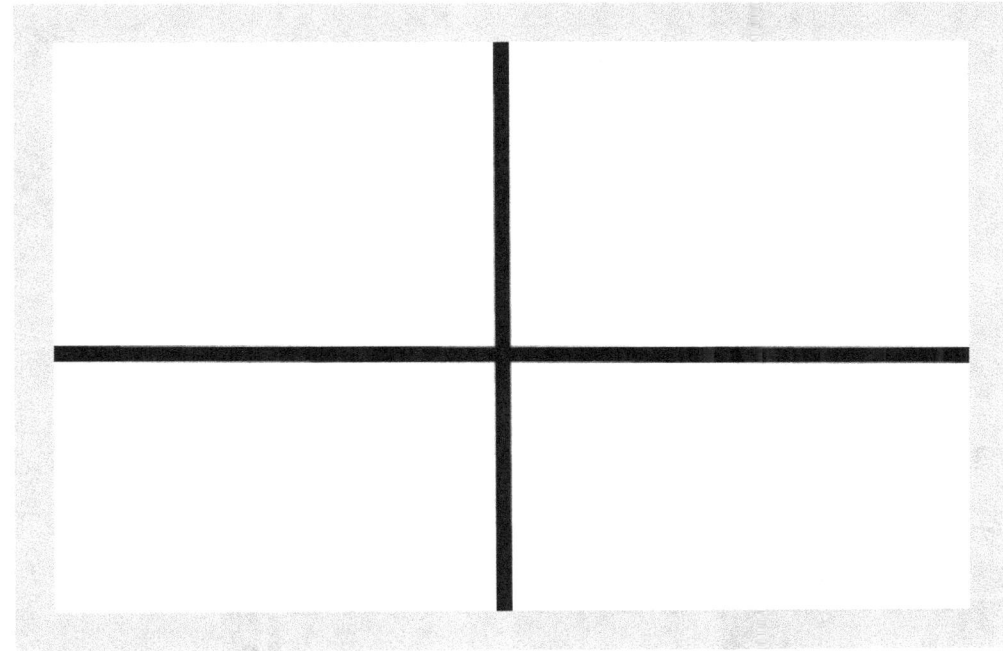

This means that the angle we are drawing is 45° in Standard position. On the next page you will be given the step-by-step process.

- Step 1: Draw the 4 quadrants of the Cartesian Plane,

- Step 2: Determine which quadrant θ lies in. θ in this case is 45°, therefore, θ lies in Quadrant 1.

- Step 3: Determine where θ lies in Quadrant 1, 45° is half of 90°. With this being said, 45° is half-way to 90°. Draw the terminal arm halfway to 90° diagonally.

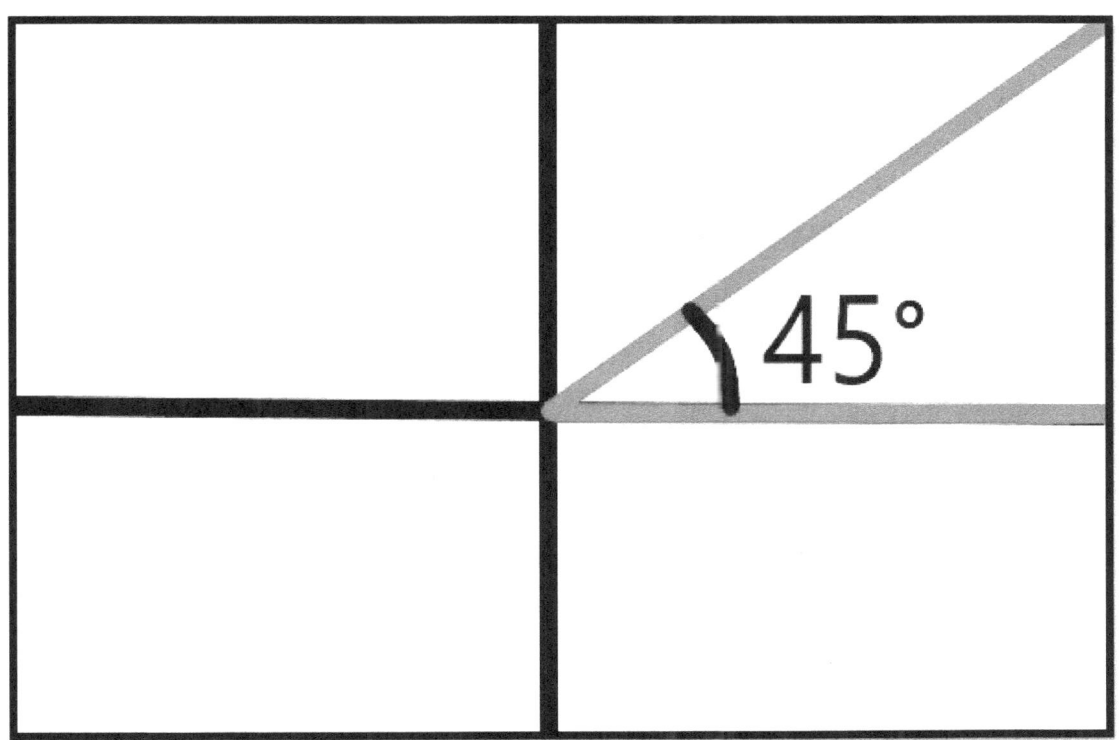

Notice how it is the Terminal arm that determines the angle of the sketch.

Example 2: $\theta = 135°$

- Step 1: Determine which quadrant 135° lies in. It is located in Quadrant 2 since any angle greater than 90° and less than 180° is found in that quadrant.

Step 3: Determine where θ lies in Quadrant 2, 135° is halfway from 90° to 180°. Draw the terminal arm diagonally in-between the 90° and 180°.

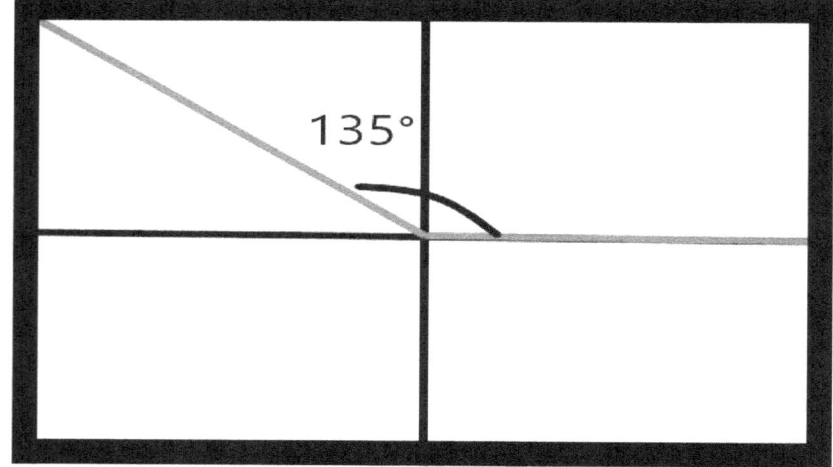

Reference Angles

A reference angle is the angle that the terminal arm makes with the x-axis. Here are two examples of reference angles. Let a represent the reference angle.

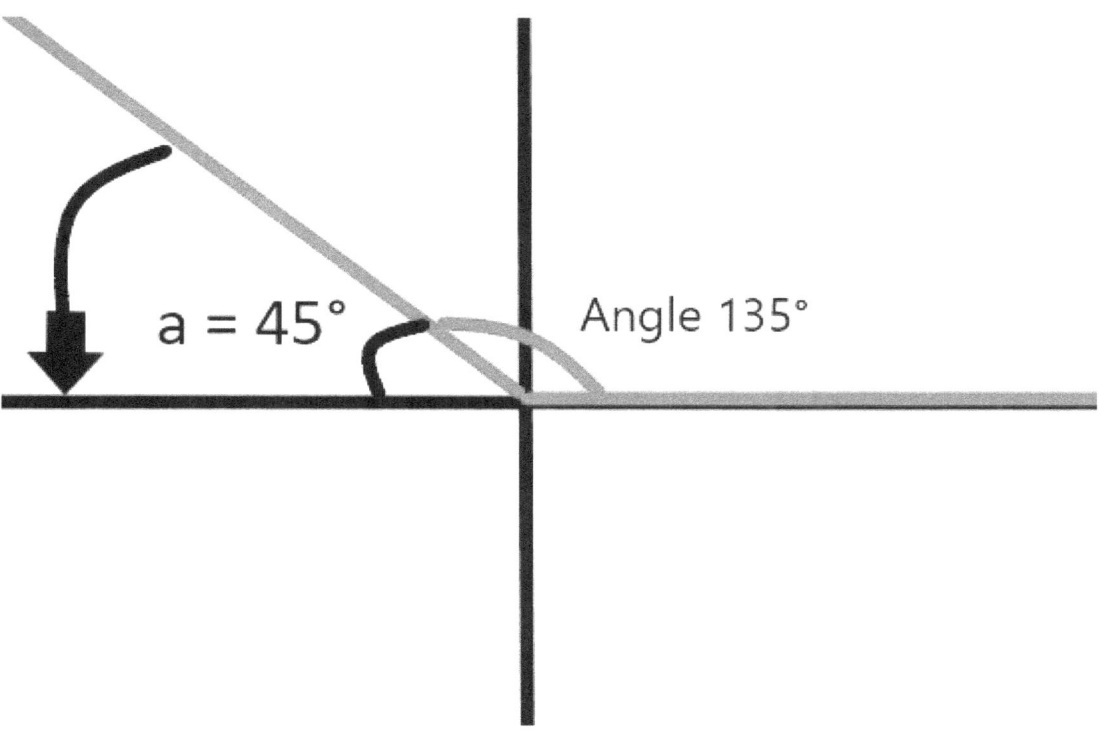

If the angle drawn is less than 180° then you basically subtract the angle drawn from 180° giving you the reference angle. Ex:

$\theta = 135°$

$a = 180° - \theta = 180° - 135° = 45°$

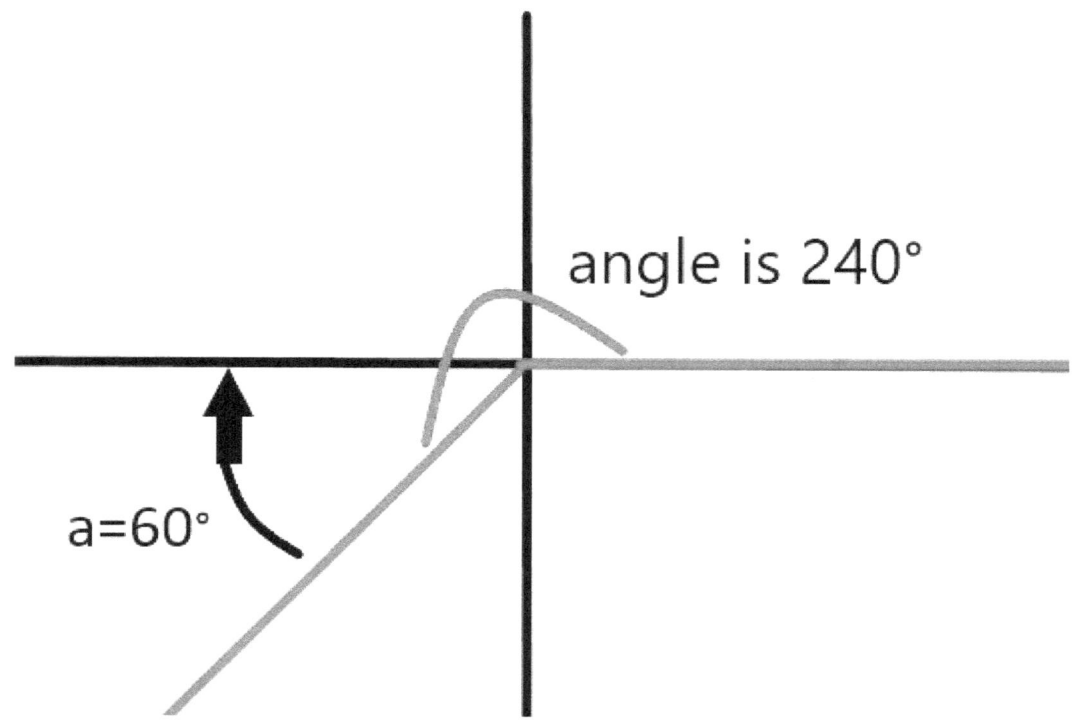

If the angle drawn is more than 180° and less than 270° then you basically subtract 180° from the angle drawn, giving you the reference angle. Ex: $\theta = 240°$

$a = \theta - 180° = 240° - 180° = 60°$

When the angle drawn is located in Quadrant 1, then the reference angle is equal to the angle of the terminal arm in Quadrant 1. For instance, if the angle drawn is 30° then the reference angle is 30°.

Sine, Cosine and Tangent Ratios

In this section we will be using angles in standard position to determine the exact value of Trigonometric ratios. Exact values mean no decimals, the exact precise value. This chapter as I have previously mentioned is very important, you will see these angles in standard position and Trigonometric ratios come back to haunt you in Grade 12.

$$\sin\theta = \frac{opp}{hyp} = \frac{y}{r}$$

$$\tan\theta = \frac{opp}{adj} = \frac{y}{x}$$

$$\cos\theta = \frac{Adj}{hyp} = \frac{x}{r}$$

$$r = \sqrt{x^2 + y^2}$$

y-coordinate is the rise, x-coordinate is the run, and r is the hypotenuse(diagonal). You can determine the r value by using the Pythagorean Theorem.

Example 1: If the coordinate is (3,4),
$$r = \sqrt{x^2 + y^2} = \sqrt{3^2 + 4^2} = \sqrt{9 + 16} = \sqrt{25} = 5$$

$r = 5$ hypotenuse is 5

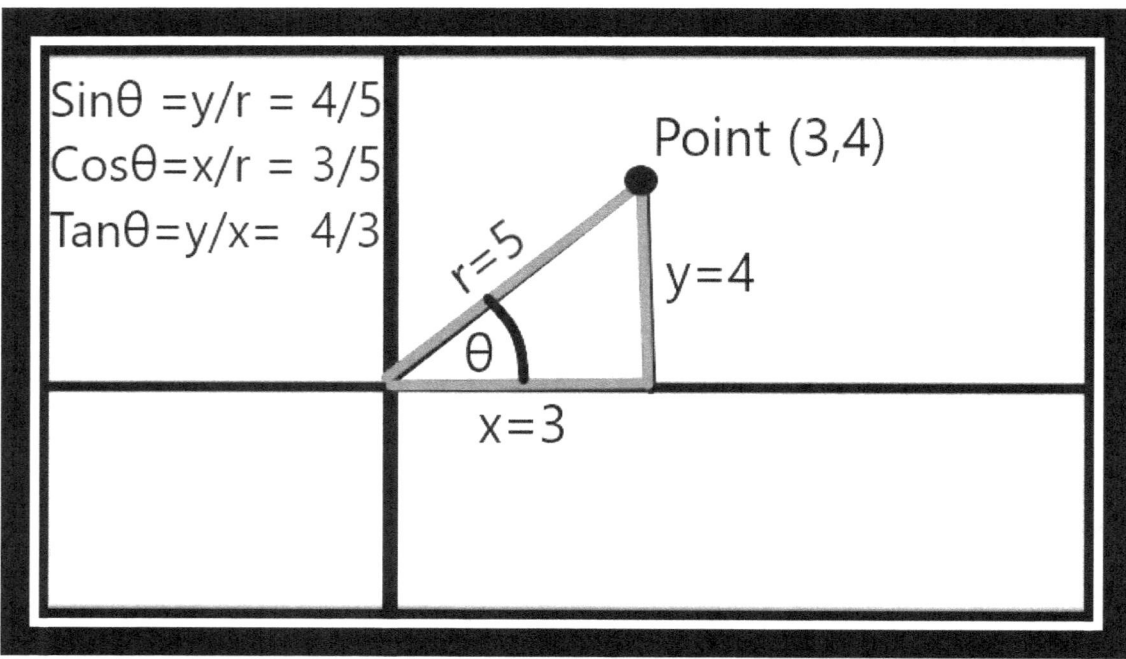

Example 2: Determine the exact value of $\sin\theta, \cos\theta, \tan\theta$, if the terminal arm of an angle in standard position passes through the Point (4,5).

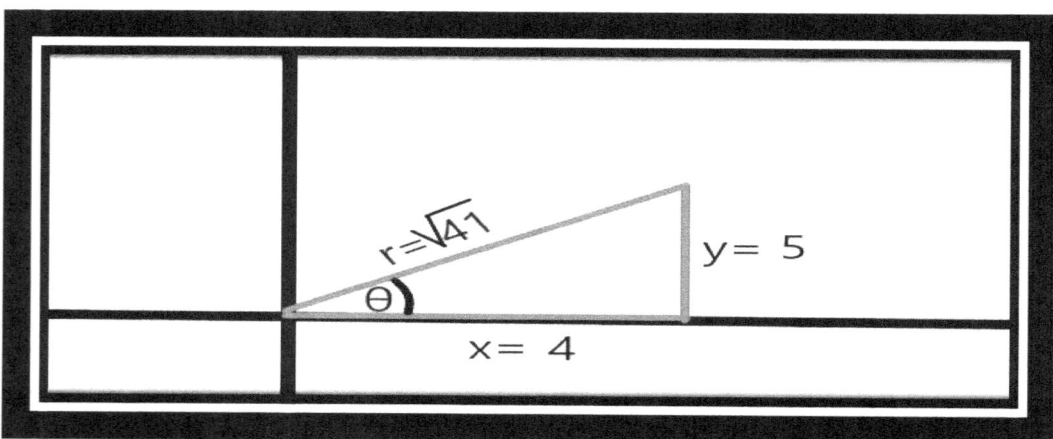

$r = \sqrt{x^2 + y^2} = \sqrt{16 + 25} = \sqrt{41}$

$r = \sqrt{41}$

Keep the "r" value as a radical, and plug the values into the corresponding ratios:

$$sin\theta = \frac{y}{r} = \frac{5}{\sqrt{41}}$$

$$cos\theta = \frac{x}{r} = \frac{4}{\sqrt{41}}$$

$$tan\theta = \frac{y}{x} = \frac{5}{4}$$

Sine Law

In this section we will learn about the Sine Law, this law is useful for finding a side or an angle of a triangle that is not a right-triangle (triangle in which one of its angles is 90°).

The Sine Law states that $\frac{a}{sinA} = \frac{b}{sinB} = \frac{c}{sinC}$

You will see a few examples that will help you understand what this means.

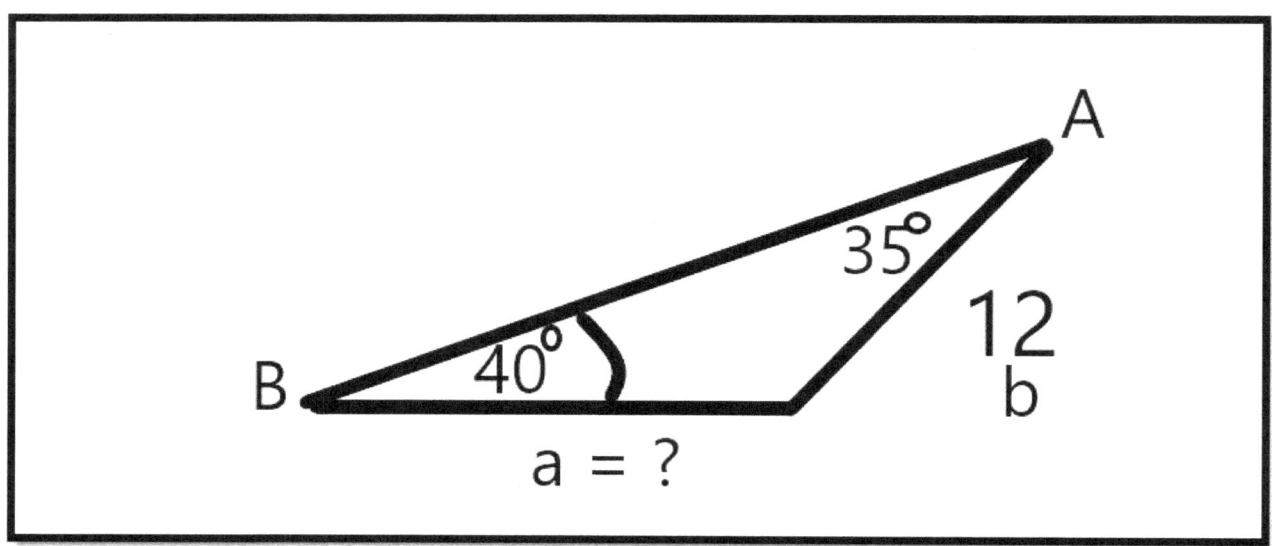

As you can see, the question is asking you to find the value of "a"

Sin B = 40°   Sin A = 35°   b = 12   a = ?

$$\frac{a}{sinA} = \frac{b}{sinB}$$

$$\frac{a}{sin35°} = \frac{12}{sin40°}$$

We now rearrange the formula to solve for "a", we multiply both sides by sin35°,

$$(sin35)\frac{a}{sin35°} = \frac{12}{sin40}(sin35)$$

$$a = \frac{12sin35}{sin40}$$

$$a = 10.7$$

The length of side "a" is 10.7, that's it we have found side "a"!
A good way to verify your answer is by substituting "a" inside of the formula and you should get the same ratio for both sides of the equation (should be very close).

Example 2: $$\frac{Sin38}{50} = \frac{SinB}{35}$$

We are looking for angle B, therefore, we will rearrange the formula to solve for Sin B, and then once you have the ratio use the Inverse Sin function(Sin^{-1}) on your calculator to solve for "Angle B"

$$\frac{sin38}{50}(35) = \frac{SinB}{35}(35)$$

$$\frac{35sin38}{50} = SinB$$

$SinB = 0.43096032...$

$B = Sin^{-1}(0.43096032...)$

$B \approx 25.53° \approx 26°$

Cosine Law

The *Law of Cosines* says that:

$$c^2 = a^2 + b^2 - 2ab\cos(C)$$

This law helps us solve triangles that do not have an angle of 90°.

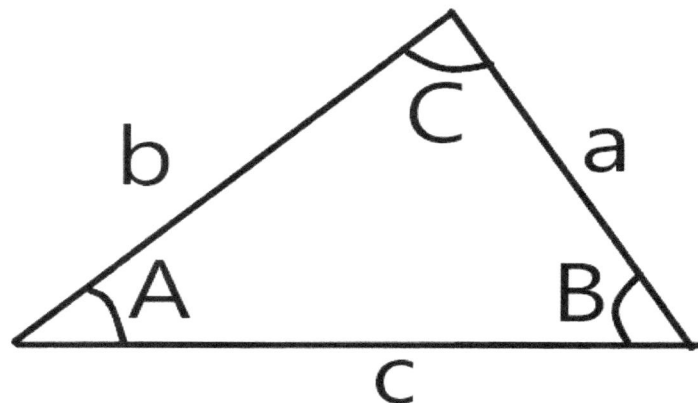

We will look at two examples of how this law can be applied:

Example 1: Find the length of side "c"

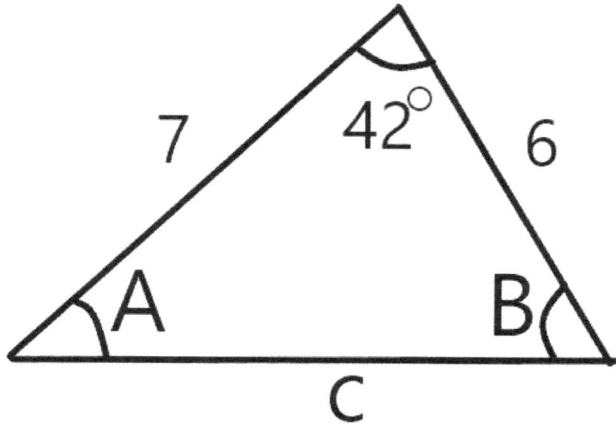

We will use the formula: $c^2 = a^2 + b^2 - 2ab\cos(C)$, since we are trying to find side "c."

Angle C = 42° $\cos(C) = \cos(42°)$

$a = 7$ $b = 6$ plugging this into our formula we get,

$c^2 = (7)^2 + (6)^2 - 2(7)(6)\cos(42°)$

$c^2 = 49 + 36 - (84)(\cos(42°))$

$c^2 = 85 - 62.42$ $c^2 = 22.58$

$c = \sqrt{22.58}$ $c \approx 4.75$

There are some other rearrangements of this formula:

$a^2 = b^2 + c^2 - 2bc\cos(A)$

$b^2 = a^2 + c^2 - 2ac\cos(B)$

Basically, the angle you use in the formula is the angle opposite to the side you are trying to find the length of. This formula can even be rearranged in order to find the measure of the angle opposite to a given side. Here are the formulas:

$$Cos(C) = \frac{a^2 + b^2 - c^2}{2ab} \qquad Cos(B) = \frac{c^2 + a^2 - b^2}{2ca}$$

$$Cos(A) = \frac{b^2 + c^2 - a^2}{2bc}$$

Example 2: Find Angle C.

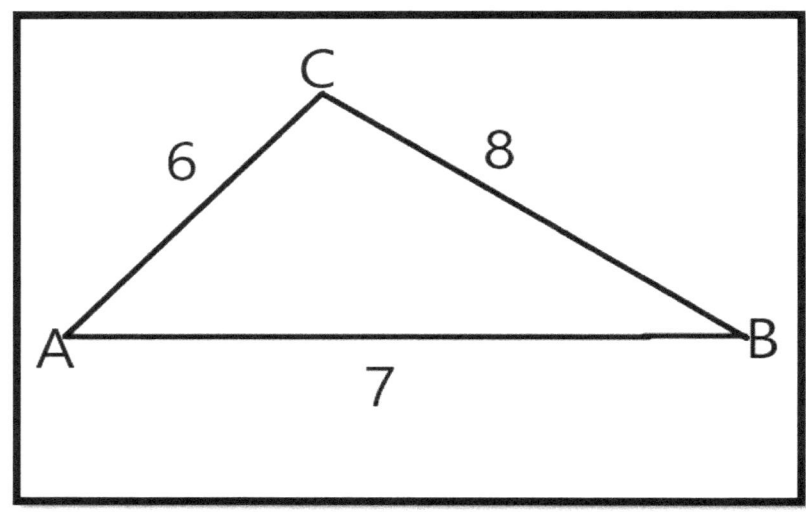

$a = 8 \quad b = 6 \quad c = 7$

Which formula should we use? We should use

$Cos(C) = \dfrac{a^2 + b^2 - c^2}{2ab}$, because we are trying to find Angle C.

Let's plug $a, b,$ and c into the formula:

$$Cos(C) = \frac{(8)^2 + (6)^2 - (7^2)}{2(8)(6)}$$

$$Cos(C) = \frac{64 + 36 - 49}{96} = \frac{100 - 49}{96} = \frac{51}{96}$$

$$Cos(C) = \frac{51}{96}$$ Use the inverse cosine function

$$C = Cos^{-1}\left(\frac{51}{96}\right) \quad C \approx 57.9°$$

Practice Questions

1. Draw an angle of 60° in standard position.

2. Draw an angle of 30° in standard position.

3. Draw an angle of 150° in standard position, what is the reference angle?

4. Determine the exact values of $sin\theta, cos\theta, tan\theta$ if the terminal arm of an angle in standard position passes through point P($1, \sqrt{3}$). Hint: $\theta = 60°$,

5. Find the length of b.

6. $\dfrac{18}{\sin 36} = \dfrac{a}{\sin 120}$ Find a.

7. What is the reference angle of $120°$?

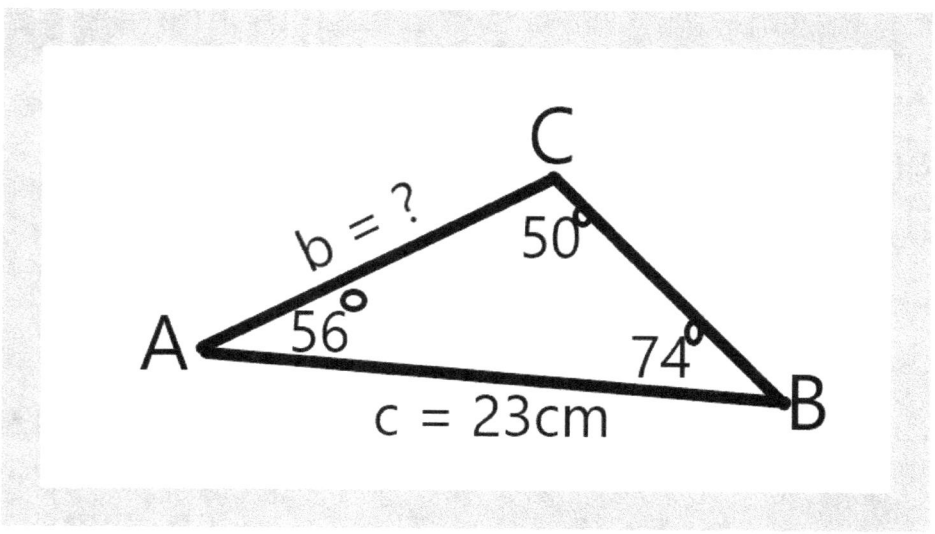

Quadratic Functions

Chapter 2

Quadratic Functions also go by another name "Parabolas" In this chapter you will learn how to graph parabolas and you will look at some of its features.

The equation of the most basic quadratic function is

$y = x^2$

What this means is that y is equal to x times itself. We will build a table of values and we will plot a few points to see the shape of a quadratic function.

x	y
-3	9
-2	4
-1	1
0	0
1	1
2	4
3	9

When graphing Quadratics DO NOT CONNECT THE POINTS WITH A STRAIGHT LINE!!! Make sure that you draw a smooth curve through the points. Plot at least 5 points before drawing a smooth curve.

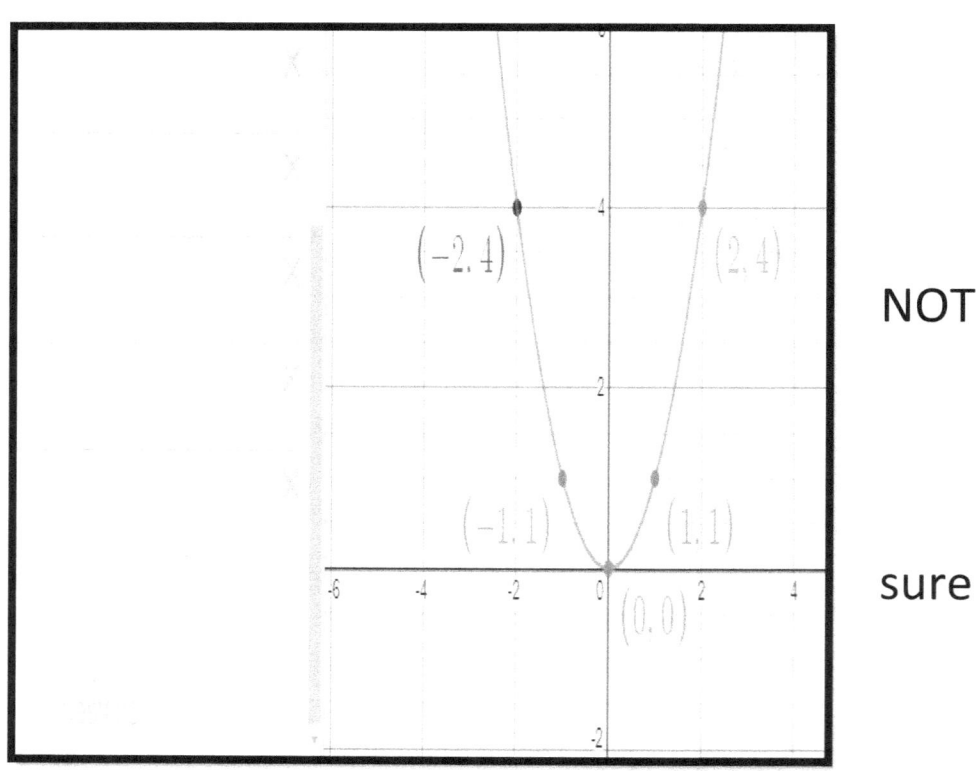

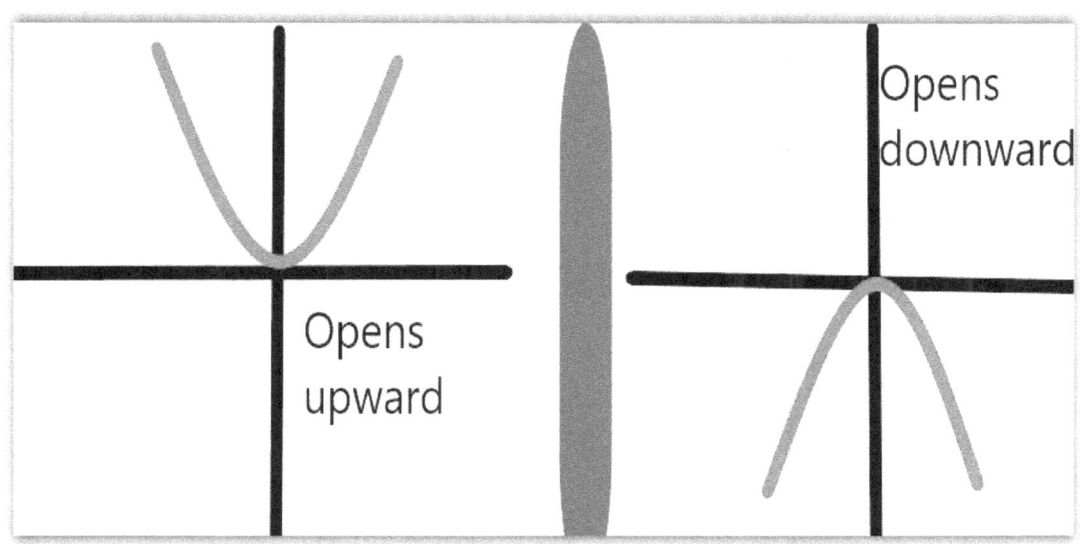

There are two different kinds of Parabolas, one that opens upwards and one that opens downwards.

If the coefficient in front of x² is negative, the parabola opens downward. If the coefficient in front of x² is positive, the parabola opens upward.

For example, in the equation $y = -x^2 - 1$ the parabola opens downward because the coefficient of x is negative.

The most common type of Quadratic equation you will see is the Vertex Form, with point (p,q) as the Vertex.

$$y = a(x - p)^2 + q$$

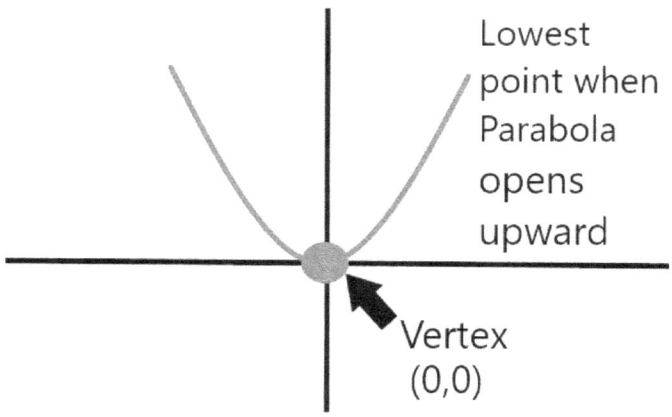

This is the Vertex of $y = x^2$

For example, in the quadratic $y = (x - 3)^2 + 2$ the vertex is point (3,2). You must be thinking, "There was a -3 in the brackets, why is point p of the vertex positive 3?" The reason for this is because in the Vertex form, we are subtracting p from x; therefore, to find p, we take the opposite of what we see. In $y = (x + 2)^2 + 1$

P = -2   q = 1

There is a fascinating concept behind this, and this concept you will learn in Grade 12, it is a concept that can be applied to all functions. I will discuss this later in this book as I want to gradually introduce you to these new subjects.

What is the vertex of, $y = (x + 7)$,

P= -7   and q = 0, since there is no "q-value"

If you see something like, $y = x^2 + 2$ the p-value is 0, and the q value is the constant being added. The vertex would be (0,2)

X-Intercepts (Zeros)

We will look at how to determine the zeros (x-intercepts) of quadratic functions. Basically, you set y to 0, you expand the equation, simplify, and then factor

$0 = (x + 3)^2 - 1$

$0 = (x + 3)(x + 3) - 1$

$0 = x^2 + 3x + 3x + 9 - 1$

$0 = x^2 + 6x + 8$

We need two numbers that multiply to 8 and add up to 2. Those two numbers are 4 and 2. 4*2=8 and 4+2=6

$0 = (x + 4)(x + 2)$

We now make use of the zero-principle, which states that A*B = 0

If B=0, or if A=0, giving two x-intercepts for this problem. We will deal with these two factors separately, and then I will show you what I mean by the zero principle.

$(x + 4) = 0$

$x + 4 - 4 = 0 - 4$

$x = -4$

-4 is one of our x-intercepts, point (-4,0)

$(x + 2) = 0$

$x + 2 - 2 = 0 - 2$

$x = -2$

-2 is the other x-intercept, point (-2,0).

$x = -4 \text{ and } x = -2$

Testing the first solution $x = -4$

$0 = ((-4) + 4)(x + 2)$

$0 = (0)(x + 2)$

$0 = 0$

Testing the second solution $x = -2$

$0 = (x + 4)((-2) + 2)$

$0 = (x + 4)(0)$

$0 = 0$

Here is a visual of this graph

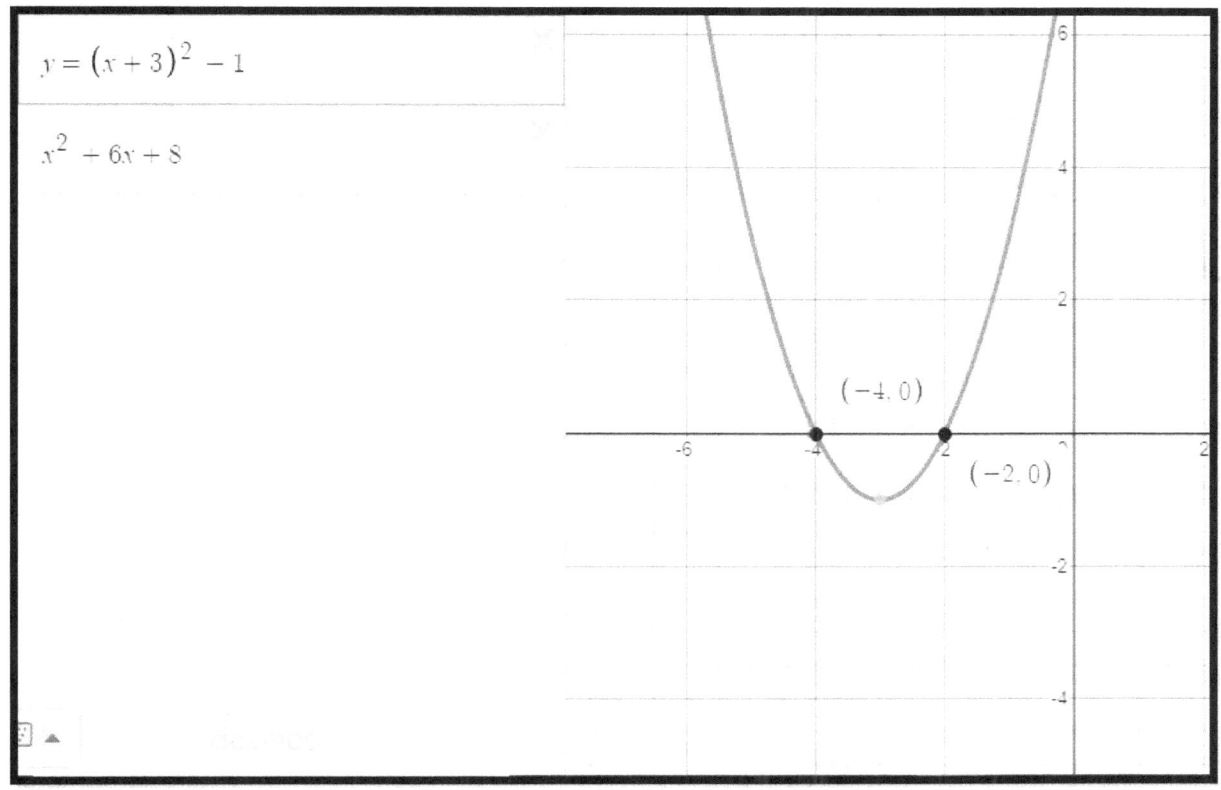

As you can see, the zeros are where the graph crosses the x-axis.

Domain and Range

As you learned in Grade 10, Domain is all the x-values (input values), and the Range is all of the y-values (Output) values.

The Domain is the easiest one to determine, there are no restrictions on x-values; therefore, the Domain is $\{x|xER\}$ which means that "X is an Element of all Real Numbers"

The range is a different story, if the parabola opens upward, then the range is any value greater than or equal to the y-value of the Vertex.

For example, in the equation $y = (x + 2)^2 + 3$

Q = 3, which is the y-value of the vertex (in this case the lowest point)

$y \geq 3$

$$Range: \{y|y \geq 3, YER\}$$

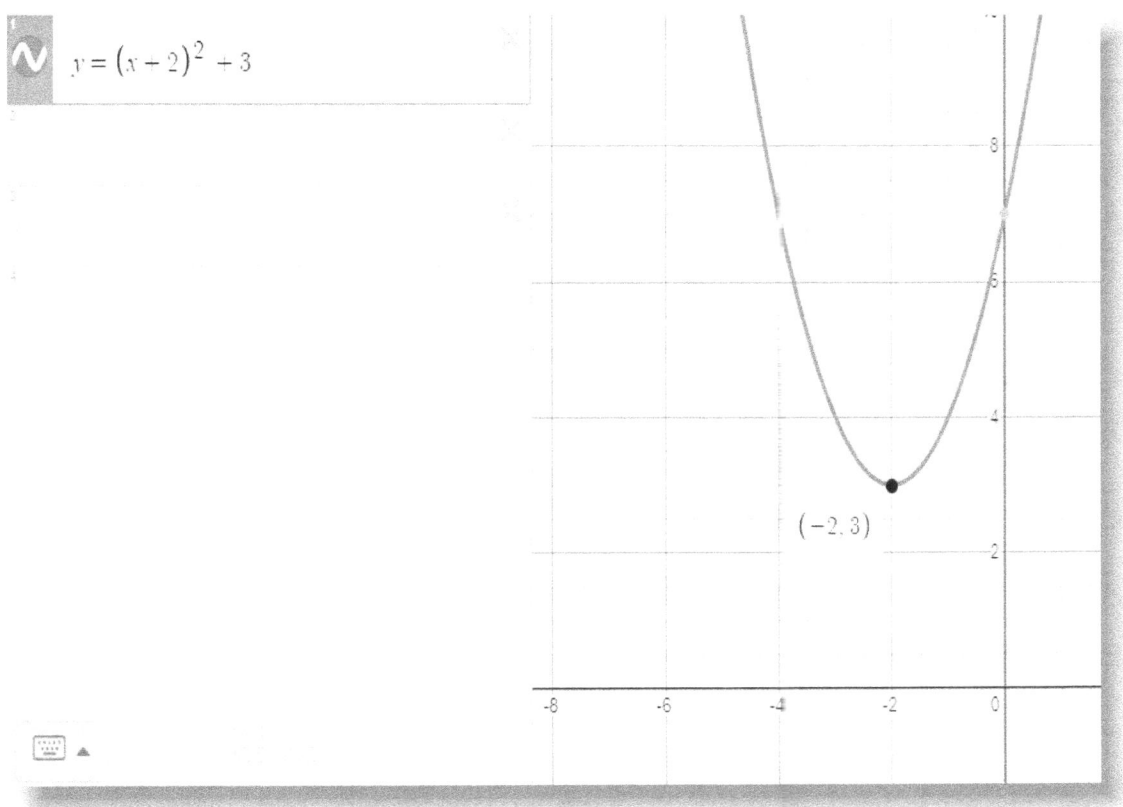

As you can see, the lowest vertical point of this graph is 3; therefore, the y-value can be anything equal to or above 3 (greater than or equal to 3).

Solving Quadratic Equations

We will learn how to solve quadratic equations in this section.

We will start by solving a few basic equations, and then we will learn about the Quadratic Formula, a formula that will prove to be very useful throughout High School.

Solve for x. $$x^2 = 81$$

- Step 1: Isolate x, we can do this by taking the square root of both sides.
$\sqrt{x^2} = \sqrt{81}$
- Step 2: Place a plus or minus sign ($\pm$) in front of the $\sqrt{81}$
$x = \pm \sqrt{81}$
$x = \pm 9$

There are two solutions to this problem, since a negative number multiplied by itself will equal a positive number as well.

Solve for x. $\quad x^2 + 3 = 19$

Subtract 3 from both sides,
$x^2 + 3 - 3 = 19 - 3$
Take the square root of both sides
$\sqrt{x^2} = \sqrt{16}$
$x = \pm \sqrt{16}$
$x = \pm 4$

Solve the factored equation, $0 = (x+7)(x-2)$
- Use the Zero-Principle $AB = 0$, if A=0 or B =0

- Solve for x in each factor separately

$$0 = (x + 7)$$
$$0 - 7 = x + 7 - 7$$
$$-7 = x$$

$$0 = (x - 2)$$
$$0 + 2 = x - 2 + 2$$
$$2 = x$$

Solutions: $x = -7$, and $x = 2$

What about factoring a difference of squares?

For example, $x^2 - 4$ as you can see there is no middle term.

To factor this, you simply use this rule,

$$a^2 - b^2 = (a + b)(a - b)$$

So, using this rule, we find that $x^2 - 4 = (x + 2)(x - 2)$

Here is another example: $x^2 - 9 = (x + 3)(x - 3)$

Since 3 times 3 is 9. You need to find the square root of x^2 and the square root of the coefficient and use the result as the "a" and "b" value.

The two solutions will always be negative b, and positive b, this only works when both terms are perfect squares, and only for a difference of squares, not a sum of squares.

Completing the Square

You will be asked to "complete the square" to basically turn an equation in the form $ax^2 + bx + c = 0$ to vertex form ($a(x-p)^2 + q = 0$) or $a(x+p)^2 = q$, this is useful because vertex form is so much easier to work with. You can see the vertex (highest or lowest point of a parabola). I will show you several examples that will help you learn how to do this. It may seem like a daunting task at first, but with practice it will become easy.

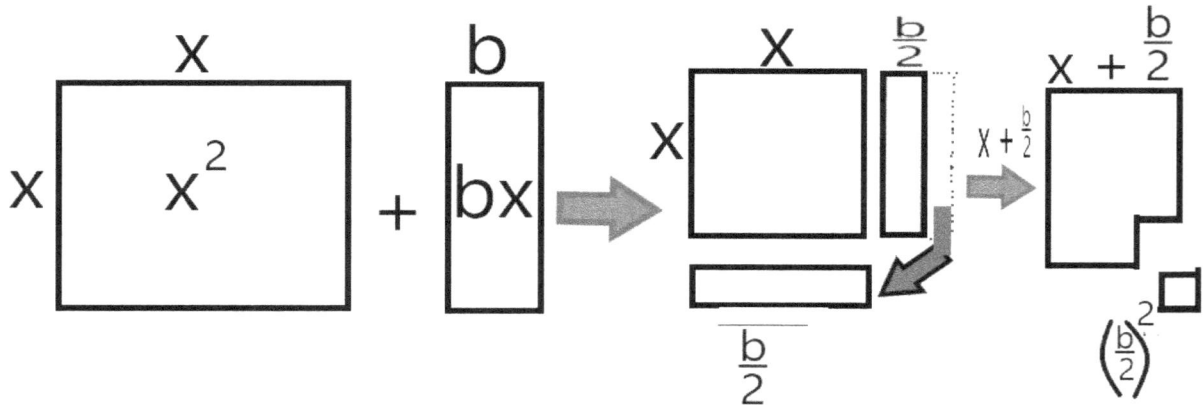

Here are the steps you should follow,

- Set up equation so that x-terms are on one side and the constant is on the other.

- Make sure coefficient of x^2 term is 1. Divide if necessary.
- ADD the SQUARE of HALF the x-term to BOTH sides.
- Factor perfect square trinomial, since it is now in the required form.

Example 1: $x^2 + 4x + 3 = 0$

- Step 1: Subtract 3 from both sides
  $x^2 + 4x = -3$
- Step 2: Coefficient of x^2 term is 1; therefore, there is no need to divide. Put $x^2 + 4x$ in brackets like this,
  $(x^2 + 4x) = -3$
- Step 3: Divide 4 by 2, giving you 2. Square 2, giving you $2^2 = 4$.
  Add 4 to both sides,
  $(x^2 + 4x + 4) = -3 + 4$
  $(x^2 + 4x + 4) = 1$

- Step 4: $(x^2 + 4x + 4)$ is now a perfect trinomial, so factor it. The trick is that we just simply use half the x-term, which is 2.

$$(x+2)(x+2) = 1$$
$$(x+2)^2 = 1$$

Example 2: $(2x^2 - 4x + 6) = 0$

- Step 1: Subtract both sides by 6,
$$2x^2 - 4x = -6$$
- Step 2: Factor 2 out of the left side,
$$2x^2 - 4x = -6$$
$$2(x^2 - 2x) = -6$$
- Step 3: Take half of 2, which is 1. Square 1, $1^2 = 1$. Be very careful here, multiply 1 by the number outside of the brackets (2). 1 multiplied by 2 is 2; therefore, add 2 to the other side.
$$2(x^2 - 2x + 1) = -6 + 2$$
- Step 4: Factor this perfect trinomial,
$$2(x-1)(x-1) = -4$$
$$2(x-1)^2 = -4$$

Quadratic Formula

Sometimes, you will not be able to factor a given quadratic expression, in a case like this, you will use something called the Quadratic Formula. Trust me, it looks scarier than it actually is,

$$x = \frac{-b \pm \sqrt{b^2 - 4ac}}{2a}$$

Quadratic equations come in the form,

$$ax^2 + bx + c = 0$$

The "a" value is the coefficient in front of x^2, the "b" value is the number in front of the "x" term. And the value of c is the constant term.

You will see a few examples of the Quadratic Formula which will help improve your understanding,

$$4x^2 - 12x - 9 = 0$$

We are unable factor this equation. With this being said, we will use the Quadratic Formula to find the two roots of this equation.

$$x = \frac{-b \pm \sqrt{b^2 - 4ac}}{2a}$$

$$4x^2 - 12x - 9 = 0$$

$$a = 4, \quad b = -12, \quad c = -9$$

$$x = \frac{-(-12) \pm \sqrt{(-12)^2 - 4(4)(-9)}}{2(4)}$$

$$x = \frac{12 \pm \sqrt{144 - 16(-9)}}{8}$$

$$x = \frac{12 \pm \sqrt{144 + 144}}{8}$$

$$x = \frac{12 \pm \sqrt{288}}{8}$$

I am already aware that I could have simplified it further, for those of you that have already noticed. I will not do that, just so I keep this introduction to the Quadratic Formula as simple as possible.

Solution 1: $$x = \frac{12 + \sqrt{288}}{8} \approx 3.62$$

Solution 2: $$x = \frac{12 - \sqrt{288}}{8} \approx -0.62$$

Just to let you know, in Grade 11 you may be asked to round the solution, like I just did; however, this is a habit you will want to drop when you get to grade 12 or University. You will lose marks if you do not write your solution as an exact value.

When I say exact value I'm referring to this, $x = \frac{12 - \sqrt{288}}{8}$ and not -0.62. -0.62 is just an approximation.

Here is another example of the use of the Quadratic Formula:

$$2x^2 + 4x - 7 = 0$$

We cannot factor this equation. With this being said, we will use the Quadratic Formula to find the two roots of this equation.

$$x = \frac{-b \pm \sqrt{b^2 - 4ac}}{2a}$$

$$2x^2 + 4x - 7 = 0$$

$$a = 2, \quad b = 4, \quad c = -7$$

$$x = \frac{-(4) \pm \sqrt{(4)^2 - 4(2)(-7)}}{2(2)}$$

$$x = \frac{-4 \pm \sqrt{16 - 8(-7)}}{4}$$

$$x = \frac{-4 \pm \sqrt{16 + 56}}{4}$$

$$x = \frac{-4 \pm \sqrt{72}}{4}$$

Solution 1: $x = \dfrac{-4 + \sqrt{72}}{4} \approx 1.12$

Solution 2: $x = \dfrac{-4 - \sqrt{72}}{4} \approx -3.12$

Quadratic Formula Proof

This chapter is intended for those who are interested in why the quadratic formula works. Do not worry if you do not understand it, it is a little bit complicated, it will not pop up on a test. The first step in proving the quadratic formula is by completing the square. The next steps are to solve for x.

- Write the standard form: $ax^2 + bx + c = 0$

- Step 2: Subtract c from both sides
$ax^2 + bx = -c$

- Step 3: Factor the a out of the left side
$a\left(x^2 + \dfrac{b}{a}x\right) = -c$

We have $\dfrac{b}{a}$ because that is the only way to factor an a out of bx since there was no a to start with.

- Step 4: Take half of $\dfrac{b}{a}$. To do this, multiply $\dfrac{b}{a}$ by $\dfrac{1}{2}$. In other words, multiply the denominator by 2.

$$\dfrac{b}{a} * \dfrac{1}{2} = \dfrac{b}{2a}$$

- Step 5: Square $\dfrac{b}{2a}$.

$$\dfrac{b}{2a} * \dfrac{b}{2a} = \dfrac{b^2}{4a^2}$$

- Step 6: Add $\dfrac{b^2}{4a^2}$ to both sides and do not forget to multiply it by a when you bring it to the other side.

$$a\left(x^2 + \dfrac{b}{a}x + \dfrac{b^2}{4a^2}\right) = -c + \dfrac{b^2}{4a}$$

- Step 7: Find the common denominator of the right side. 4a

$$a\left(x^2 + \dfrac{b}{a}x + \dfrac{b^2}{4a^2}\right) = -\dfrac{(4a)c}{4a} + \dfrac{b^2}{4a}$$

$$a\left(x^2 + \dfrac{b}{a}x + \dfrac{b^2}{4a^2}\right) = \dfrac{b^2 - 4ac}{4a}$$

- Step 8: Divide both sides by a.

$$\frac{a\left(x^2 + \frac{b}{a}x + \frac{b^2}{4a^2}\right)}{a} = \frac{b^2 - 4ac}{4a} * \frac{1}{a}$$

$$\left(x^2 + \frac{b}{a}x + \frac{b^2}{4a^2}\right) = \frac{b^2 - 4ac}{4a^2}$$

- Step 9: Factor this perfect trinomial.

$$\left(x + \frac{b}{2a}\right)^2 = \frac{b^2 - 4ac}{4a^2}$$

- Step 10: Take the square root of both sides.

$$\sqrt{\left(x + \frac{b}{2a}\right)^2} = \sqrt{\frac{b^2 - 4ac}{4a^2}}$$

$$x + \frac{b}{2a} = \pm \frac{\sqrt{b^2 - 4ac}}{2a}$$

- Final Step: Subtract $\frac{b}{2a}$ from both sides.

$$x = \frac{-b \pm \sqrt{b^2 - 4ac}}{2a}$$

That is one of the ways that the quadratic formula is proven.

Practice Questions

1. Sketch the graph of $y = x^2 + 3$. Hint: Make a table of values and plot at least 5 points before drawing a smooth curve.

2. Determine the vertex of each of the following equations
   a) $y = (x + 2)^2 - 3$
   b) $y = x^2 + 4$
   c) $y = (x + w)^2 - h$
   d) $y = (x - 1)^2 + 80$
   e) $y = (x - 200)^2 - 300$

3. Solve each of the following equations by factoring. Find the zeros.
   a) $x^2 + 4x + 3$
   b) $x^2 + 4x - 21$

4. What is the range of $y = (x-2)^2 + 8$?

5. Find the Zeros using the Quadratic Formula,
   a) $2x^2 + 3x - 7 = 0$
   b) $-6x^2 + 17x + 5$

6. Complete the square, $x^2 + 4x + 5 = 0$

7. Challenge question, sketch the graph of $y = (x+2)^2 - 1$.

   Hint: First build a table of values, then determine the vertex.

Radical Expressions & Equations

Chapter 3

In Grade 10, you learned to convert mixed radicals to entire radicals and vice versa. In this chapter, we will be learning how to simplify radicals when multiplying or adding two radicals, we will also learn how to rationalize a denominator (remove a radical expression from the denominator). Before we do all of this, we will go over a brief review of converting radicals to either mixed or entire form.

When you convert mixed radicals to entire radicals, you multiply the number or variable in front of the radical by itself the number of times indicated in the index. In the first example the index is 2, in the second example the index is 3.

$5\sqrt{6} = \sqrt{(5)(5)(6)} = \sqrt{150}$

$x\sqrt[3]{2} = \sqrt[3]{(x)(x)(x)(2)} = \sqrt[3]{2x^3}$

When you convert entire radicals to mixed radicals, you write out the numbers and variables inside the radicand as a product of all of its prime factors, and you look for factors that appear the number of times indicated by the index. For example, if the index is 2, then you look for factors that appear twice and take the square root of that number or variable and then move it front of the radical.

Here are a few examples:
- $\sqrt{72} = \sqrt{(6)(6)(2)} = 6\sqrt{2}$ there is a 6 in front because $\sqrt{(6)(6)} = 6$
- $\sqrt[3]{5y^3} = \sqrt[3]{5(y)(y)(y)} = y\sqrt[3]{5}$
- $\sqrt{80} = \sqrt{(2)(2)(2)(2)(5)} = 4\sqrt{5}$ 2 appears 4 times therefore, $2^4 = 16$ $\sqrt{16} = 4$ that is why there is a 4 in front of the radical.

Adding and Subtracting radical expressions

In order to simplify radical expressions, you combine the skill of collecting "Like-terms" in this case, "Like-radicals" with the skill of converting between mixed and entire radicals. If the radicand and index of both terms is the same, then you add the coefficient in front of both radicals together.

$a\sqrt{x} + b\sqrt{x} = (a+b)\sqrt{x}$

For example:
- $5\sqrt{2} + 7\sqrt{2} = (5+7)\sqrt{2} = 12\sqrt{2}$
- $\sqrt{3} + 2\sqrt{3} = (1+2)\sqrt{3} = 3\sqrt{3}$
- $2\sqrt{5} + x\sqrt{5} = (2+x)\sqrt{5}$

Sometimes, it will not always be that easy. There are some case in which you must try to find a way to make both radicals have the same radicand, so you can simplify.

$\sqrt{18} + \sqrt{50}$

As you can see, they do not have the same radicand. In order to simplify, we must write all of the factors of both numbers and compare.

$\sqrt{18} = \sqrt{(3)(3)(2)}$
$\sqrt{50} = \sqrt{(5)(5)(2)}$

For $\sqrt{18}$ we can write it as $3\sqrt{2}$
$\sqrt{50}$ can be written as $5\sqrt{2}$
$\sqrt{18} + \sqrt{50} = 3\sqrt{2} + 5\sqrt{2} = (3+5)\sqrt{2} = 8\sqrt{2}$

And then when you subtract radicals, it is the same concept, except you subtract instead of adding.

$$a\sqrt{x} - b\sqrt{x} = (a - b)\sqrt{x}$$

For example:
- $8\sqrt[3]{7} - 2\sqrt[3]{7} = (8 - 2)\sqrt[3]{7} = 6\sqrt[3]{7}$
- $22\sqrt{xy} - 3\sqrt{xy} = (22 - 3)\sqrt{xy} = 19\sqrt{xy}$
- $6\sqrt{5} - 2\sqrt{5} = (6 - 2)\sqrt{5} = 4\sqrt{5}$

Multiplying radical expressions

To multiply radical expressions, you simply multiply the coefficient in front of both radicals together, and you multiply both radicands together when they have the same index. There is a way to simplify two radicals of different indexes being multiplied by each other; however, to keep this topic as simple as possible it will be discussed later on in this book.

$$a\sqrt{x} * b\sqrt{y} = (a * b)\sqrt{x * y} = ab\sqrt{xy}$$

For example:
- $5\sqrt{2} * 6\sqrt{3} = (5 * 6)\sqrt{2 * 3} = 30\sqrt{6}$

- $3\sqrt{5} * a\sqrt{x} = (3*a)\sqrt{5*x} = 3a\sqrt{5x}$
- $8\sqrt{6} * 4\sqrt{5} = 32\sqrt{30}$
- $3\sqrt[3]{6} * 5\sqrt[3]{3} = 15\sqrt[3]{18}$

Be careful, sometimes you will be multiplying two expressions with the same radicand, in this case, you are cancelling out the radical. For instance, $2\sqrt{3} * 5\sqrt{3}$

$= (2*5)\sqrt{3*3} = 10\sqrt{9} = 10 * \sqrt{9} = 10 * 3 = 30$

Rationalizing Denominator

In order to rationalize the denominator, which is the process of removing a radical expression from the denominator, we multiply the denominator and the numerator by its **conjugate**. For example, the conjugate of $(a + b)$ is $(a - b)$ and vice versa.

Example 1: Simplify $\dfrac{3}{\sqrt{2}}$ in this case we simply multiply the denominator and numerator by $\sqrt{2}$.

$$\frac{3(\sqrt{2})}{\sqrt{2}(\sqrt{2})} = \frac{3\sqrt{2}}{2}$$

Example 2: $\dfrac{3}{\sqrt{5} + 2}$ First, we need to find the conjugate of $(\sqrt{5} + 2)$ which is $(\sqrt{5} - 2)$. We then multiply the numerator and denominator by $(\sqrt{5} - 2)$ removing the radical from the denominator.

$$\frac{3(\sqrt{5} - 2)}{(\sqrt{5} + 2)(\sqrt{5} - 2)} = \frac{3\sqrt{5} - 6}{5 + 2\sqrt{5} - 2\sqrt{5} - 4}$$

$$= \frac{3\sqrt{5} - 6}{1} = 3\sqrt{5} - 6$$

Example 3: $\dfrac{a + \sqrt{b}}{a - \sqrt{b}}$

Find the conjugate of $(a - \sqrt{b})$ which is $(a + \sqrt{b})$. Multiply the denominator and the numerator by $(a + \sqrt{b})$.

$$\frac{a + \sqrt{b}}{a - \sqrt{b}} = \frac{(a + \sqrt{b})(a + \sqrt{b})}{(a - \sqrt{b})(a + \sqrt{b})} = \frac{a^2 + a\sqrt{b} + a\sqrt{b} + b}{a^2 - a\sqrt{b} + a\sqrt{b} - b}$$

$$= \frac{a^2 + 2a\sqrt{b} + b}{a^2 - b}$$

Practice Questions

1. Simplify, the following radicals by using addition,

   a) $3\sqrt{7} + 92\sqrt{7}$

   b) $2\sqrt{5} + 3\sqrt{5}$

   c) $\sqrt{28} + \sqrt{63}$

   d) $\sqrt{xz^2} + \sqrt{xy^2}$

2. Simplify the following radicals by using subtraction,

   a) $9\sqrt{3} - 2\sqrt{12}$

   b) $25\sqrt{7} - 16\sqrt{7}$

3. Simplify the following radicals by using the property of multiplication,
   a) $2\sqrt{3} * 4\sqrt{2}$
   b) $3\sqrt{5} * 5\sqrt{3}$

4. Rationalize the denominator of each of the following expressions,
   a) $\dfrac{a}{\sqrt{b}}$
   b) $\dfrac{a}{y + \sqrt{x}}$

Rational Expressions & Equations

Chapter 4

We will learn how to simplify rational expressions in this chapter. A rational expression is an expression where there is a polynomial expression in both the numerator and in the denominator. When simplifying these expressions one thing to keep in mind is that there are values that are **Non-Permissible** (not allowed because it would put a 0 in the denominator making the expression undefined). For example, in the expression $\dfrac{x+2}{a}$ $a \neq 0$ because it would zero the denominator, which would cause the expression to be undefined. I know I have already said this the phrase before, I just want to make sure you know this, as this will be something to be very careful with when you graph rational functions.

The non-permissible value will not always be zero, it all depends on the expression in the denominator. You look only at the denominator to determine the non-permissible value(s). It does not matter what is in the numerator, it does not make a difference.

Now we will look at how to determine the non-permissible value of some other rational expressions.

- $\dfrac{x+2}{y-2}$ we use the fact that the denominator cannot equal 0, and we solve for the non-permissible value like this,
  $y - 2 \neq 0$
  $y - 2 + 2 \neq 0 + 2$
  $y \neq 2$

- $\dfrac{5}{b^2 + 4}$
  $b^2 + 4 \neq 0$
  $b^2 + 4 - 4 \neq 0 - 4$
  $b^2 \neq -4$

$b = \sqrt{-4}$ This is impossible, you cannot take the square root of a negative number; therefore, there are no restrictions on b.

- $\dfrac{5}{y^2 - 4}$

  $y^2 - 4 \neq 0$

  $y^2 - 4 + 4 \neq 0 + 4$

  $y^2 \neq +4$

  $y \neq \pm\sqrt{4}$

  $y \neq \pm 2$

  2 restrictions since a negative number multiplied by itself will result in a positive number.

What about simplifying expressions and determining non-permissible values.

Before you simplify, look for a non-permissible value. After that, simplify, and then check for another non-permissible value with what is left in the denominator.

For example,

$\dfrac{3x^2 + 9}{3y + 12}$

First thing we see is that $3y + 12 \neq 0$

$3y + 12 - 12 \neq 0 - 12$

$3y \neq -12$

$\dfrac{3y}{3} \neq -\dfrac{12}{3}$

$y \neq -4$

Now we look for the GCF in both the numerator and the denominator.

$$\dfrac{3x^2 + 9}{3y + 12} = \dfrac{3(x^2 + 3)}{3(y + 4)} = \dfrac{x^2 + 3}{y + 4}$$

The non permissible value is still $y \neq -4$

Sometimes there will be more than 1 non-permissible value, what you do is you factor the denominator, and then use the zero principle to find restrictions. Here are a few examples,

$$\dfrac{z}{x^2 - 3x - 4}$$

Which two numbers multiply to -4 and add up to -3? These two numbers are -4 and 1.

$$\dfrac{z}{(x - 4)(x + 1)}$$

Treat each factor separately,

$x - 4 \neq 0$ $\qquad\qquad\qquad\qquad$ $x + 1 \neq 0$

$x - 4 + 4 \neq 0 + 4$ $\quad$ $x + 1 - 1 \neq 0 - 1$

$x \neq 4$ $\qquad\qquad\qquad\qquad$ $x \neq -1$

These are both of the non-permissible values.

Here are some examples of rational expressions being simplified:

- $\dfrac{j^2 + 5j + 6}{j^2 + 2j - 3} = \dfrac{(j + 2)(j + 3)}{(j + 3)(j - 1)}$

See how (j+3) appears in both the numerator and the denominator? Now we cancel this factor and remove it from the top and bottom of this expression like this,

$\dfrac{(j + 2)(j + 3)}{(j + 3)(j - 1)} = \dfrac{j + 2}{j - 1}$

Restrictions: $j \neq -3$ and $j \neq 1$

- $\dfrac{(a - 1)(a - 2)}{(a - 1)(a - 2)} = \dfrac{(a - 1)(a - 2)}{(a - 1)(a - 2)} = 1$

- $\dfrac{w^2 + 3w}{w^2 + w} = \dfrac{w(w + 3)}{w(w + 1)} = \dfrac{w(w + 3)}{w(w + 1)} = \dfrac{w + 3}{w + 1}$

$w \neq 0 \qquad w \neq -1$

- $\dfrac{(k+2)(k-3)(k+4)}{(k+4)(k-1)(k-3)} = \dfrac{(k+2)(k-3)(k+4)}{(k+4)(k-1)(k-3)} = \dfrac{k+2}{k-1}$
  Restrictions: $k \neq -4 \qquad k \neq 3 \qquad k \neq 1$

- $\dfrac{x^2 + x - 2}{x^2 - x - 6} = \dfrac{(x-1)(x+2)}{(x-3)(x+2)} = \dfrac{(x-1)(x+2)}{(x-3)(x+2)} = \dfrac{x-1}{x-3}$
  Restrictions: $x \neq 3 \qquad x \neq -2$

Practice Questions

1. Identify any non-permissible value(s),

   a) $\dfrac{1}{j+7}$

   b) $\dfrac{1}{x}$

   c) $\dfrac{x}{y^2 - 4}$

2. Simplify the following expressions and identify any non-permissible values before and after the simplification.

   a) $\dfrac{3x+6}{3y+9}$

b) $\dfrac{x^2 + x - 2}{x^2 + 2x - 3}$

c) $\dfrac{a^2 b + ab + b}{b^2 + b}$

d) $\dfrac{4xy + 8x}{12x^2 + 4x}$

Absolute Value & Reciprocal Functions
Chapter 5

An absolute value is basically the non-negative value of a number without regard to its sign. In other words, if you put a positive number inside of the absolute value brackets, you get a positive number. If you put a negative number inside of the absolute value brackets you get a positive number. For example, $|3| = 3$ and $|-3| = 3$. Absolute values are used to do things such as, calculating difference between a negative temperature and a positive temperature. Another way to think of the Absolute value of a number, is as the distance between 0 and the number inside the brackets. Here is an example:

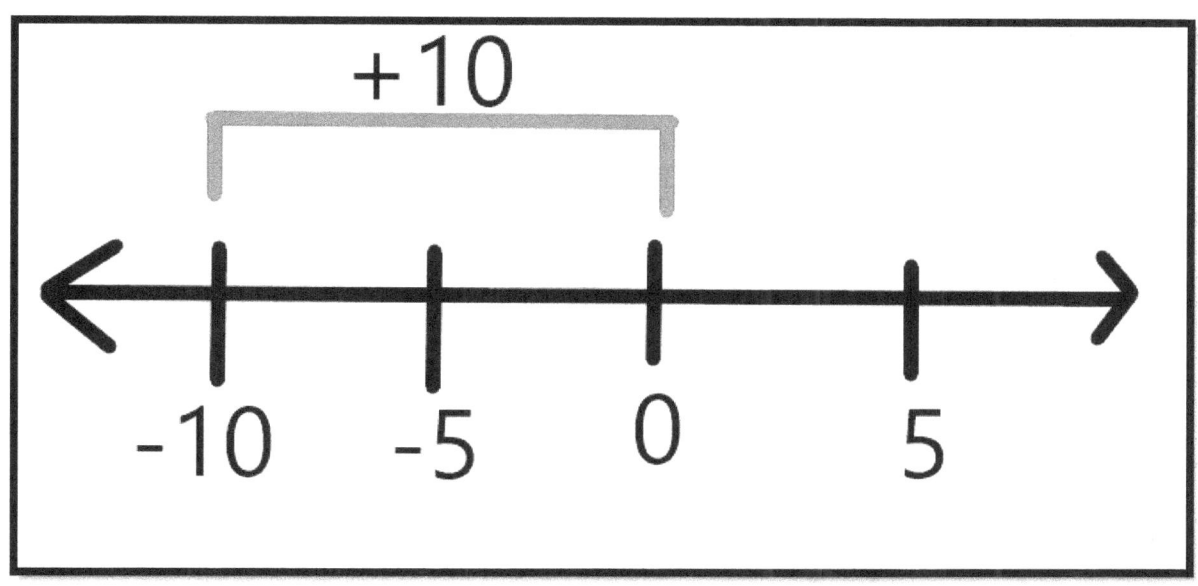

Example 1: Determine the value of $|5-7|$

$|5-7| = |-2| = 2$

Example 2: Determine the value of $|-(-4)^2|$

$|-(16)| = |-16| = 16$

As you can see whether there is a positive number or a negative number inside the absolute value brackets, the result will always be a positive number. Think about it this way, when a positive number goes inside, nothing happens to the number. When a negative number goes inside, something does happen, what happens? The negative number becomes a positive number. How do we mathematically change a negative number to a positive number? The only way we can do this is by either multiplying the negative number by -1, or we can divide it by -1.

The most basic Absolute Value function is

$y = |x|$

This means that "y" is equal to the absolute value of "x"

In order to graph an Absolute value function, we use piecewise functions. Piecewise functions are basically a function composed of multiple functions, in this chapter, we will only look at piecewise functions composed of two functions.

$y = |x|$

Is composed of two functions.

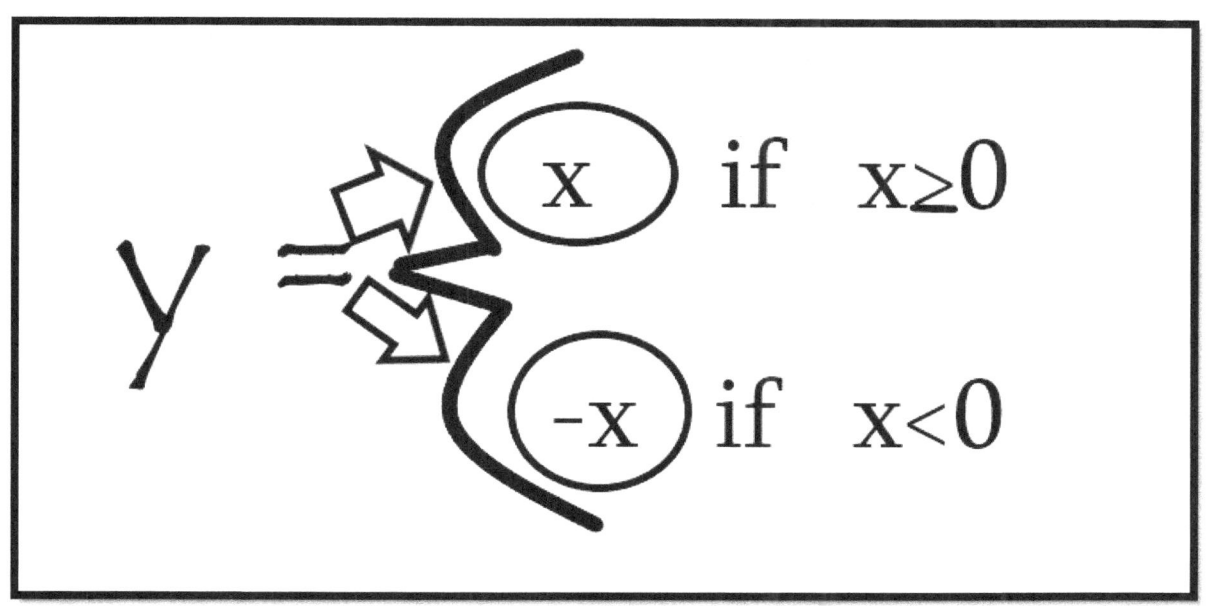

What this is saying is that, "For any x-value less than 0, draw the line y= -x, and when x is greater than or equal to 0, you draw the line y=x."

Basically, you draw $y = -x$ until you hit y=0, and then after that, you draw the line $y = x$

Here is a picture of this graph:

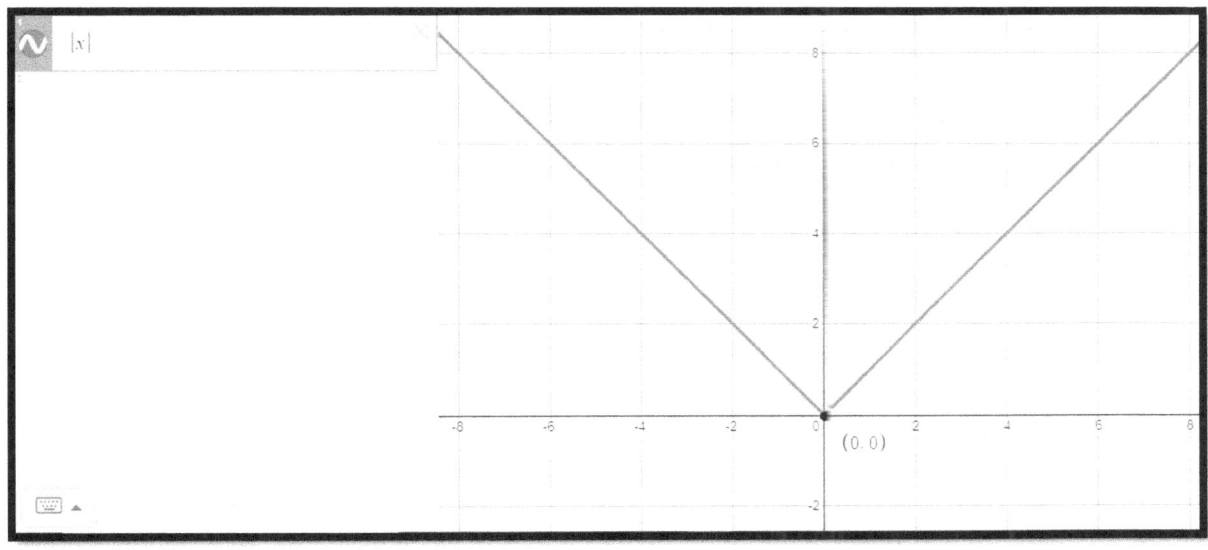

The quickest way to sketch these functions is like this:

- Step 1: Draw the original function
- Step 2: Reflect everything that is below the x-axis, above the x-axis.

Here is an example:
$|x^2 - 4|$

Step 1: Sketch $y = x^2 - 4$

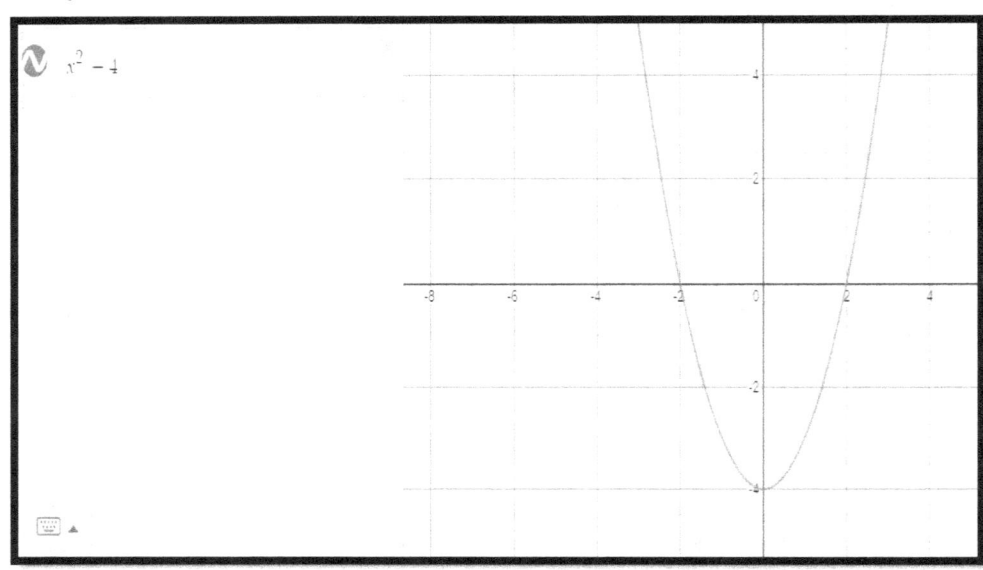

Step 2: Take the section below the x-axis and reflect it above and erase the bottom section.

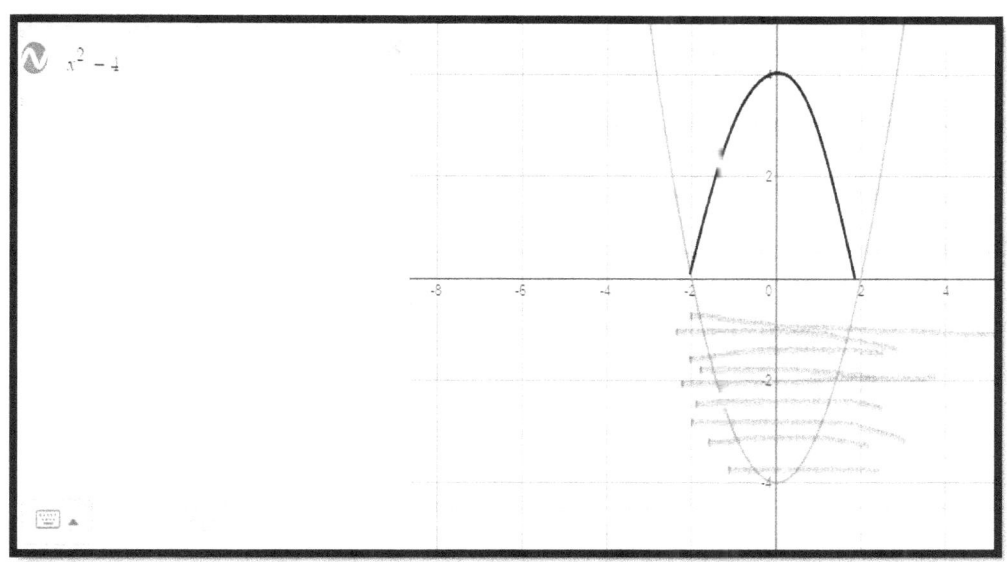

This is what the final graph looks like:

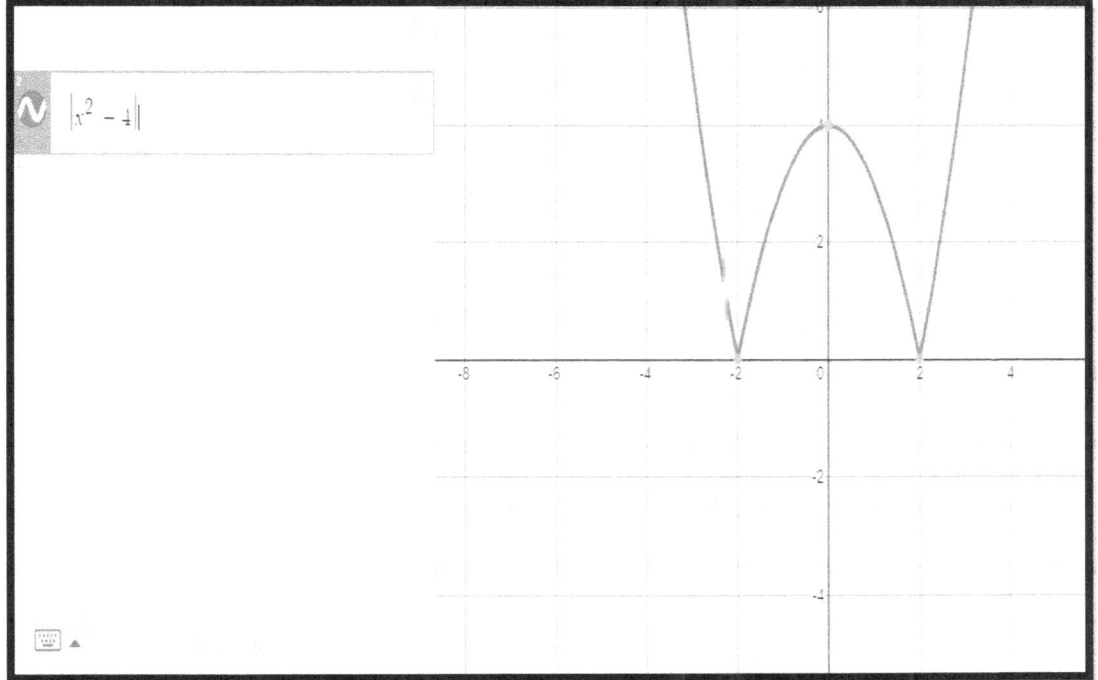

It may look complicated; however, if you really take the time to think about it, it actually makes sense. The portion below the x-axis was negative, and the absolute value function makes all negative y-values positive. That is why all the negative y-values became positive y-values.

The output values cannot be negative, unless you have a function with other operations outside of the absolute value function like this, $y = |x| - 2$

The graph is on the next page,

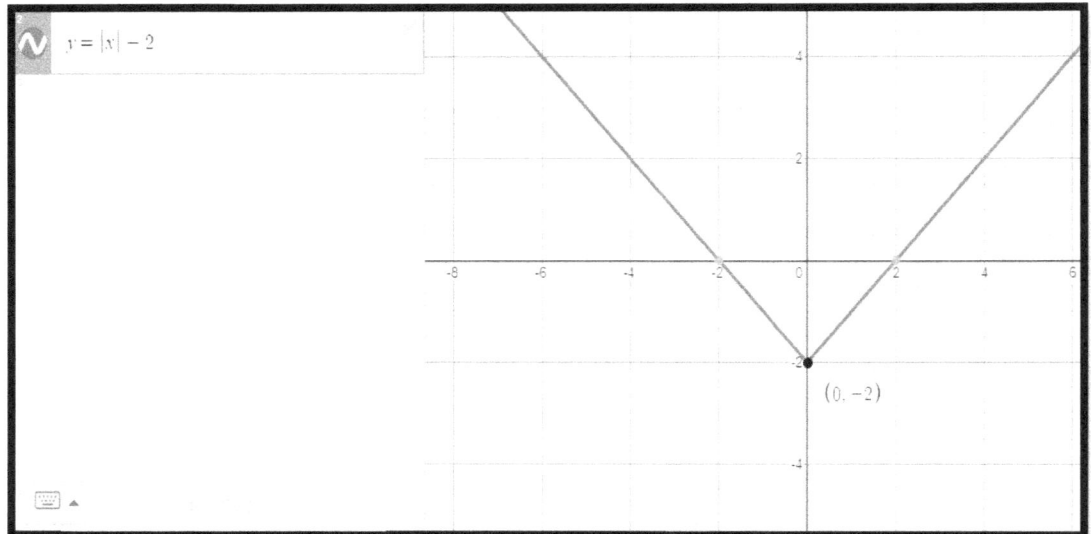

The graphing method I have shown you will not work in a case like this. If you want to use this graphing method I have shown, you will need to reflect the graph above the y=-2 part of the graph.

- Step 1: Draw the original function, $y = |x|$

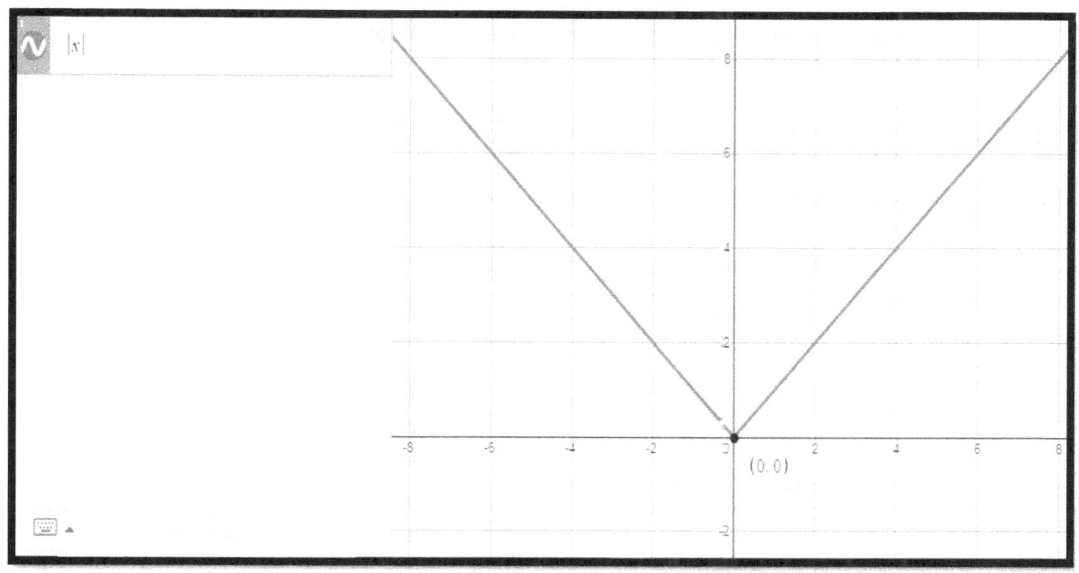

- Step 2: shift it 2 units down

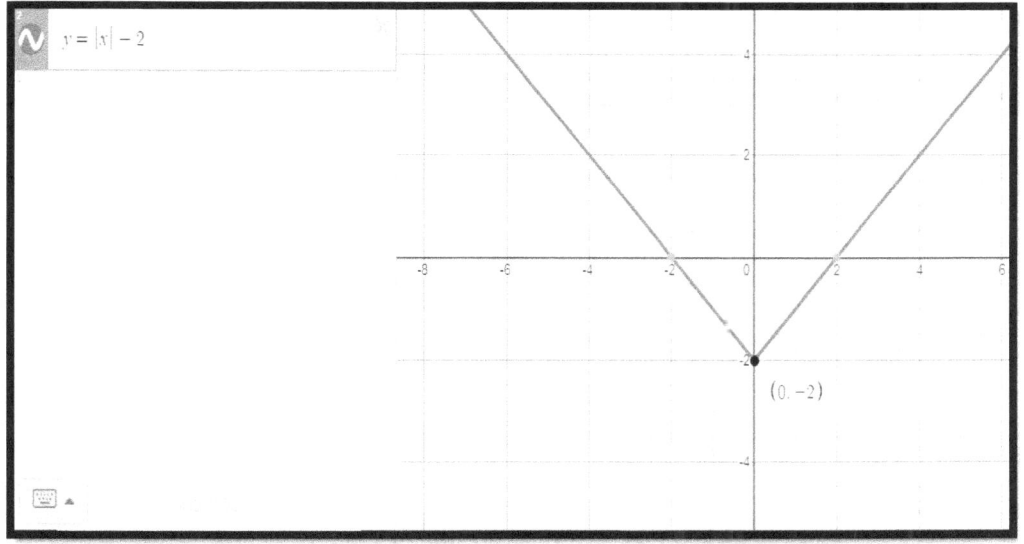

Solving Absolute Value Expressions

When you solve Absolute Value expressions, you solve it just like you would for a normal equation; however, you solve it twice. First time in the positive version, and the second time in the negative version.

For example, $|x + 2| = 4$

Two cases,

- Case 1: $x + 2 = 4$
- Case 2: $-(x + 2) = 4$
  $-x - 2 = 4$

We will start by solving case 1:
$x + 2 = 4$
$x = 4 - 2$
$x = 2$

Now we will solve case 2:
$-x - 2 = 4$
$-x = 4 + 2$
$-\dfrac{x}{-1} = \dfrac{6}{-1}$
$x = -6$

We will now test out the solutions,
$x = 2$
$|x + 2| = 4$
$|2 + 2| = 4$
$4 = 4$

$x = -6$
$|-6 + 2| = 4$
$|-4| = 4$
$4 = 4$

Both of the solutions work. Here is another example,
$|x - 1| = 2$
Case 1: $x - 1 = 2$
Case 2: $-(x - 1) = 2$
$-x + 1 = 2$

We will start by solving case 1,
$x - 1 = 2$
$x = 2 + 1$
$x = 3$

Now we will solve case 2,
$-x + 1 = 2$
$-x = 2 - 1$

$$-x = 1$$
$$-\frac{x}{-1} = \frac{1}{-1}$$
$$x = -1$$

Test out the solutions and see if they are both correct,
$$x = 3$$
$$|3 - 1| = 2$$
$$|2| = 2$$
$$2 = 2$$
$$x = -1$$
$$|-1 - 1| = 2$$
$$|-2| = 2$$
$$2 = 2$$
Both of the solutions work.

Reciprocal Functions

We will now look at reciprocal functions. Reciprocal functions are in the form $y = \dfrac{1}{x}$

These functions are complicated to graph for many students, my goal in this section is to help you understand how to graph these functions in the simplest way possible.

This is what the most basic reciprocal function looks like,

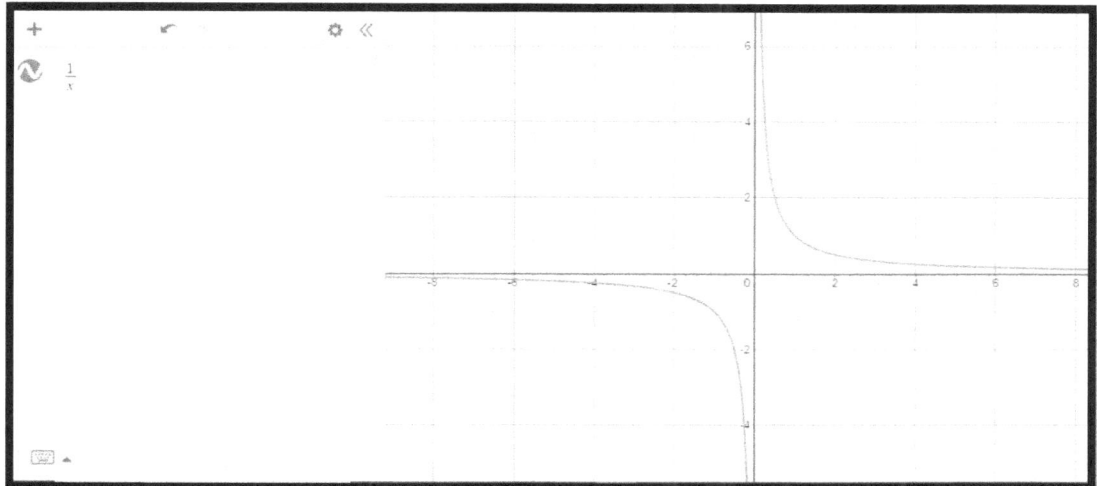

This probably seems very scary to you right now, trust me, it is not as scary as you think it is. The reason the graph takes that shape is because in the equation

$$y = \frac{1}{x}$$

X cannot equal 0 ($x \neq 0$). Why? Because the function is undefined at 0, you cannot divide a number by 0. You will now take a look at a table of values, that will help you understand what happens when the number gets close to 0.

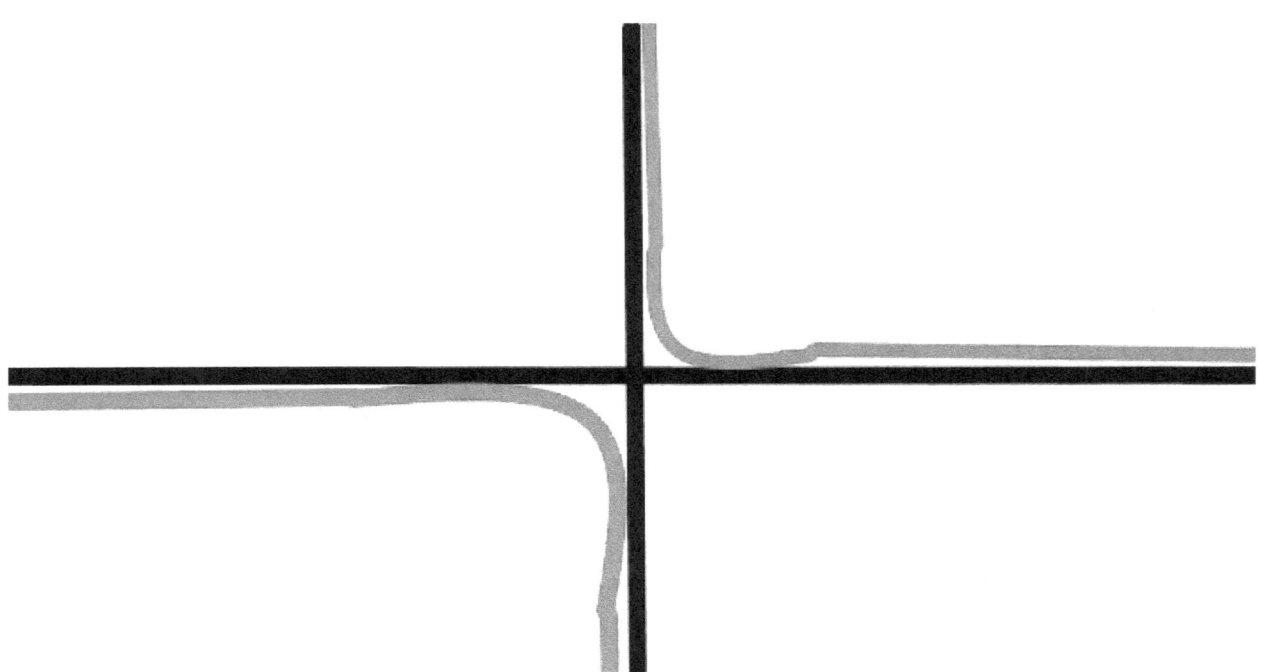

We will start with the part in Quadrant 1.

As you can see, the graph never actually reaches $x=0$. Why is that? The reason for that is because as I have previously mentioned, you cannot divide anything by zero.

Why does the graph become so steep as it approaches zero from the right? The reason for that is because the graph reaches for positive infinity in the y-axis. It goes very high, enter these numbers on your calculator and see what happens,

$$\frac{1}{0.1} \quad \frac{1}{0.01} \quad \frac{1}{0.001} \quad \frac{1}{0.0001}$$

Did you notice how as the denominator gets smaller, the result gets larger and larger, this is what the whole idea is.

And for the part in Quadrant 3, as the denominator gets smaller and smaller in terms of negative x-values less than 1 and greater than 0, the resulting value gets smaller and smaller.

$$\frac{1}{-0.1} = -10 \qquad \frac{1}{-0.001} = -1000$$

In Quadrant 3, as the x-values approach zero, y-values reach for negative infinity, they go very low on the y-axis forever.

0 is the **Vertical Asymptote**, a vertical asymptote is a non-permissible x-value. You sketch a vertical asymptote by basically drawing a dashed vertical line on the non-permissible value, I will show you a reciprocal function with a vertical asymptote at x=2,

So you see what I'm talking about.

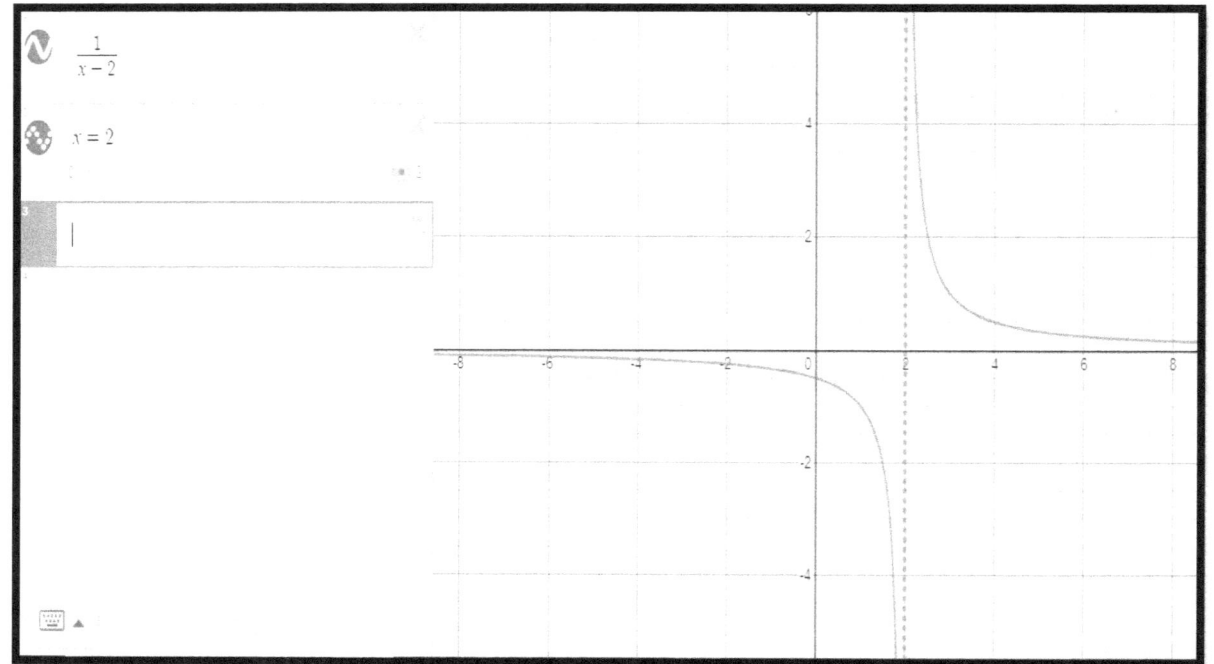

Do you see that dotted line at $x = 2$ going up and down? That is the vertical asymptote.

We will now graph $y = \dfrac{1}{x + 1}$

- Step 1: Identify the non-permissible value,
$x + 1 \neq 0$
$x \neq -1$
- Step 2: Draw the vertical asymptote at $x = -1$,

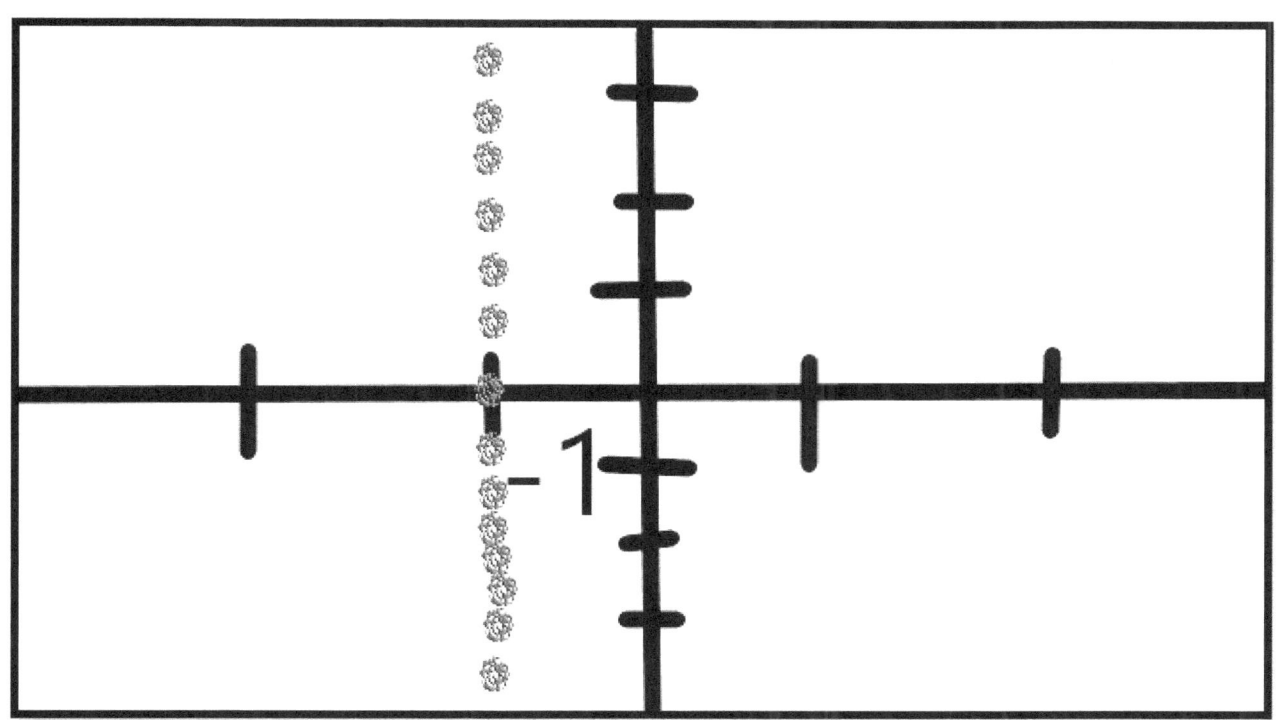

- Step 3: Now that we have drawn the vertical asymptote, we will now finish drawing the graph.

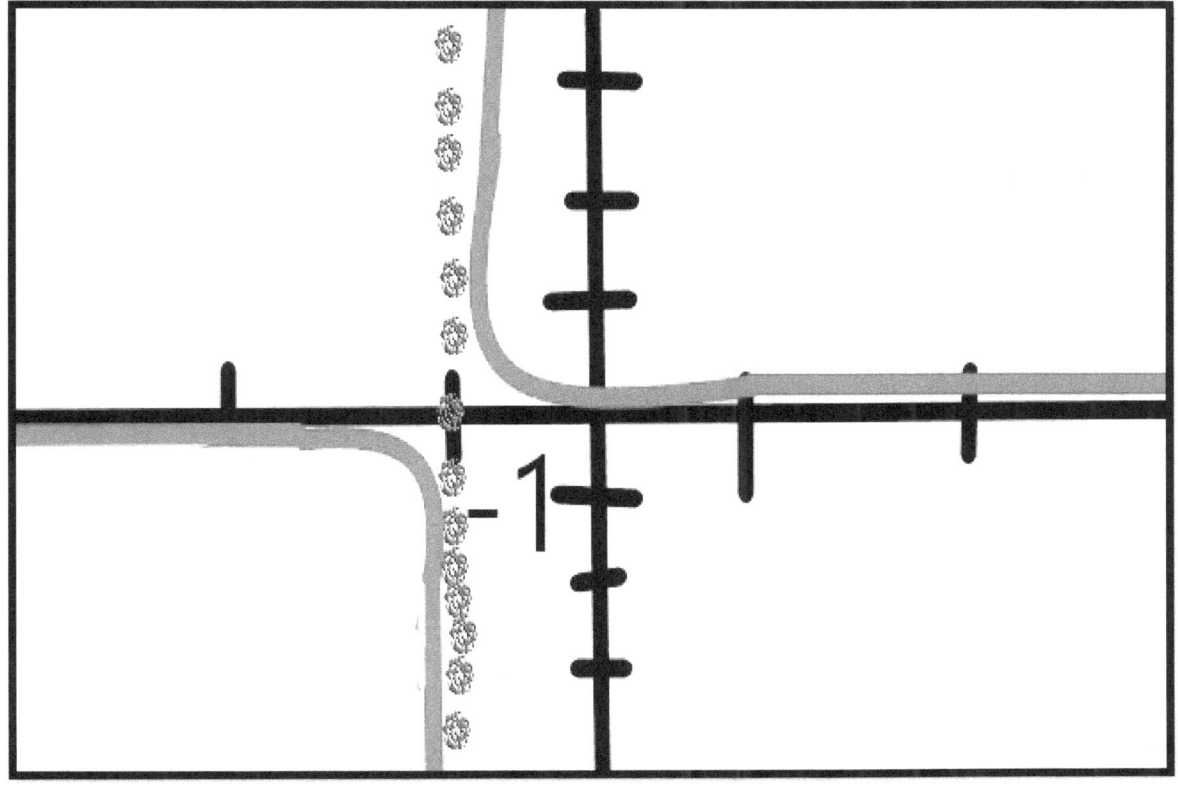

It is just like the first graph you saw, except it has moved one unit to the left because the vertical asymptote has moved to $x = -1$.

If the function is positive, it will always take that form.

If the function is in the form of, $y = -\dfrac{1}{x}$ then it will take this form instead:

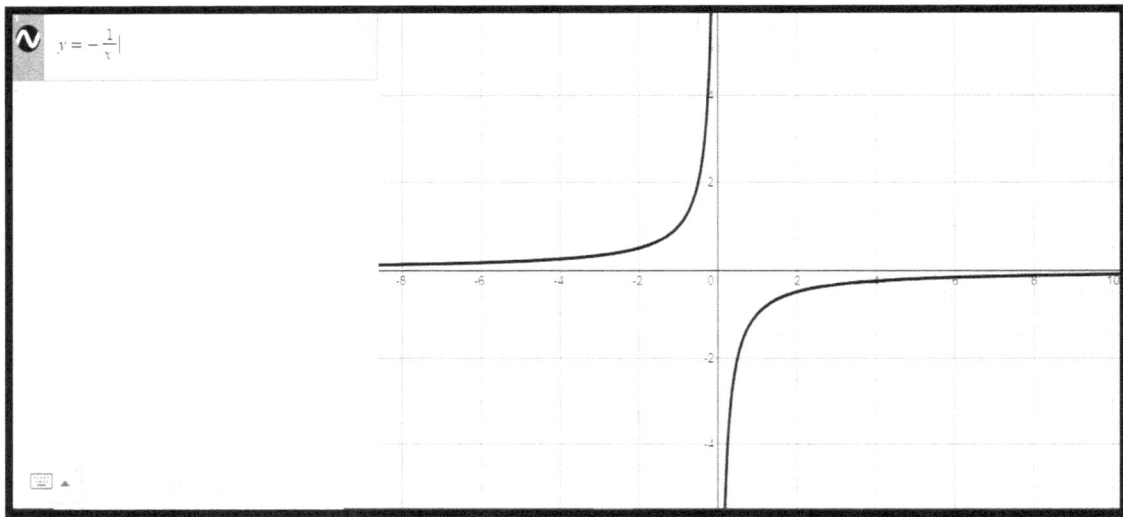

As you can see, it has basically been reflected in the x-axis.

What was in Quadrant 3 is now in Quadrant 2,

and what was in Quadrant 1 is now in Quadrant 4.

If you are unsure of its shape, draw the vertical asymptote, and plot several points on the graph. This is a good method to use in high school; however, it is not a good method to use in University Calculus, things will be different, and you need to actually know several features of the graph before drawing, as you might miss an important feature if you just plot points and guess what it looks like. For now, plot points if you are unsure, you are in High School right now, take your time and try your best. Do not put pressure on yourself, you do not need extra stress as a teen. High School should be a fun part of your life, before you start to see reality (lots of bills, responsibilities, etc.). Enjoy time while you are young because next thing you know, you will be all grown up. Trust my words, I may only be 18 years old, but what I have just told you is very true. Next thing you know you will be in University or College. I am currently enrolled in *Athabasca University's* ***Bachelor of Science in Computing and Information Systems with a Minors in Applied Mathematics***. I am planning on becoming a Computer Programmer. Not putting stress on yourself is the best thing. When I was in my Junior years in High School, I use to be very hard on myself. And then in Grade 12, I stopped being hard on myself and all of my grades went up, and so did my happiness. I finished high school with an excellent average. Being hard on yourself in High School and University will only make your years miserable. What matters at the end of the day is trying your best and being yourself, it does not matter what other people think about you. That is one of the biggest problems in High School, most of the kids are pretending to be someone else, they are not being their true selves. They are playing this whole act just to impress their friends, be cool, and fit in. I am going to tell you this right now, there is no such thing as the "cool kids" in High School. They are not better than anyone; we are all equal. If everyone would realise this, the world would be a better place. This now concludes this chapter. You can try the practice questions, and then you can verify your answers by looking at the answer key at the end of this book.

Practice Questions

1. Solve for x,
   a) $|x + 2| = 3$
   b) $|x - 4| = 2$
   c) $|x - 1| = 5$

2. Graph $y = |x| - 1$
3. Graph $y = |x^2 - 3|$

4. Write a piecewise function for $y = |x - 2|$

5. Graph $y = \dfrac{1}{x + 3}$ sketch the vertical asymptote with a dotted or dashed line.

Linear & Quadratic Inequalities

Chapter 6

In this chapter, you will learn about Linear and Quadratic Inequalities. What is an inequality? An inequality is a relation that makes a comparison between two numbers or expressions. For example: $x > 2$, x is greater than 2. What satisfies this inequality? Any value greater than 2. This is what that would look like on the number line,

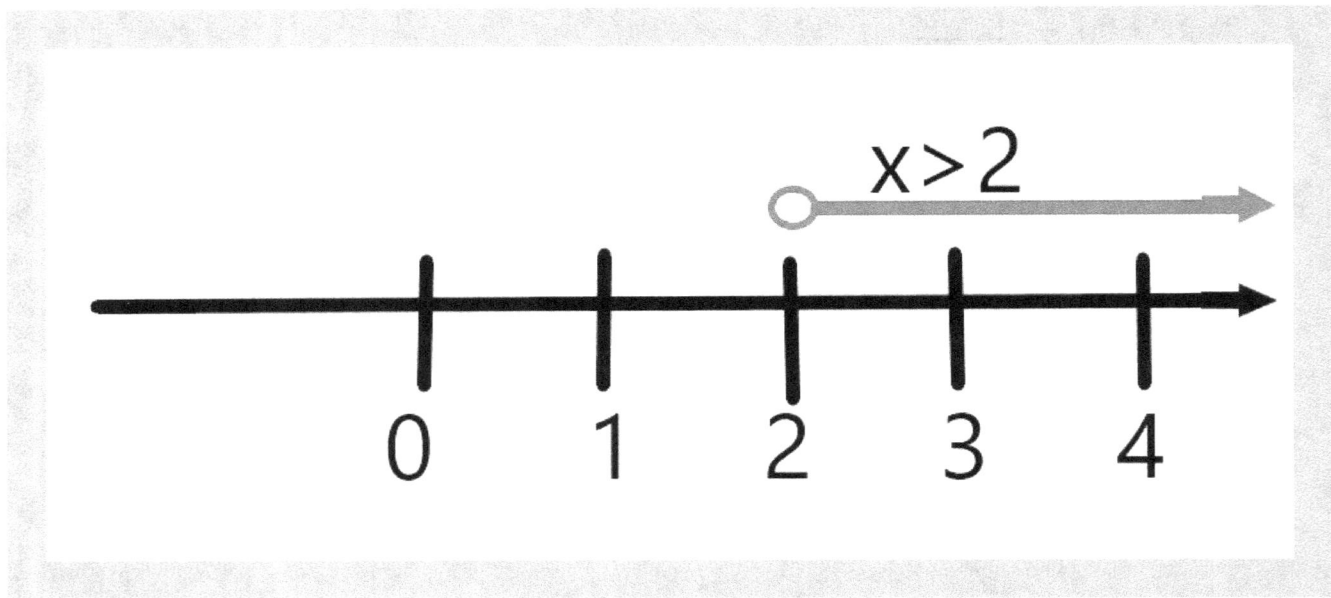

Do you see the open circle on top of the 2? The reason there is an open circle is because the inequality means anything greater than 2 but not equal to 2; therefore, the open circle means not equal to 2. What about when $x \geq 2$? This means that x is greater than or equal to 2, with this being said, we can draw a closed circle on top of the 2.

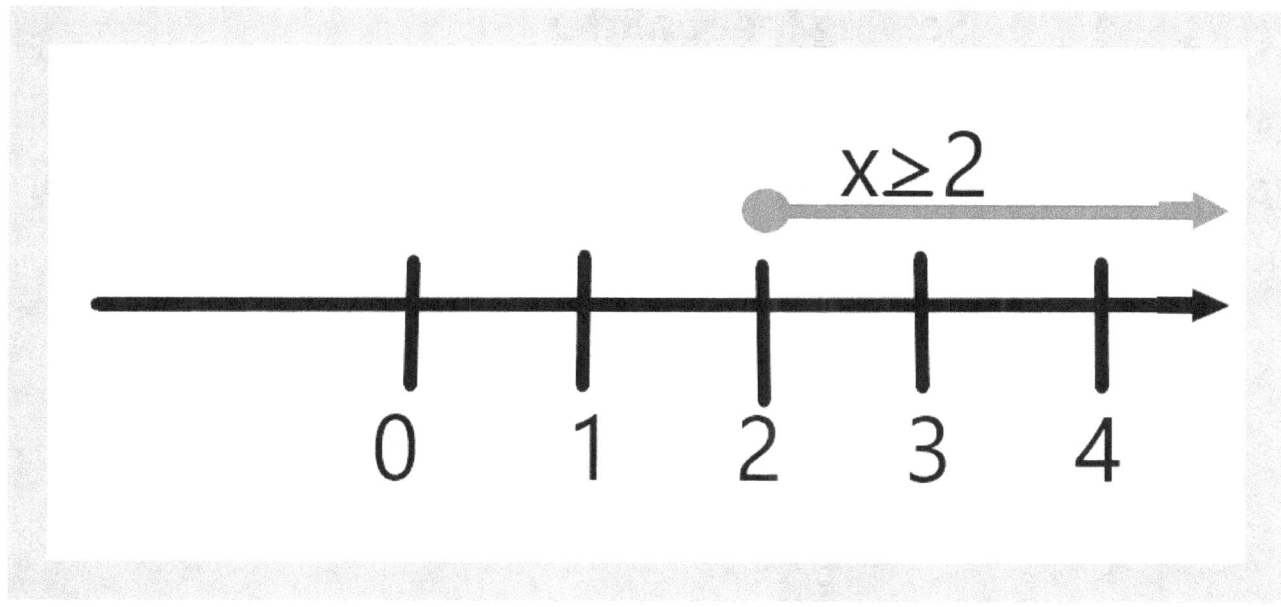

$a > b$	a is greater than b
$a \geq b$	a is greater than or equal to b
$a < b$	a is less than b
$a \leq b$	a is less than or equal to b

This table is useful whenever you are unsure of how to read the inequality. The next section will show you how to solve inequalities, there are a few rules that you should make note of.

Solving Inequalities

We will now learn how to solve inequalities. Linear and quadratic inequalities. We will start by solving inequalities for x, and representing the inequality in terms of x.

Example 1: $3x - 1 > 8$

- Step 1: Add 1 to both sides to isolate the $3x$
  $3x > 8 + 1$
  $3x > 9$
- Step 2: Divide both sides by 3
  $$\frac{3x}{3} > \frac{9}{3}$$
  $x > 3$
  In order to satisfy this inequality x needs to be greater than 3. $x = 3, 3.00001, 3.7, 4, 4.00001822,$ 38288282828, 999944440443, etc. The important thing to grasp is that there are several solutions that can satisfy an inequality (too many to count), there is not just one value.

Example 2: $-x + 3 > 5$

- Step 1: Subtract 3 from both sides
  $-x > 5 - 3$

  $-x > 2$

- Step 2: Divide both sides by -1. Be very careful here, when you divide or multiply by a negative number, you need to flip the inequality symbo the other way.

$$-\frac{x}{-1} < \frac{2}{-1}$$

$x < -2$

X is less than -2. We will now test out a number less than -2, to verify this solution.

$-(-3) + 3 > 5$

$3 + 3 > 5$

$6 > 5$

This is true, because 6 is greater than 5.

Remember this, when you multiply or divide by a negative you need to FLIP the inequality around. Do not forget this, you must remember to FLIP it around, or else you will get the wrong answer.

Quadratic inequalities are more difficult to solve than linear inequalities.

Example 1: $x^2 + 4x + 3 < 0$

You may be tempted to say, "Okay, let's factor this and find the roots." That is not the final solution, you can do this, but that is only the first step in solving quadratic inequalities.

- Step 1: Factor $x^2 + 4x + 3 < 0$
$(x+1)(x+3) < 0$
Zeros are $x = -1$ and $x = -3$

- Step 2: $x^2 + 4x + 3 < 0$ means that the solution will be when the graph of $x^2 + 4x + 3$ is below 0 on the y-axis. In other words, this is the section where the graph lies below the x-axis.

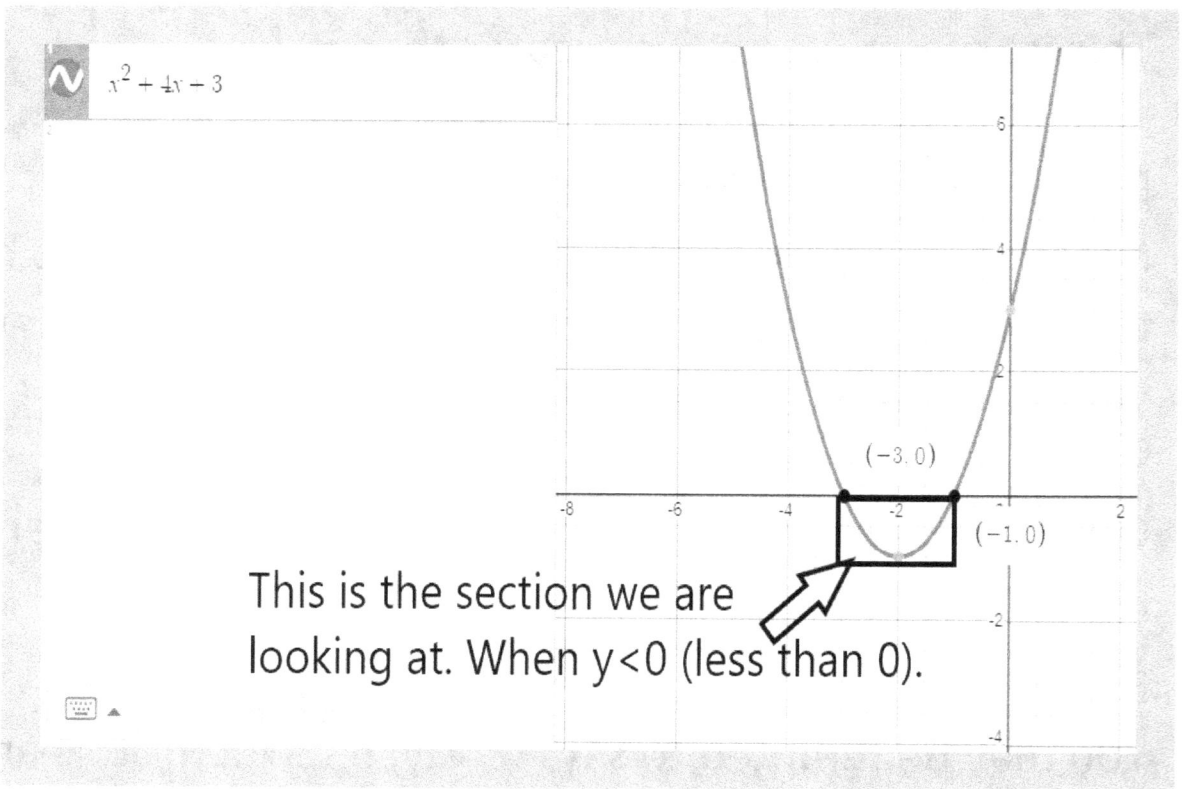

This is the section we are looking at. When y<0 (less than 0).

The solution of this inequality as you can see is the interval in-between -3 and -1; therefore, $-3 < x < -1$

This means x is greater than -3 and less than -1, which makes sense, since you can see that it is only in this interval y is less than 0. The solution is any number that is greater than -3 and less than -1.

Graphing Inequalities

In this section we will graph inequalities. In order to graph inequalities properly there are a few rules that you should take not of.

$y > x$ shade above the graph and sketch the graph with a broken line.

$y \geq x$ shade above the graph and sketch the graph with a solid line. A way to remember this is that as you can see there is a solid line below the > symbol.

$y < x$ shade below the graph and sketch the graph with a broken line.

$y \leq x$ shade below the graph and sketch the graph with a solid line. A way to remember this is that as you can see there is a solid line below the < symbol.

Linear inequality example 1: $y > x + 2$

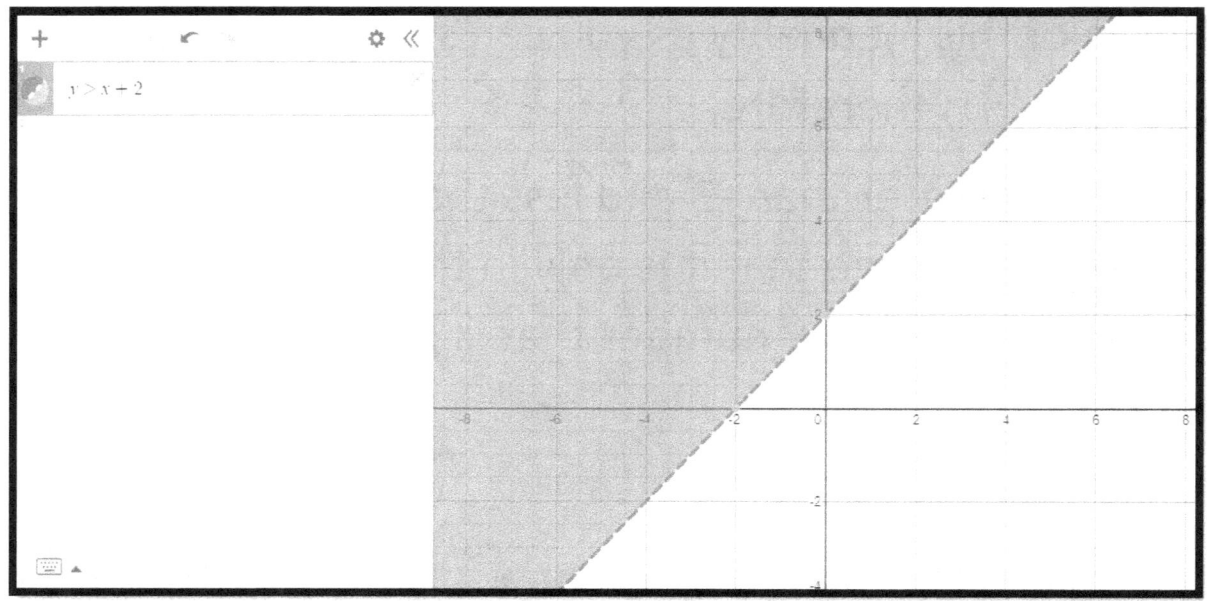

Linear inequality example 2: $y \geq x + 2$

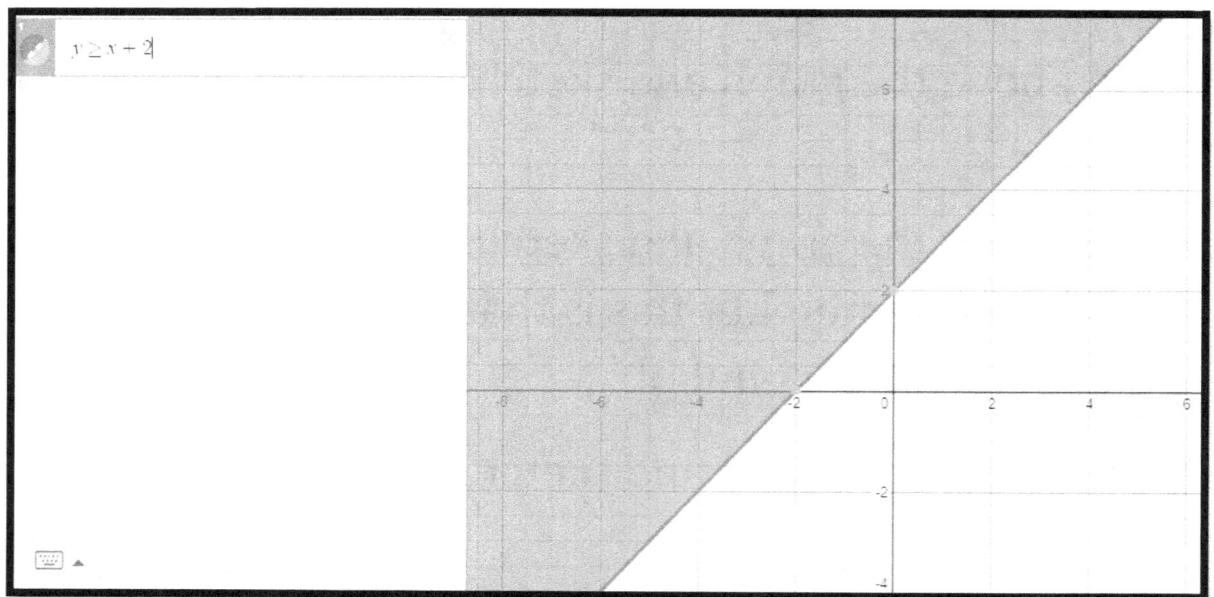

Linear inequality example 3: $y < x + 2$

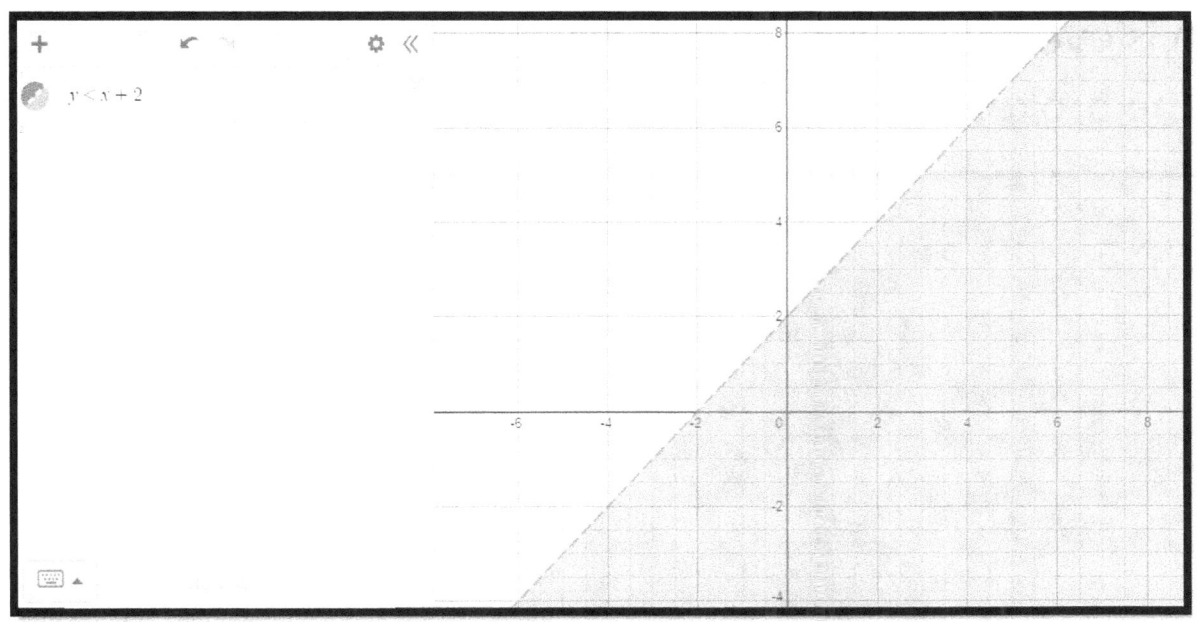

Linear inequality example 4: $y \leq x + 2$

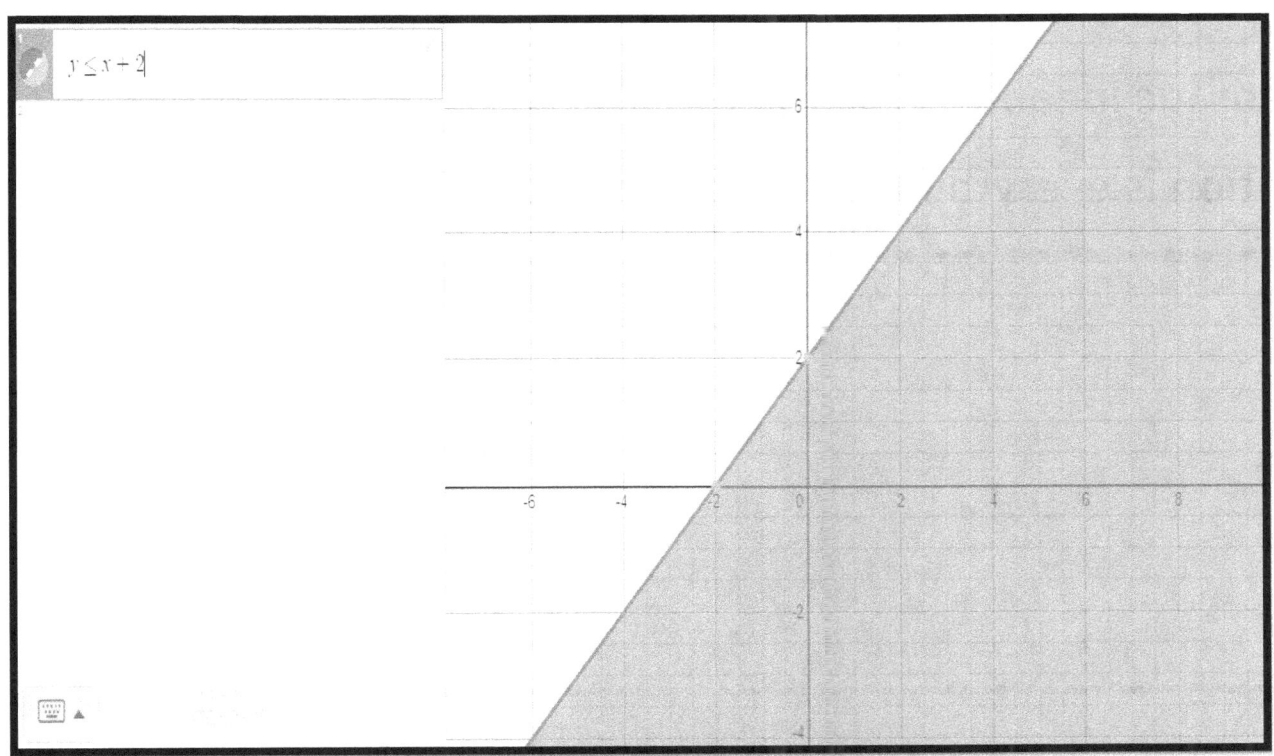

Quadratic inequality example 1: $y > (x - 1)^2 + 1$,

Solution is everything inside of the graph except anything that lies on the broken line

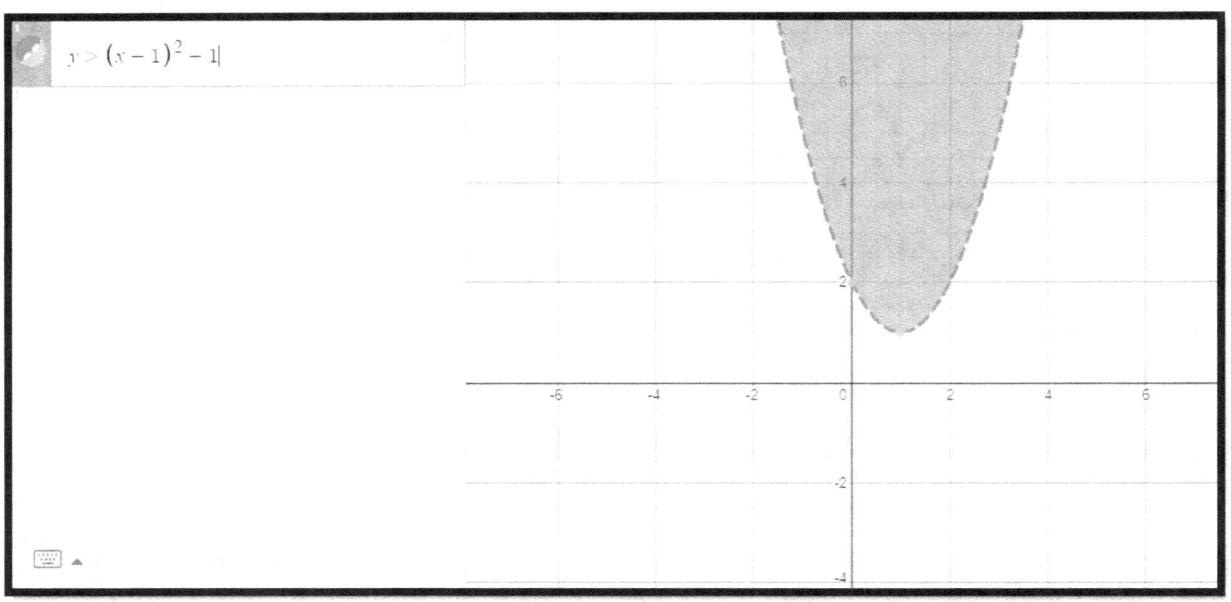

Quadratic inequality example 2: $y \geq (x-1)^2 + 1$

Solution is everything inside of the graph

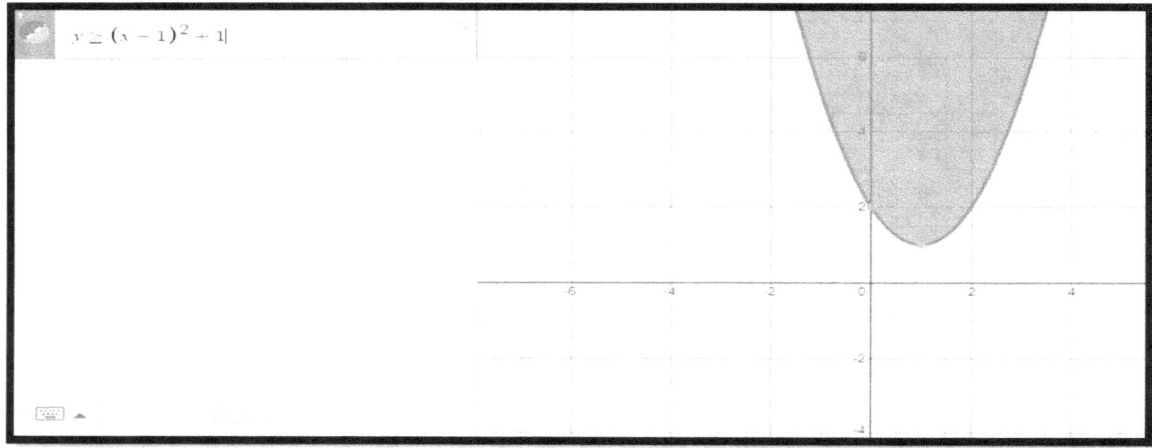

Quadratic inequality example 3: $y < (x-1)^2 + 1$,

Solution is everything outside of graph except for anything that lies on the broken line

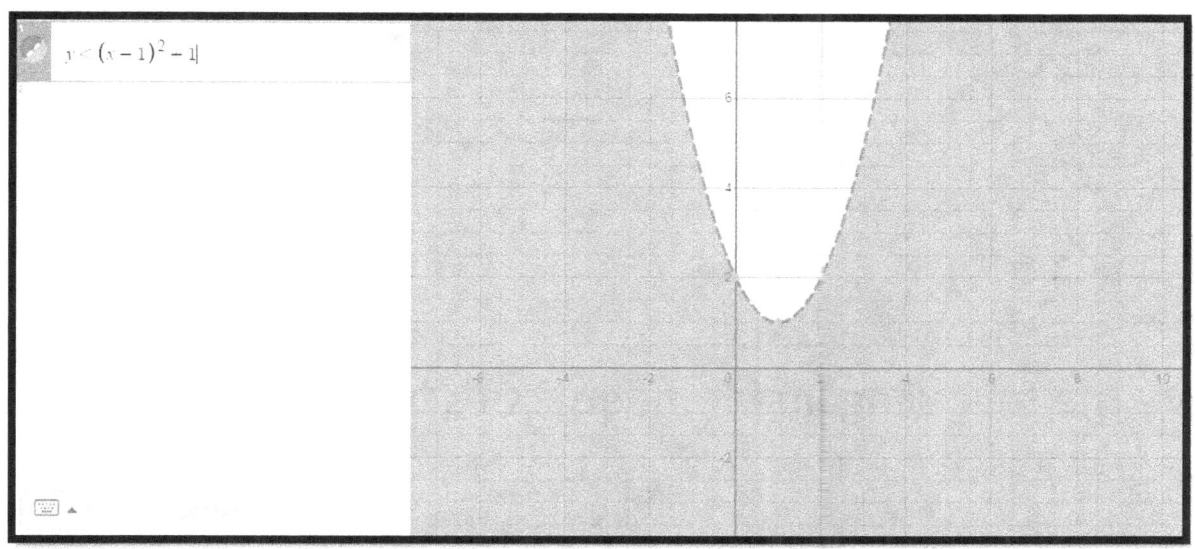

Quadratic inequality example 4: $y \leq (x-1)^2 + 1$,

Solution is everything outside of graph

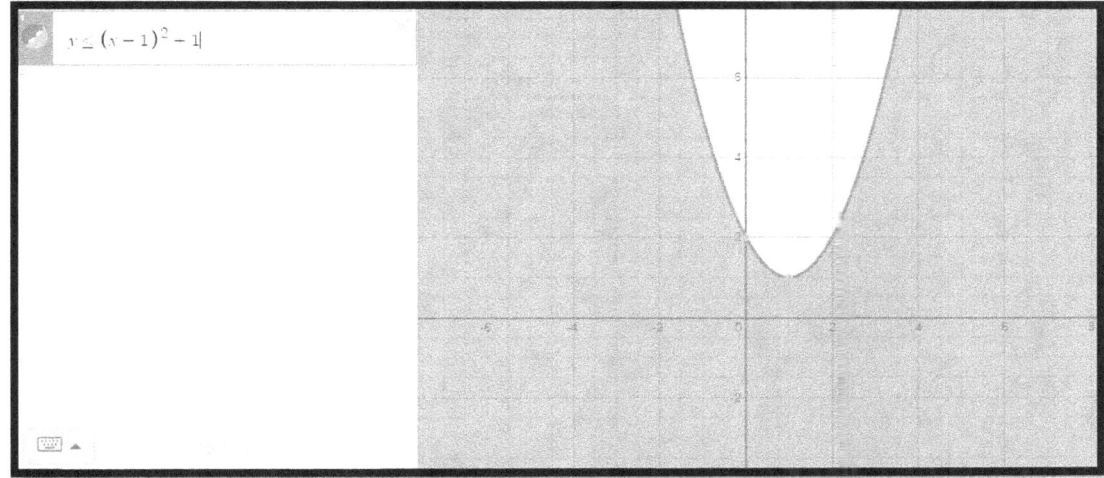

Practice Questions

1. Solve the following inequalities in terms of x.
   a) $-x \geq 2$
   b) $x - 3 < 5$
   c) $4 - x < 2$
   d) $-3x \leq -6$

2. Solve the following linear inequalities in terms of y
   a) $y - 2 > x$
   b) $-2y \geq 4x + 6$

3. Graph $y < x - 1$

4. Graph $y \geq (x + 2)^2 - 1$

PART 2: MATH 12

Transformations Of Functions

Chapter 7

In this chapter, you will be learning about a fascinating concept that can be applied to all kinds of functions, the concept of transformations. Before we learn this, I will give you a brief review on Functions. In this example, I will be putting fish on the scale, and the output will be the weight.

Fish name	Weight(kg)
Fish 1	1.9
Fish 2	2.3
Fish 3	1.7
Fish 4	1.9

See how Fish 1 and Fish 4 have the same weight? This is fine, it is still considered a function. The only thing that would matter is if one of the Fish would have more than one weight. That would not make sense, would it? How can you have more than one weight possible per fish? The important thing to note is that each INPUT has only **ONE** OUTPUT. When we are looking at graphs to determine if it is a function or not, the best thing to use is the vertical line test. You draw a straight vertical line through several sections of the graph. If it only intersects with the graph once, then it is a function. If it crosses through the graph more than once, then it is NOT a function.

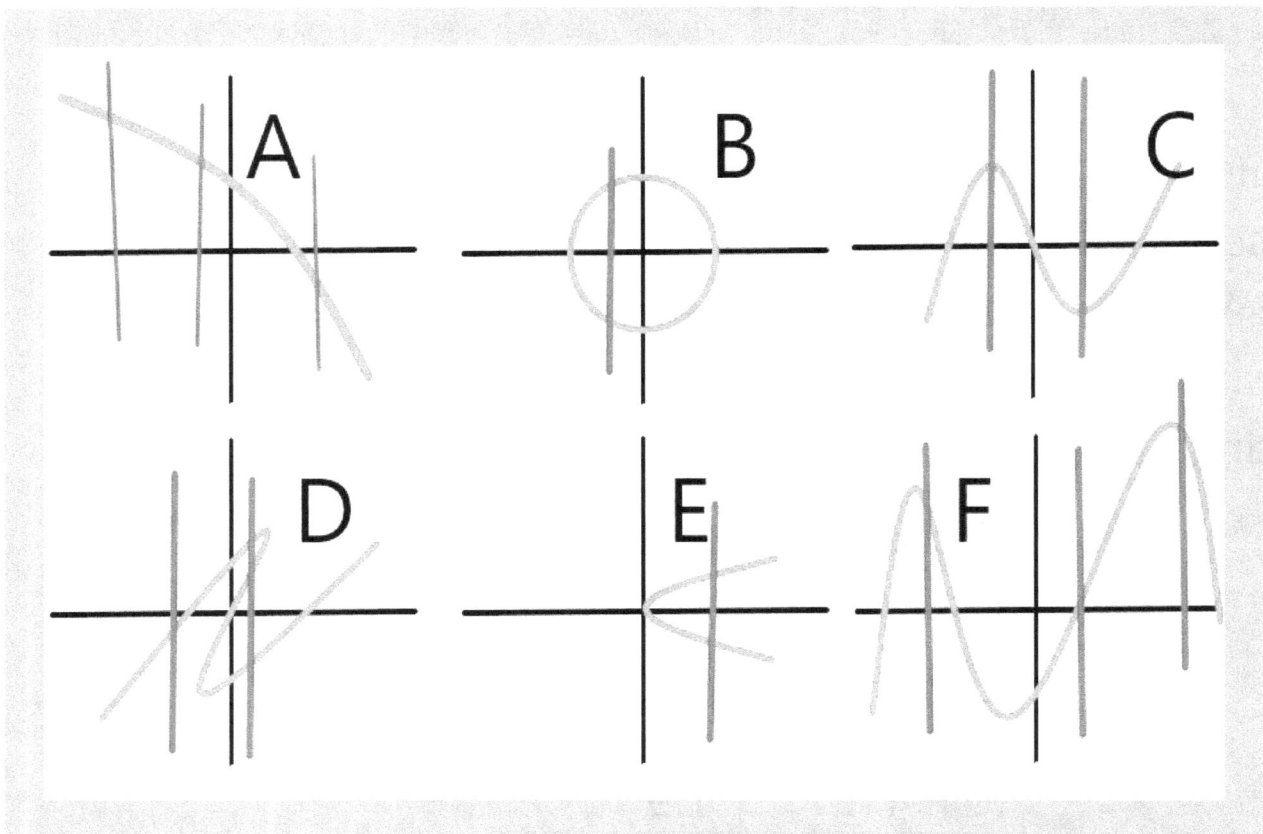

Only the graphs A, C, and F are functions. As you can see, the vertical line only passes through the graph once, this means that it is a function. So whenever you are in doubt if it is a function or not, the vertical line test will be your best bet.

Translations

We will start by looking at vertical translations, that is equations in the form, $y = f(x) + d$, where d denotes the vertical translation, the number of units up or down.

First, we replace y with $y - d$, and then we solve the equation for y.

Example 1: $d = 3$,

$y - d = |x|$

$y - (3) = |x|$

$y - 3 = |x|$

$y - 3 + 3 = |x| + 3$

$y = |x| + 3$

The graph is translated 3 units up.

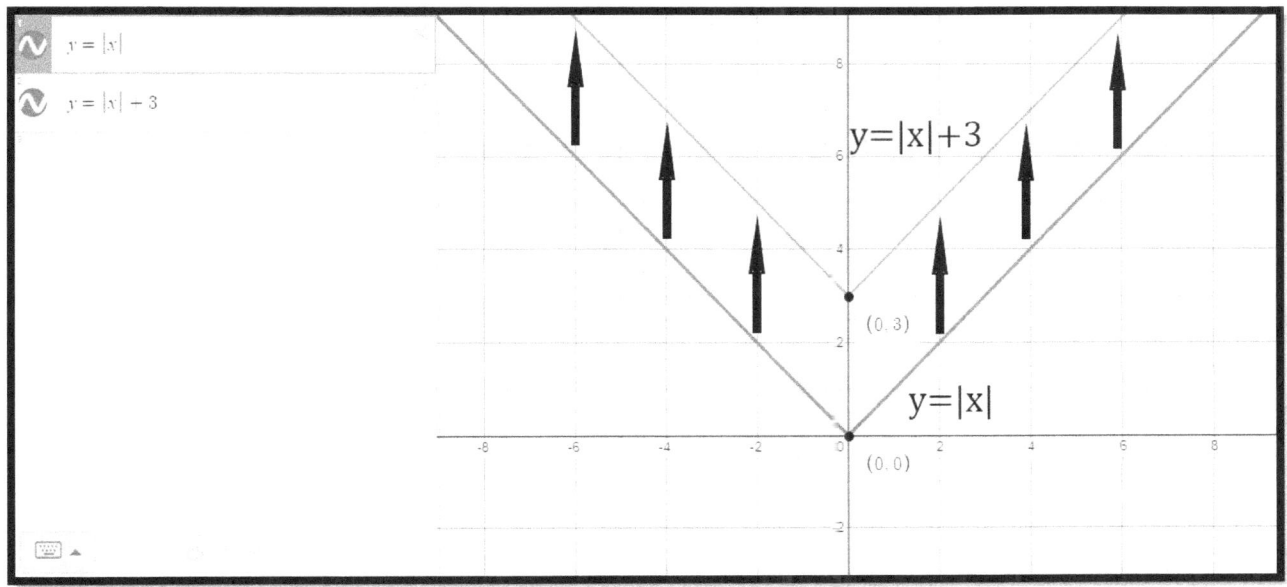

Example 1: $d = -3$,

$y - d = |x|$

$y - (-3) = |x|$

$y + 3 = |x|$

$y - 3 + 3 = |x| - 3$

$y = |x| - 3$

The graph is translated 3 units down.

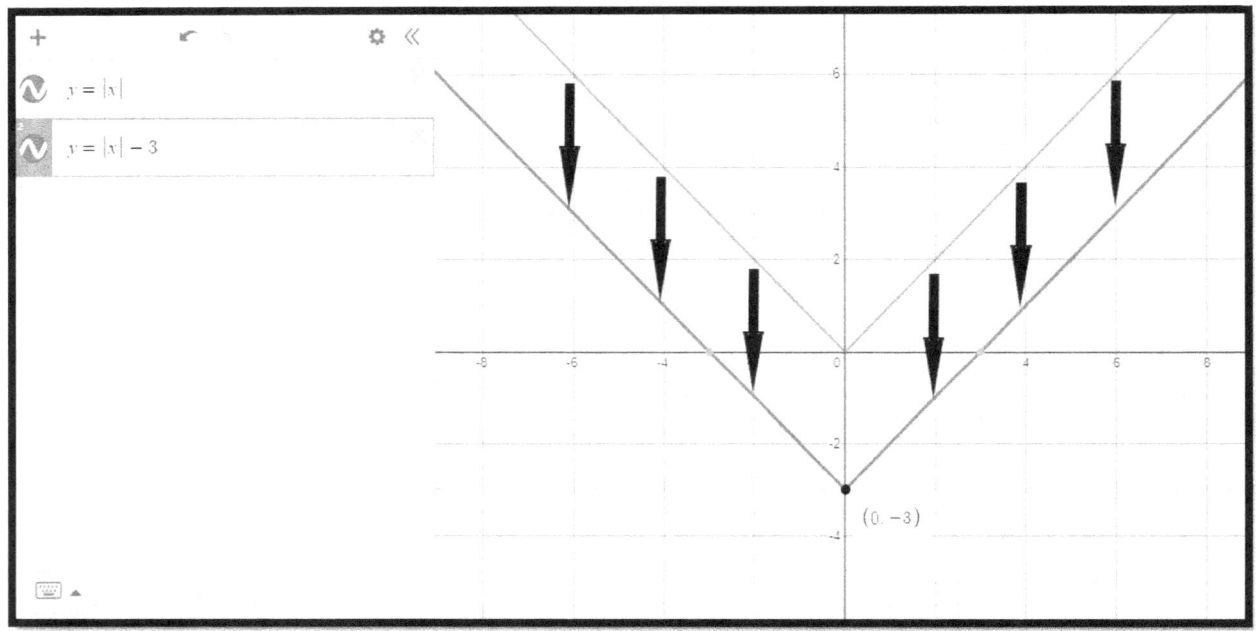

When $d > 0$, vertical translation up

When $d < 0$, vertical translation down

There are two other examples on the next page.

Vertical Translation of 2 units up.

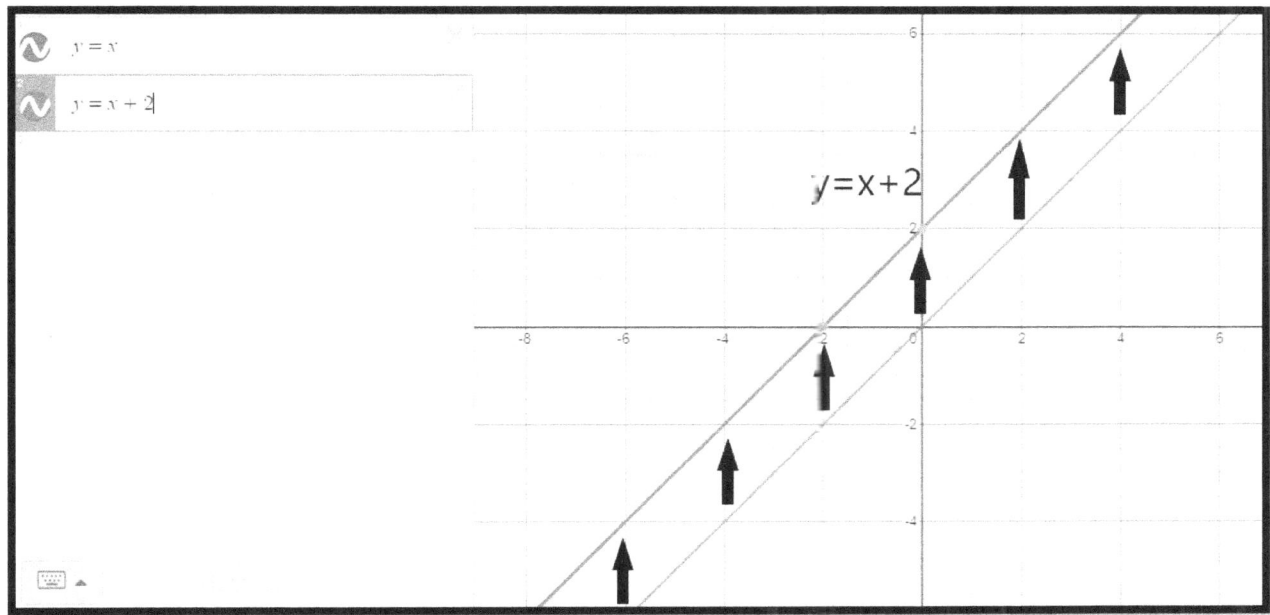

Vertical Translation of 3 units down

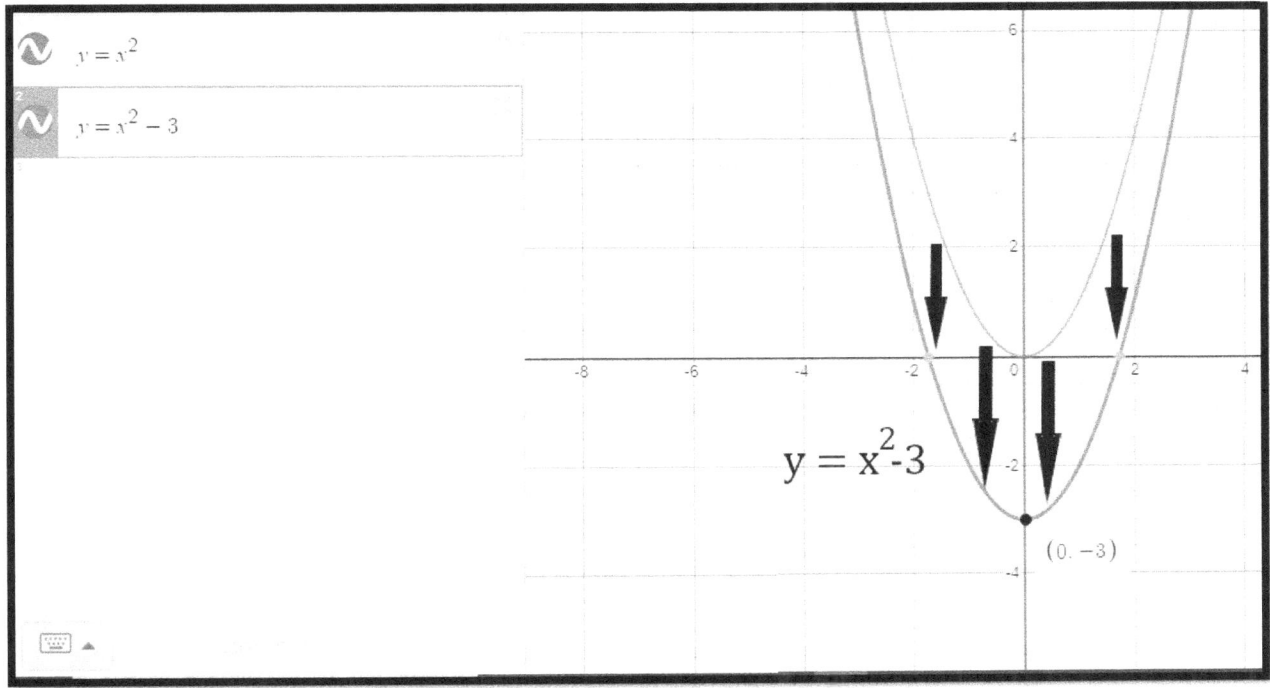

We will now look at horizontal translations in the form of

$y = f(x - c)$

A value of $c > 0$, means a horizontal translation of c units right.

A value of $c < 0$, means a horizontal translation of c units left. Pay very close attention, because a lot of students get messed up with the horizontal translations because they will see the opposite value of c inside of the brackets. For example, in the equation $y = |x - 3|$ there is a horizontal translation of 3 units right. The reason for this is because you are subtracting c from x; therefore, you get the opposite of c inside of the brackets.

$y = (x + 2)^2$ would mean a horizontal translation of 2 units left.

I will use w to represent a number inside of the brackets.

If you see $f(x + w)$, think left.

If you see $f(x - w)$ think right.

Here are some examples:

Horizontal translation of 2 units right.

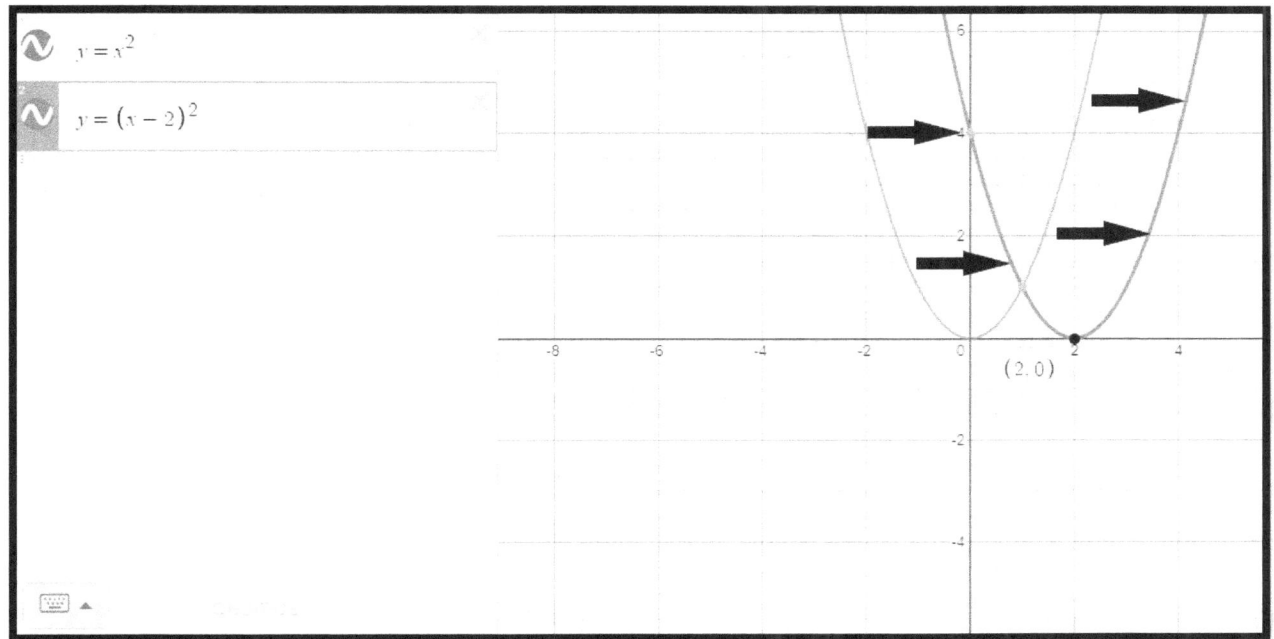

Horizontal translation of 2 units left.

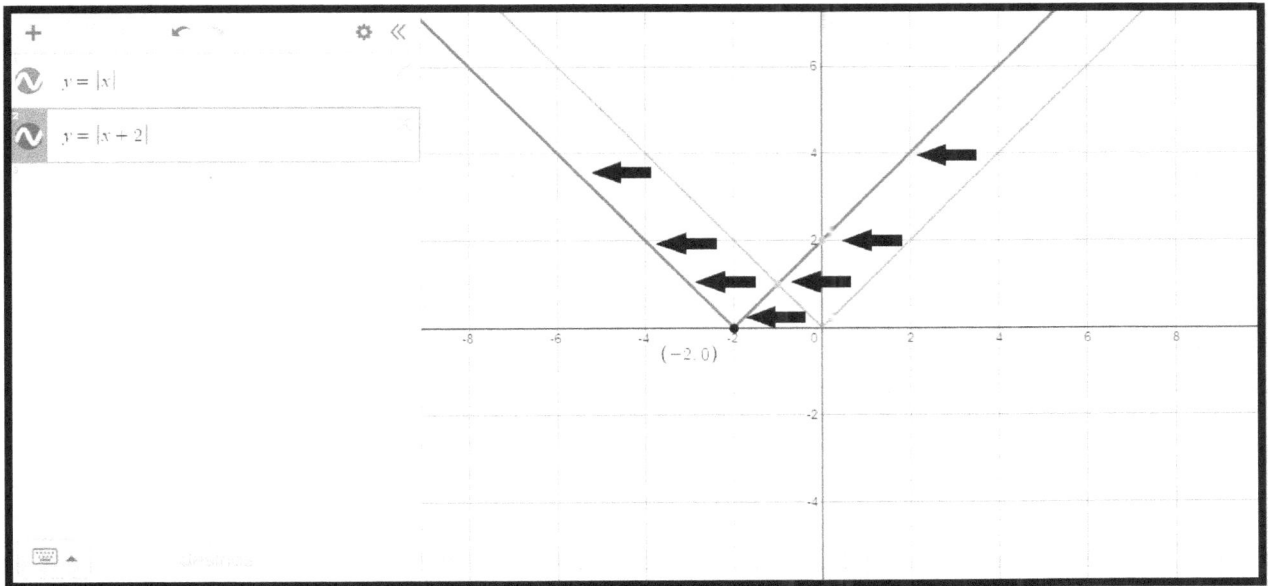

Reflections

In this section you will learn about reflections. There are two kinds of reflections, reflections in the y-axis, and reflections in the x-axis. Reflections in the y-axis occur when x is replaced with $-x$, causing all of the x-coordinates to be multiplied by -1 ($y = f(-x)$). Reflections in the x-axis occur when y is replaced with $-y$, causing all of the y-coordinates to be multiplied by -1. We will start by looking at reflections in the x-axis ($y\ becomes\ -y$). Another way to represent this is as, $y = -f(x)$.

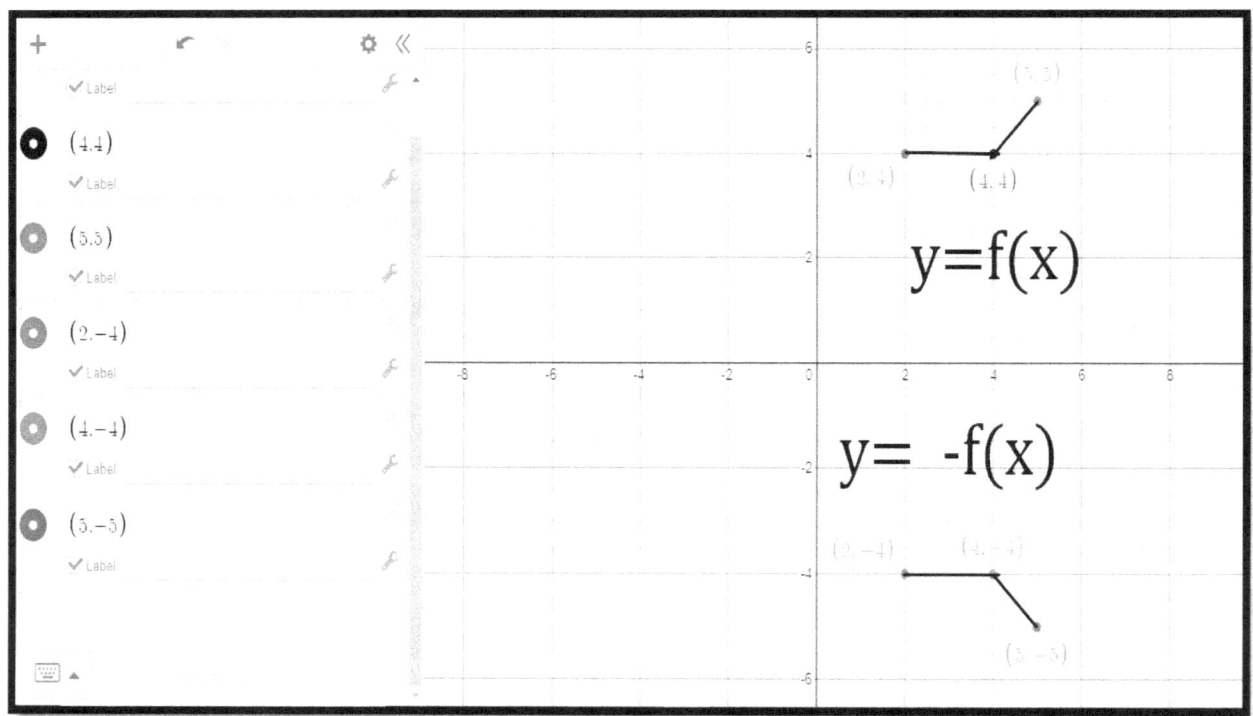

As you can see, all of the y-coordinates were multiplied by negative 1. This is known as a vertical reflection.

I will now use the same graph of $f(x)$ to show you what happens when x is replaced with $-x$.

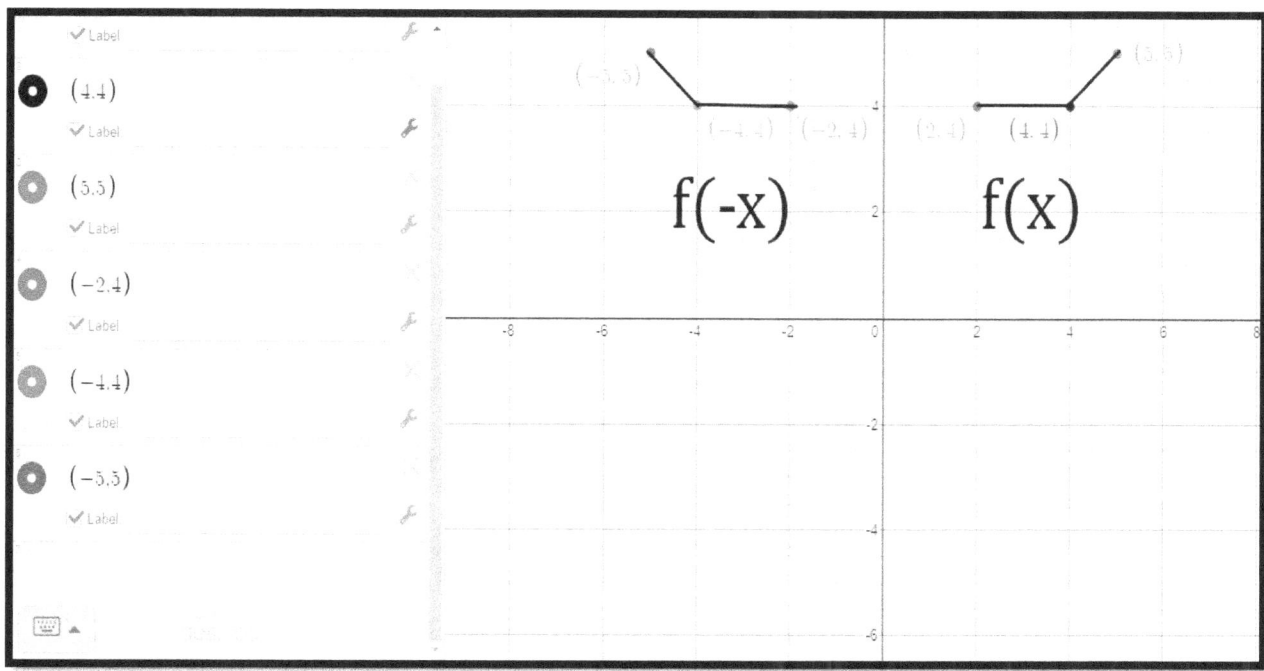

I hope that you noticed that all of the x-coordinates were multiplied by negative 1. If $f(x)$ would have been on the left side (Quadrant 2) then $f(-x)$ would have been on the right side (Quadrant 1).

To sum up, when $y \text{ is replaced with } -y$ it results in a reflection in the x-axis. $y = -f(x)$

When x is replaced with $-x$ it results in a reflection in the y-axis. $y = f(-x)$

Expansions and Compressions

We will begin this section by looking at vertical expansions and compressions, and then we will move on to look at horizontal expansions and compressions.

We will compare $y = f(x)$ to $y = af(x)$

There is a vertical expansion when the value of $|a| > 1$

The reason why I put the absolute value brackets is because sometimes you will see something like this,

$y = -2f(x)$ this means a reflection in the x-axis and a vertical expansion by 2.

In the equation $y = 3(x + 1)^2$ there is a vertical expansion by a factor of 3.

There is a vertical compression when $0 < |a| < 1$.

For example, in the equation $y = \frac{1}{3}|x|$ there is a vertical compression by a factor of $\frac{1}{3}$.

Try to determine if the following graphs have a vertical expansion or compression, and by which factor.

a) $y = \frac{1}{2}f(x)$

b) $y = 7|x|$

c) $\frac{1}{5}y = f(x)$

Answers

a) Vertical compression by a factor of ½
b) Vertical expansion by a factor of 7
c) Vertical expansion by a factor of 5. You need to solve for y. You multiply both sides by 5,

$$5\left(\frac{1}{5}y\right) = 5(f(x))$$

$$y = 5f(x)$$

Here is what a vertical expansion looks like graphically: $y = 2|x|$ this means a vertical expansion by a factor of 2.

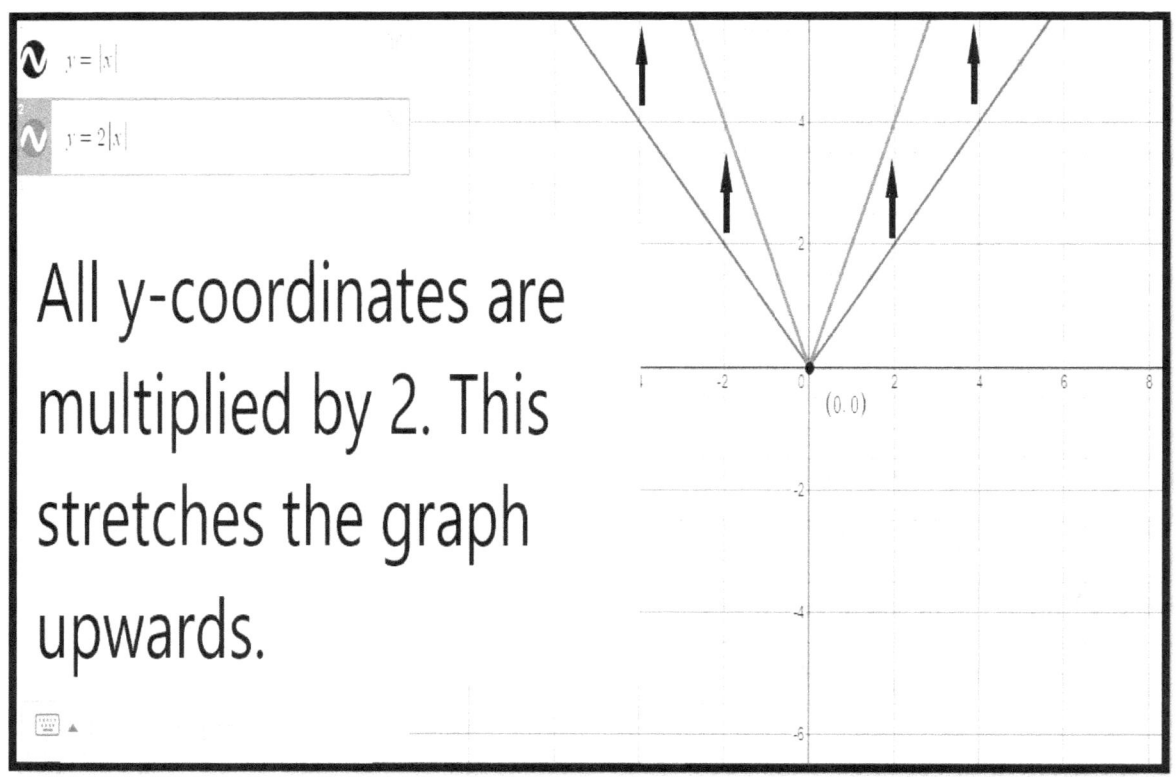

Here is what a vertical compression looks like graphically:

$y = \frac{1}{2}|x|$ this means a vertical compression by a factor of 1/2.

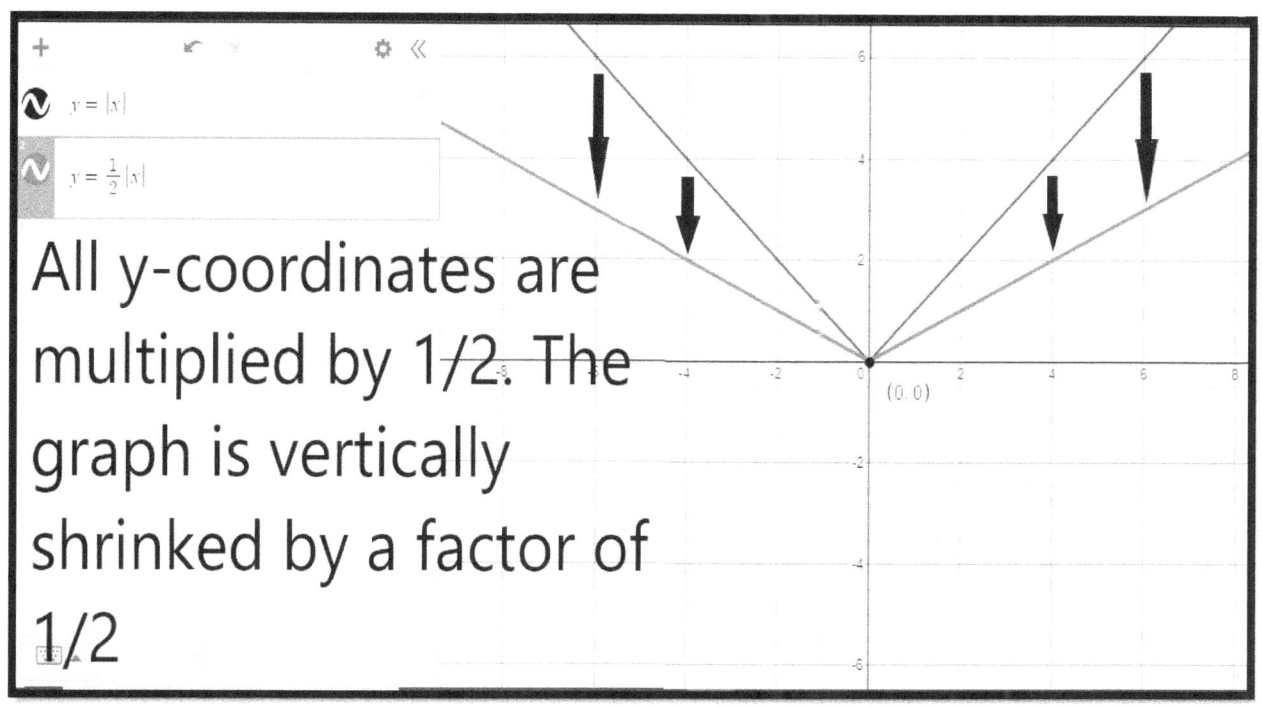

When you graph these functions, first graph the original function, and then determine if it is being vertically expanded or compressed and then multiply all of the y-coordinates by a.

A horizontal expansion or compression occurs in the form, $y = f(bx)$ where 1/b is the horizontal expansion or compression, depending on the value of b.

If $|b| > 1$ then this means that there is a horizontal compression by a factor of 1/b.

If $0 < |b| < 1$ then there is a horizontal expansion by a factor of 1/b.

$y = f(3x)$, $b = 3$ horizontal compression by a factor of $\frac{1}{3}$

$y = f\left(\frac{1}{5}x\right), b = \frac{1}{5}$ horizontal expansion by a factor of 5

$y = f(-2x)$ reflection in the y-axis, and a horizontal compression by a factor of $\frac{1}{2}$.

$y = f(-\frac{1}{2}x)$ reflection in the y-axis, and a horizontal expansion by a factor of 2.

I will now explain why $y = f\left(\frac{1}{2}x\right)$ is a horizontal expansion by a factor of 2

Here is a table of values:

$f(x)$

x	y
0	0
1	1
2	2
3	3
4	4

$f(\frac{1}{2}x)$

x	y
0	0
1	0.5
2	1
3	1.5
4	2

Here are the graphs of $f(x)$ and $f(\frac{1}{2}x)$

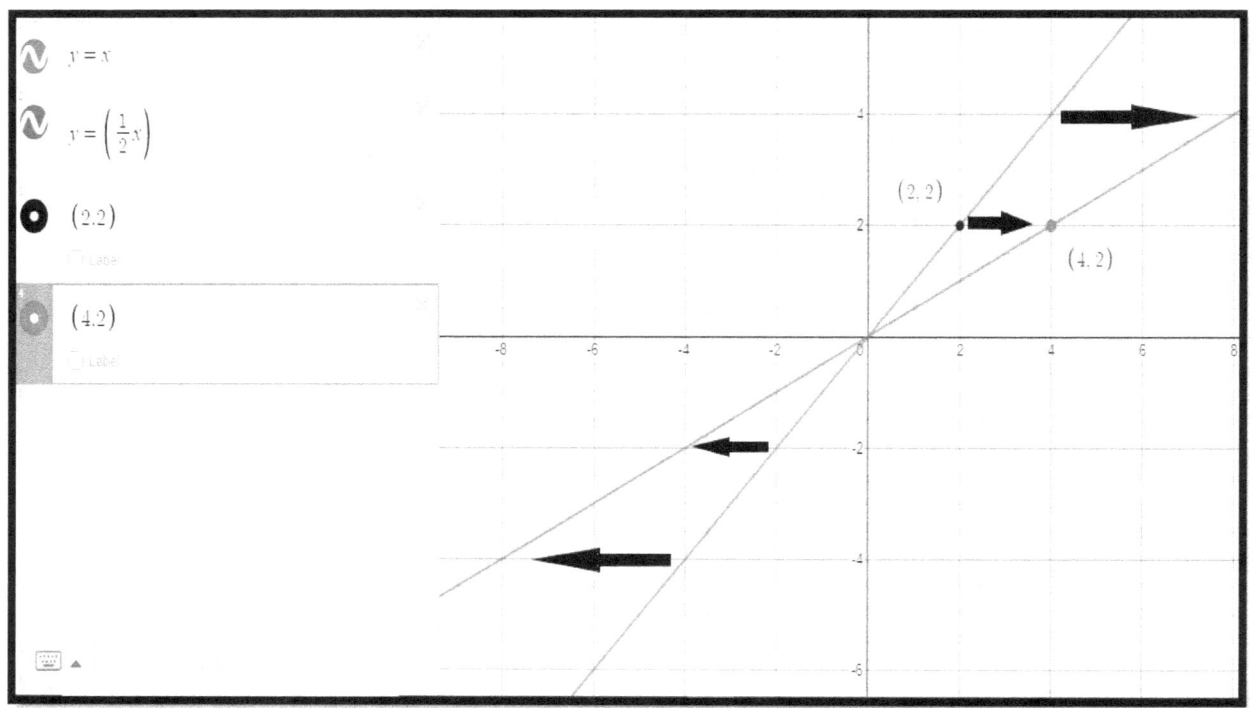

As you can see, the graph gets wider, that is why it is a horizontal expansion, it stretches sideways. It takes double the original input value to get the same output value in the function, $f(\frac{1}{2}x)$. Instead of plugging in 2 to get an output of 2, you now need to plug in a 4 to get the same output value of 2.

Here is the graph of $f(2x)$ (horizontal compression by a factor of 2):

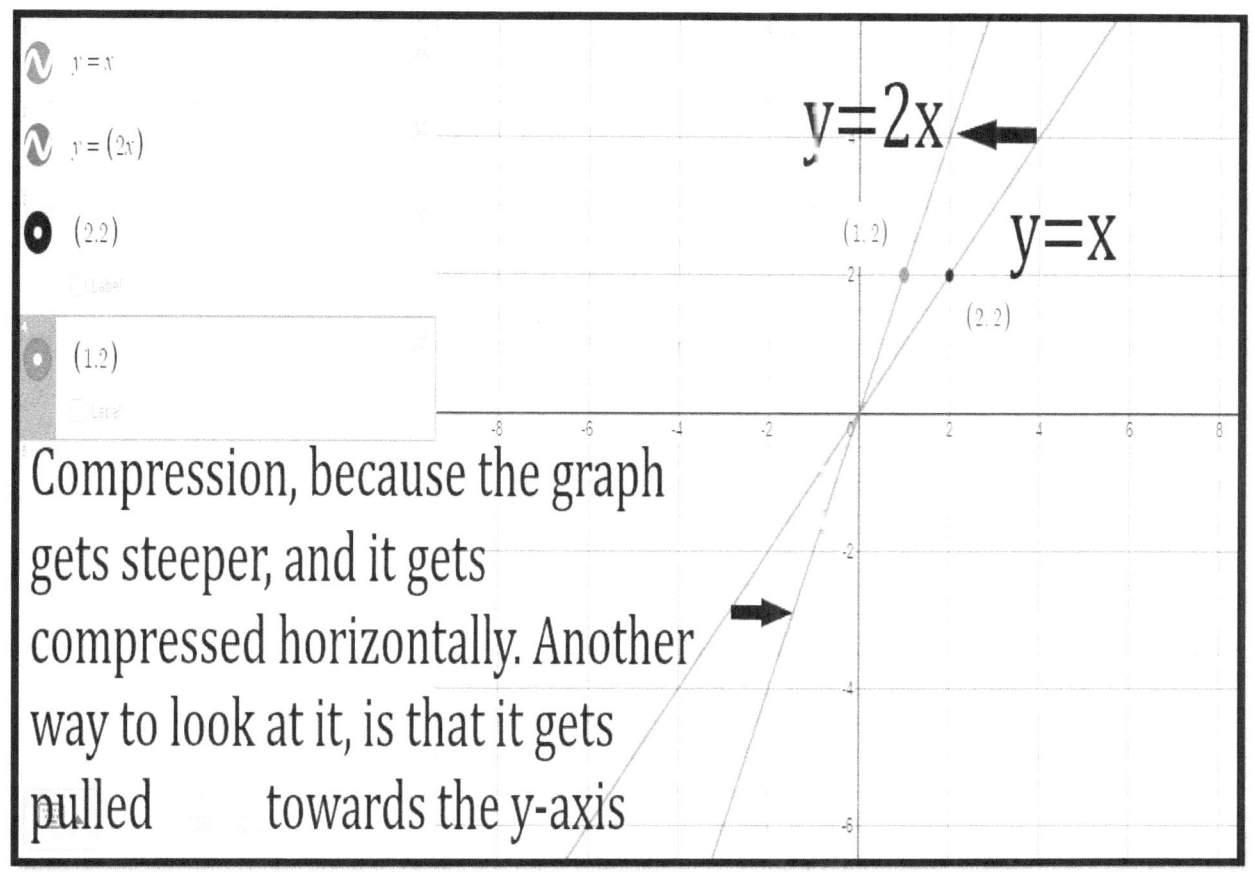

Combination of Transformations

In this section, we will put everything we have learned so far in this chapter to use. We will look an example of a graph, and then after that you will learn about inverse functions. Transformations come in the form of

$y = af(b(x - c)) + d$

a is the vertical expansion or compression,

b is the horizontal expansion or compression,

c is the horizontal translation (left or right)

d is the vertical translation (up or down)

One thing that is very important to take note of is when you are trying to determine the amount of horizontal translation, you must factor out the b-value. For example, $y = f(2x + 8)$ you must factor out the 2, giving $y = f(2(x + 4))$. This means that there is a horizontal translation of 4 units left, and a horizontal compression by a factor of ½. Be very careful with this.

Here are a few other examples:

$y = 2f(\frac{1}{3}x - 3)$ factor out 1/3 giving, $y = 2f(\frac{1}{3}(x - 9))$ which means a horizontal translation of 9 units right and a horizontal expansion by a factor of 3.

$y = f(5x + 10)$ factor out the 5 giving, $y = f(5(x + 2))$ which means a horizontal translation of 2 units left and a horizontal compression by a factor of 1/5.

Always make sure that when you are trying to determine a horizontal translation, that your final answer is in the form $y = f(b(x - c))$ and not in the form $y = f(bx - bc)$ If you forget to factor, your final answer will be wrong.

We will now look at some questions you may face on an assignment or test.

Example 1: Consider $f(x)$, Replace x with $3x$ then replace x with $x-2$. Describe the transformations.

Step 1: Replace x with 3x,

$$y = f(3x)$$

Step 2: Replace x with x-2,

$$y = f(3(x-2))$$

The graph is horizontally compressed by a factor of 1/3, and there is a horizontal translation of 2 units right.

When you combine transformations, it is very important that you apply them in a certain order to your graph.

Here is the order you should follow:

1. Expansion or Compression
2. Reflection

3. Vertical or Horizontal Translation

Example 1: $y = -2(x-1)^2$:

- Sketch its parent function first, $y = x^2$

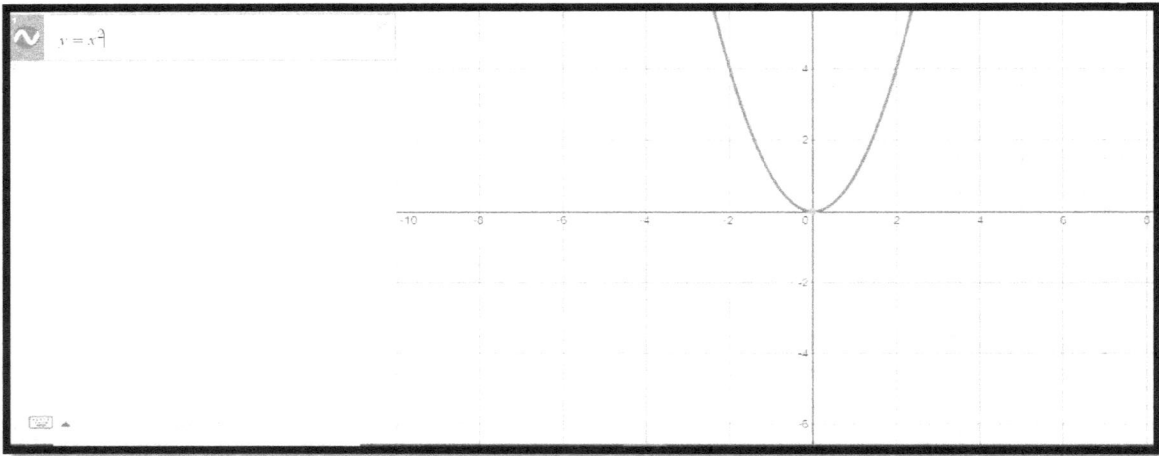

- Expand it vertically by a factor of 2, $y = 2x^2$

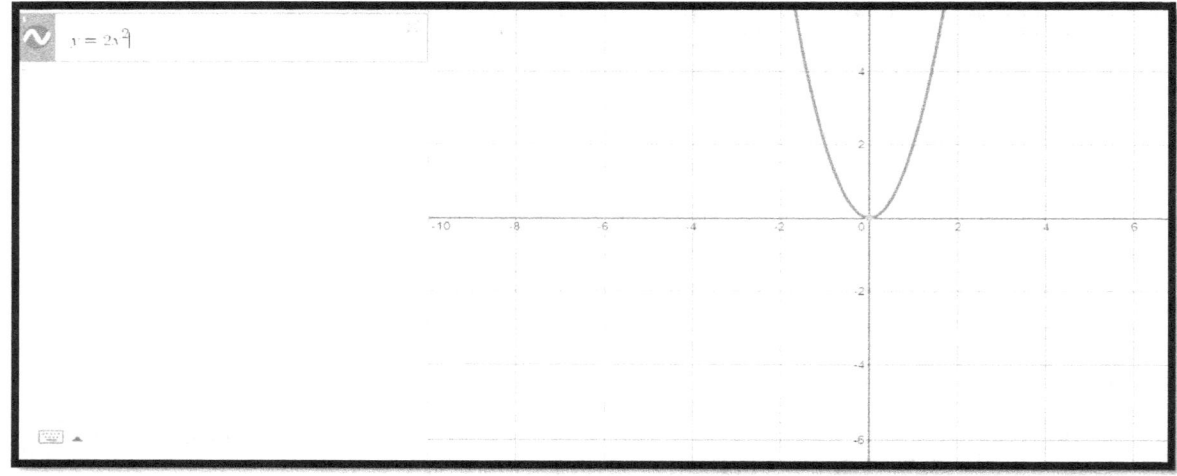

- Reflect it in the x-axis, $y = -2x^2$

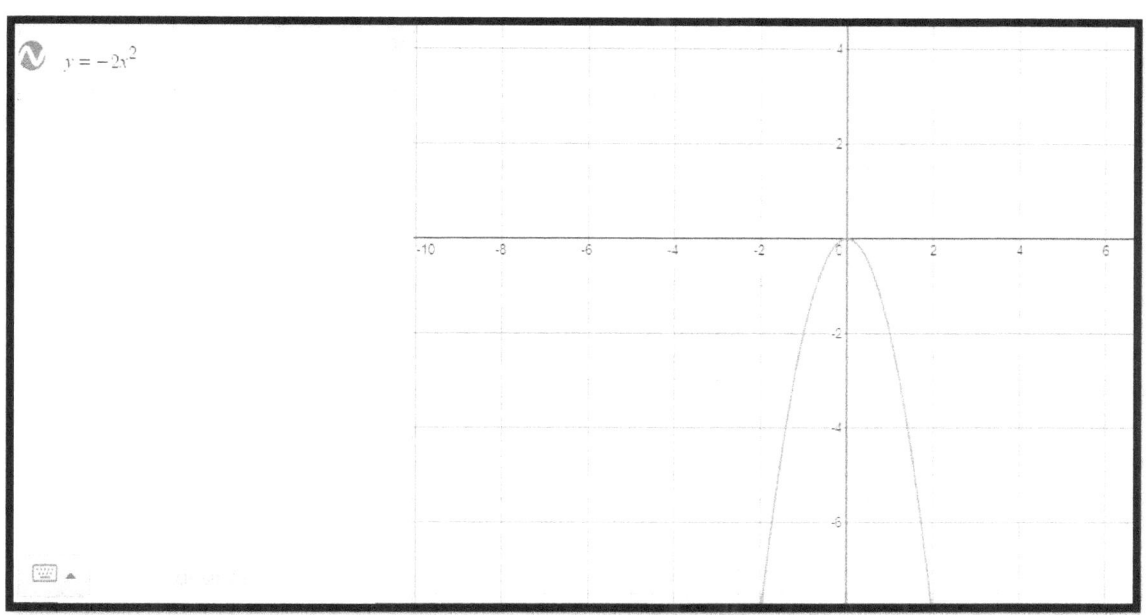

- Translate the graph 1 unit right, $y = -2(x-1)^2$

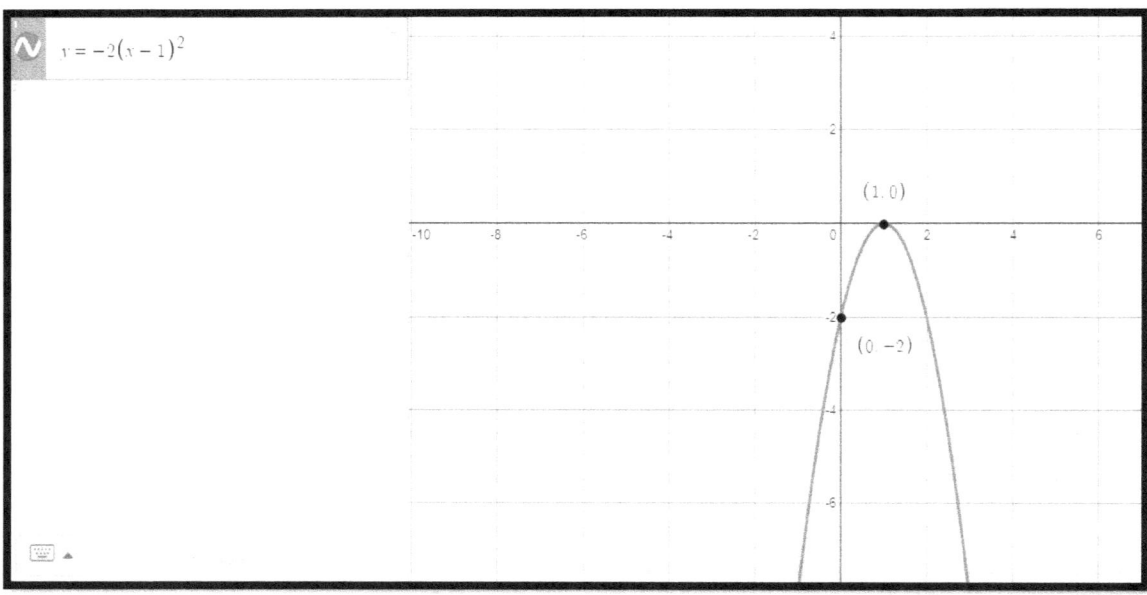

Inverse Functions

We will now compare $y = f(x)$ to $x = f(y)$,

as you can see, the x and y have been switched. This is known as an inverse function.

Note: $y = f^{-1}(x)$ is sometimes used to represent the equation of an inverse. The f^{-1} is not a negative exponent, it simply means inverse.

The x-values become the y-values and the y-values become the x-values. Another interesting property to not is that the two functions (the original and the inverse) will be a perfect reflection on the line $y = x$. If given an equation, the first step in finding the inverse is to switch the x with the y, and the y with the x.

Example 1: f(x) is $y = 5x$ find the equation of $f^{-1}(x)$.

- Step 1: Swap x and y
$$x = 5y$$
- Step 2: Solve for y
$$\frac{5y}{5} = \frac{x}{5}$$
$$y = \frac{1}{5}x$$

Example 2: $f(x) = x^2 - 3$ sketch the inverse.
- Step 1: Swap x and y.
$$x = y^2 - 3$$

- Step 2: Solve for y
$$x + 3 = y^2$$
$$y^2 = x + 3$$
$$y = \pm \sqrt{x + 3}$$

- Step 3: Graph it

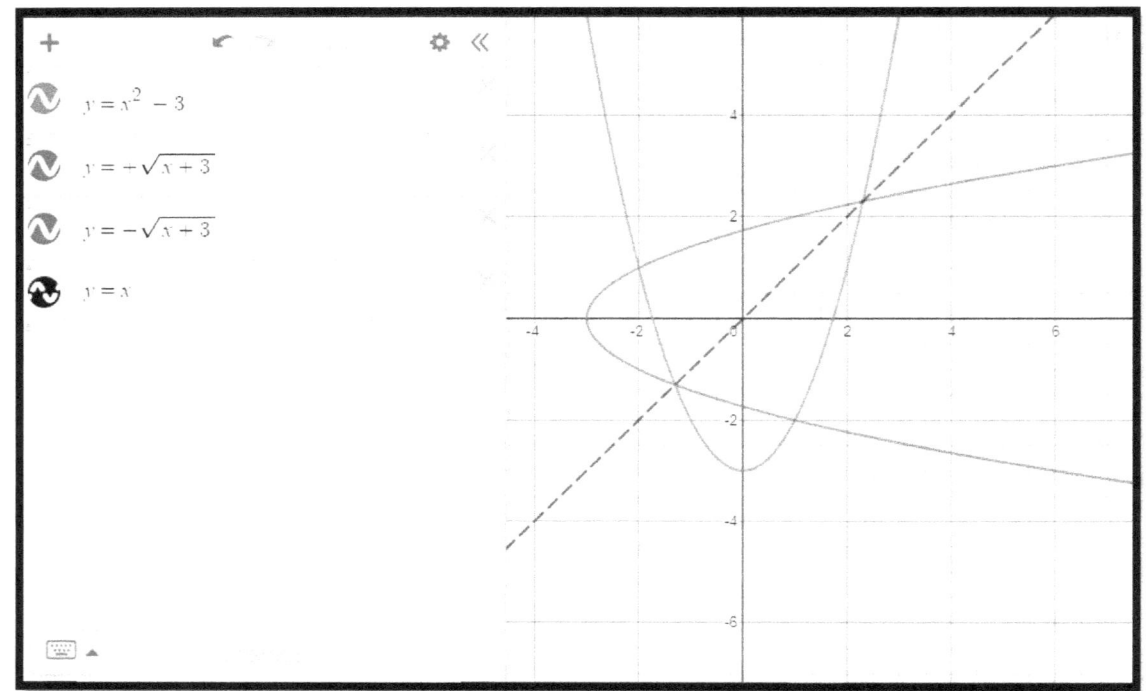

Example 3: Sketch the inverse of $f(x) = 3x$

- Step 1: Swap x and y

  $x = 3y$

- Step 2: Solve for y

  $3y = x$

  $\dfrac{3y}{3} = \dfrac{x}{3}$

$$y = \frac{1}{3}x$$

- **Step 3: Graph it**

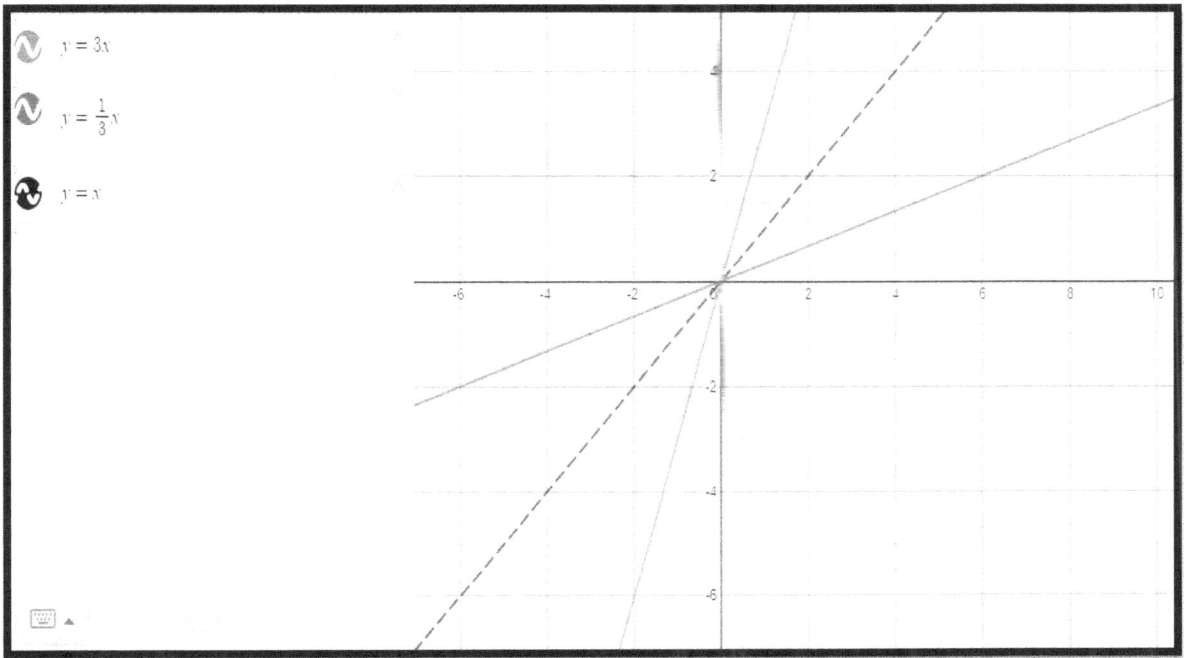

If you are given a graph without the equation, make note of a few base points, and swap the x and y values of these base points, and graph this inverse function.

For example,
$f(x) = $ (2,3) (4,5) (6,3)
$f^{-1}(x) = $ (3,2) (5,4) (3,6)

This is the end of Chapter 7; I hope this chapter made sense to you. It may seem difficult at first, but with practice it will become easier.

Practice Questions

1. Consider $f(x)$ determine the transformations that have been applied to the following:
   a) $3f(6x - 12)$
   b) $-f(x + 3)$
   c) $f(-7x) - 1$
   d) $3f\left(\frac{1}{3}x\right) + 2$

2. Sketch the graph of $y = 2|x - 3| + 1$ using transformations

3. $y = \left(\frac{1}{2}x\right)^2 - 2$

4. Consider the graph of $f(x)$ (the graph below)

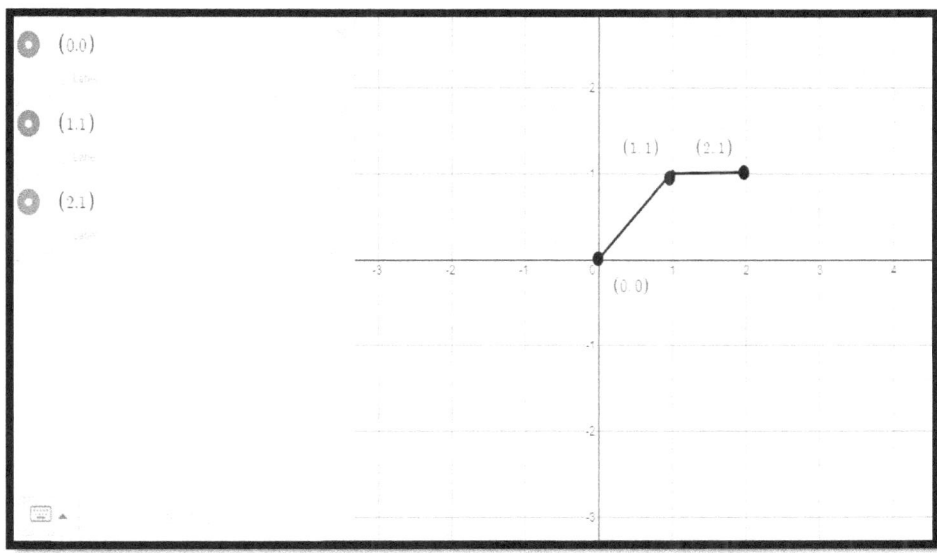

a) Sketch $y = 2f(x)$

b) Sketch $y = f^{-1}(x)$

Radical & Rational Functions

Chapter 8

What is a radical function? A radical function is a function that has a radical with a variable in the radicand. For example, $f(x) = \sqrt{x-2}$

The simplest radical function is $y = \sqrt{x}$

The domain of this function is $x \geq 0$ since you cannot take the square root of a negative number.

The range of this function is $y \geq 0$ since it does not make sense to have a negative output in this function.

This is the most basic graph of a square root function:

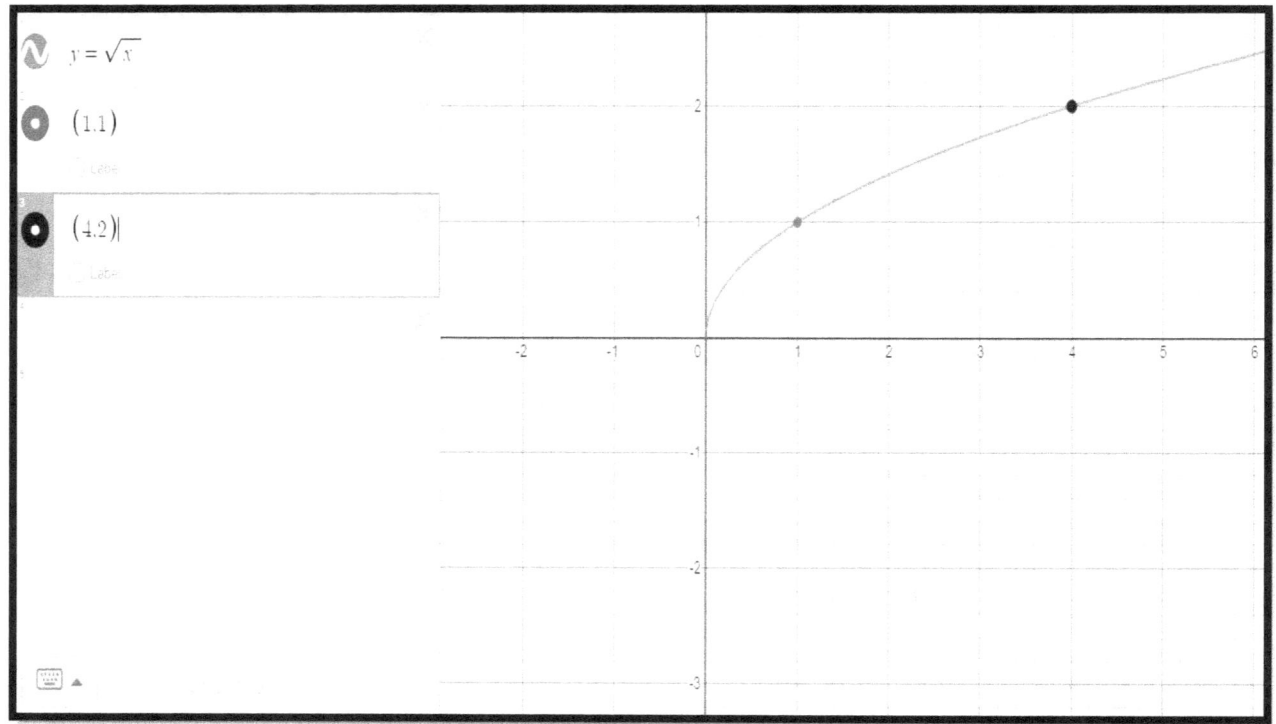

Domain and Range

To determine the domain, we are only concerned with what is inside of the radicand. For example, in the radical function $y = 3\sqrt{x+2} + 1$ we are only concerned with the $\sqrt{x+2}$ portion.

$x + 2 \geq 0$

$x \geq -2$

What is the range? We need to use the fact that the smallest possible input in this function is -2, and we need to plug in this number to find the smallest possible output.

$y \geq smallest\ possible\ output$

And of course, if the graph is reflected in the x-axis then,

$y \leq Largest\ possible\ output$

Range of $y = 3\sqrt{x+2} + 1$,

$y = 3\sqrt{-2+2} + 1$

$y = 3(0) + 1$

$y = 1$

$y \geq 1$

What about the domain and range of $y = -\sqrt{-x+1} + 6$

Domain:

Reflected in y-axis,

$-x + 1 \geq 0$

$-x \geq -1$

$x \leq 1$

Range:

Reflected in x-axis; therefore,

Largest possible input is 1,

$y \leq largest\ possible\ output$

$y \leq 6$

Find the domain and range of the following radical functions, and verify your answers on the next page:

a) $y = \sqrt{x + 3} - 2$
b) $y = -\sqrt{x}$
c) $y = \sqrt{-x}$

Answers:

a) Domain: $x \geq -3$ Range: $y \geq -2$
b) Domain: $x \geq 0$ Range: $y \leq 0$
c) Domain: $x \leq 0$ Range: $y \geq 0$

Transformations of Radical Functions

Transformations of radical functions occur in the form of:

$y = a\sqrt{b(x - c)} + d$, $y = a\sqrt{(x - c)} + d$, or $y = \sqrt{b(x - c)} + d$ where,

- a is the vertical expansion or compression,
- b is the horizontal expansion or compression,
- c is the horizontal translation (left or right),

- d is the vertical translation (up or down)

Example 1: Write the equation of a radical function after a horizontal expansion by 5, a horizontal translation of 7 units left, a reflection in the x-axis, and a vertical translation of 4 units up.

$a = -1, \quad b = \dfrac{1}{5}, \quad c = -7, \quad d = 4$

$y = -\sqrt{\dfrac{1}{5}(x+7)} + 4$

Example 2: Write the equation of a radical function after a vertical expansion by 3, reflection in the y-axis, horizontal compression by 1/7, and a horizontal translation of 2 units right.

$a = 3, \quad b = -7, \quad c = 2, \quad d = 0$

$y = 3\sqrt{-7(x-2)}$

You will also sometimes be asked to identify transformations. For example, $y = 2\sqrt{4-x}$

There is a vertical expansion by 2, a reflection in the y-axis, and a horizontal translation of 4 units right. What is the domain? The domain is,

$4 - x \geq 0$

$-x \geq -4$

$x \leq 4$

Example 3: Find the equation of a radical function with a starting point of (-3,7) and passes through (-2,2).

The starting point of a basic square root point is (0,0), therefore, we can assume there was a horizontal translation of 3 units left, and a vertical translation of 7 units up. We can plug this into the equation,
$y = a\sqrt{(x+3)} + 7$.

Now we can plug the point (-2, -2) and solve for a to determine if there is a vertical expansion or compression.

$-2 = a\sqrt{-2+3} + 7$

$-2 - 7 = a\sqrt{1} + 7 - 7$

$-9 = a$

This means that there is a vertical expansion by 9, and a reflection in the x-axis.

$y = -9\sqrt{x+3} + 7$

Example 4: Find the equation of a radical function with a starting point of (5,-1) and passes through (21,2).

Step 1: Write the equation.

The function was translated 5 units to the right and was translated 1 unit down.

$$y = a\sqrt{x-5} - 1$$

Step 2: Plug the second point into the equation and solve for a.

$$2 = a\sqrt{21-5} - 1$$
$$2 + 1 = a\sqrt{16} - 1 + 1$$
$$3 = a(4)$$
$$\frac{4a}{4} = \frac{3}{4}$$
$$a = \frac{3}{4}$$

Final equation: $y = \frac{3}{4}\sqrt{x-5} - 1$

When you sketch transformations of radical functions, you follow the same order as for any other function transformation:

1. Expansion or Compression
2. Reflection
3. Vertical or Horizontal Translation

Solving Radical Equations

You will now learn how to solve radical equations for "roots" (x-intercepts). Something important to keep in mind is that sometimes you may get more than one solution, and sometimes only one of those solutions actually work, these are known as **Extraneous Roots**. These are roots that you find from solving the equation; however, they do not actually work when you test out the root.

Example 1: Solve $\sqrt{x+9} = 7$

$(\sqrt{x+9})^2 = (7)^2$

$x + 9 = 49$

$x = 49 - 9$

$x = 40$ *this is called the "root" of the equation*

Example 2: Solve $\sqrt{x-4} + 4 = 0$

$\sqrt{x-4} + 4 - 4 = 0 - 4$

$\sqrt{x-4} = -4$

$(\sqrt{x-4})^2 = (-4)^2$

$x - 4 = 16$

$x = 16 + 4$

$x = 20$

Don't celebrate yet! We need to test this solution to see if it actually works.

$\sqrt{20-4}+4=0$
$\sqrt{16}+4=0$
$4+4=0$
$8=0$ This is not a solution, 8 does not equal 0; therefore, $x=20$ is an extraneous root and this equation does not have a real solution.

Example 3: $\sqrt{x-5}+7=x$
$\sqrt{x-5}=x-7$
$(\sqrt{x-5})^2=(x-7)^2$
$x-5=(x-7)(x-7)$
$x-5=x^2-7x-7x+49$
$x-5=x^2-14x+49$ Set the left side of the equation to 0, and solve for x.
$0=x^2-14x+49+5-x$
$0=x^2-15x+54$ Which two numbers multiply to 54 and add up to -15. -6 and -9.
$0=(x-6)(x-9)$
$x=6, x=9$

Now we need to test out both solutions to see if they both work.

<u>Verification of Solutions</u>

$x=6$ $\qquad\qquad$ $x=9$

$$\sqrt{6-5}+7=6 \qquad \sqrt{9-5}+7=9$$
$$\sqrt{1}+7=6 \qquad \sqrt{4}+7=9$$
$$8 \neq 6 \qquad 2+7=9$$
$$9=9$$

$x=6$ is an extraneous solution

$x=9$ is the only real solution

Solving Radical Equations Graphically

There are two ways of solving Radical equations graphically.

Method 1: enter the left and right side of the equation as two different functions, and then look for points of intersection.

Method 2: Write the equation as a single function, and look for "zeros" where the graph crosses the x-axis.

Example 1: $\sqrt{3x^2-4}=x+5$

Method 1 (Two Separate Functions on Graphing Calculator):
$$y_1 = \sqrt{3x^2 - 4}$$
$$y_2 = x + 5$$

Graph it on your graphing calculator. The instructions I will give you are based on the TI-83 Plus Graphing Calculator that I am familiar with.

Once you have graphed it and the graph is displayed on your screen, you will see that the graphs intersect twice. To find the x-coordinate of the intersection, click the 2nd function key on your calculator, and then press the "TRACE" button. A menu named "CALCULATE" will appear and will ask you to pick an option. You will pick option 5, "Intersect." The first thing it will say is "First Curve?" pick a point near the intersection, and press "Enter". After that it will say, "Second Curve?" again you will pick a point near that same point of intersection, and click "Enter". It will then say, "Guess?", click "Enter" once again and it will find the x-coordinate of the solution. Repeat the same process for the other point of intersection. These are the two solutions you should get:
$x = 7.06 \quad x = -2.06$

Method 2 (One Function on Graphing Calculator):

$y_1 = \sqrt{3x^2 - 4} - x - 5$

Graph it on your graphing calculator. The instructions I will give you are based on the TI-83 Plus Graphing Calculator that I am familiar with.

Once you have graphed it and the graph is displayed on your screen, you will see that there are two x-intercepts. To find the roots click the 2nd function key on your calculator, and then press the "TRACE" button. A menu named "CALCULATE" will appear and will ask you to pick an option. You will pick option 2, "Zero" The first thing it will say is "Left Bound?" pick a point that is to the left of the first x-intercept, and press "Enter". After that it will say, "Right Bound?" Pick a point that is right of the same x-intercept and click "Enter". It will then say, "Guess?", click "Enter" once again and it will find the root. Repeat the same process for the other root. These are the two solutions you should get:

$x = 7.06 \quad x = -2.06$

Method 3 (Solving $\sqrt{3x^2 - 4} = x + 5$ Algebraically):

$\sqrt{3x^2 - 4} = x + 5$

$(\sqrt{3x^2 - 4})^2 = (x + 5)^2$

$3x^2 - 4 = x^2 + 5x + 5x + 25$

$0 = x^2 + 10x + 25 + 4 - 3x^2$

$0 = -2x^2 + 10x + 29$ It doesn't look like we can factor, what about when we divide the equation by -1, to remove the negative from the x^2 term?

$0 = 2x^2 - 10x - 29$ We can not factor, do we stop here? NO! You can use your favorite formula of all time to solve this equation, the QUADRATIC FORMULA!

$$x = \frac{-b \pm \sqrt{b^2 - 4ac}}{2a}$$

$$x = \frac{-(-10) \pm \sqrt{(-10)^2 - 4(2)(-29)}}{2(2)}$$

$$x = \frac{10 \pm \sqrt{100 - 8(-29)}}{4}$$

$$x = \frac{10 \pm \sqrt{100 + 232}}{4}$$

$$x = \frac{10 \pm \sqrt{332}}{4}$$
We need to simplify this radical,

$$x = \frac{10 \pm \sqrt{(2)(2)(83)}}{4}$$

$$x = \frac{10 \pm 2\sqrt{83}}{4}$$ We can actually simplify this even further by noticing that there is a common factor in both the numerator and the denominator,

$$x = \frac{2(5 \pm \sqrt{83})}{2(2)}$$

$$x = \frac{5 \pm \sqrt{83}}{2}$$

$$x = \frac{5 + \sqrt{83}}{2} \qquad x = \frac{5 - \sqrt{83}}{2}$$

$$x = 7.06 \qquad x = -2.06$$

Rational Functions

As you probably know, a rational function is a function with a variable in the denominator, and sometimes with a variable in the numerator as well.

As mentioned in Chapter 12, a vertical asymptote results from a restriction in the denominator, creating a restriction in the domain. For example, in the function

$$y = \frac{1}{x+2},$$

$x + 2 \neq 0$

$x \neq -2$

This would mean that there is a vertical asymptote at $x = -2$.

Four important points to mention about a vertical asymptote is that:
- The graph will never cross a vertical asymptote
- A vertical asymptote will always come from the restriction on the domain
- Vertical asymptotes stand for x-values not allowed.
- When a graph approaches a vertical asymptote from the right or left, the y-values go to positive or negative infinity.

Here is a table of values for the function $y = \frac{1}{x}$.

x	y

We will consider the behaviour of the y-values as x approaches 0 from the right.

x	y
1	1
0.1	10
0.01	100
0.001	1000
0.0001	10000

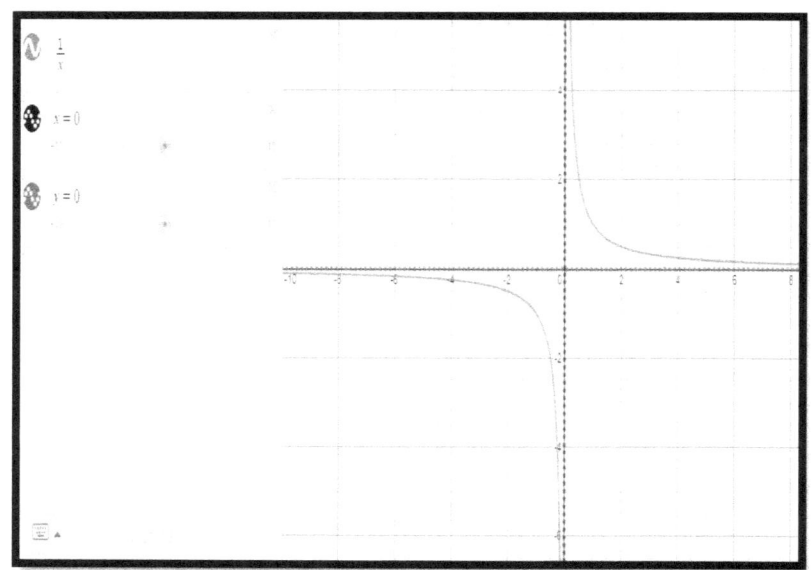

As x gets infinitely small, what happens to the y-values? The y-values get infinitely large.

There is also a horizontal asymptote. A horizontal asymptote is a non-permissible y-value that results from restrictions on the domain.

10	0.1
10000	0.0001
100000	0.00001
1000000000	0.00000001

x	y
-10	-0.1
-10000	-0.0001
-100000	-0.00001
-1000000000	-0.00000001

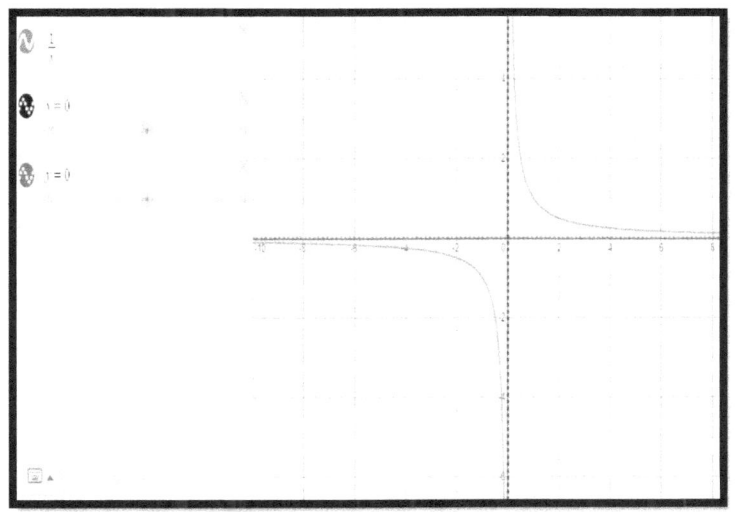

As x gets infinitely large, y approaches what number? 0

Will y ever reach zero? NO!

Equation of the Horizontal Asymptote: $y = 0$

Horizontal asymptotes can be identified when x is really, really big positive and negative values.

Unlike a vertical asymptote, a Horizontal asymptote can sometimes be crossed by the graph, not in this case, however. You will now see the four different kinds of rational functions you will come across in this course.

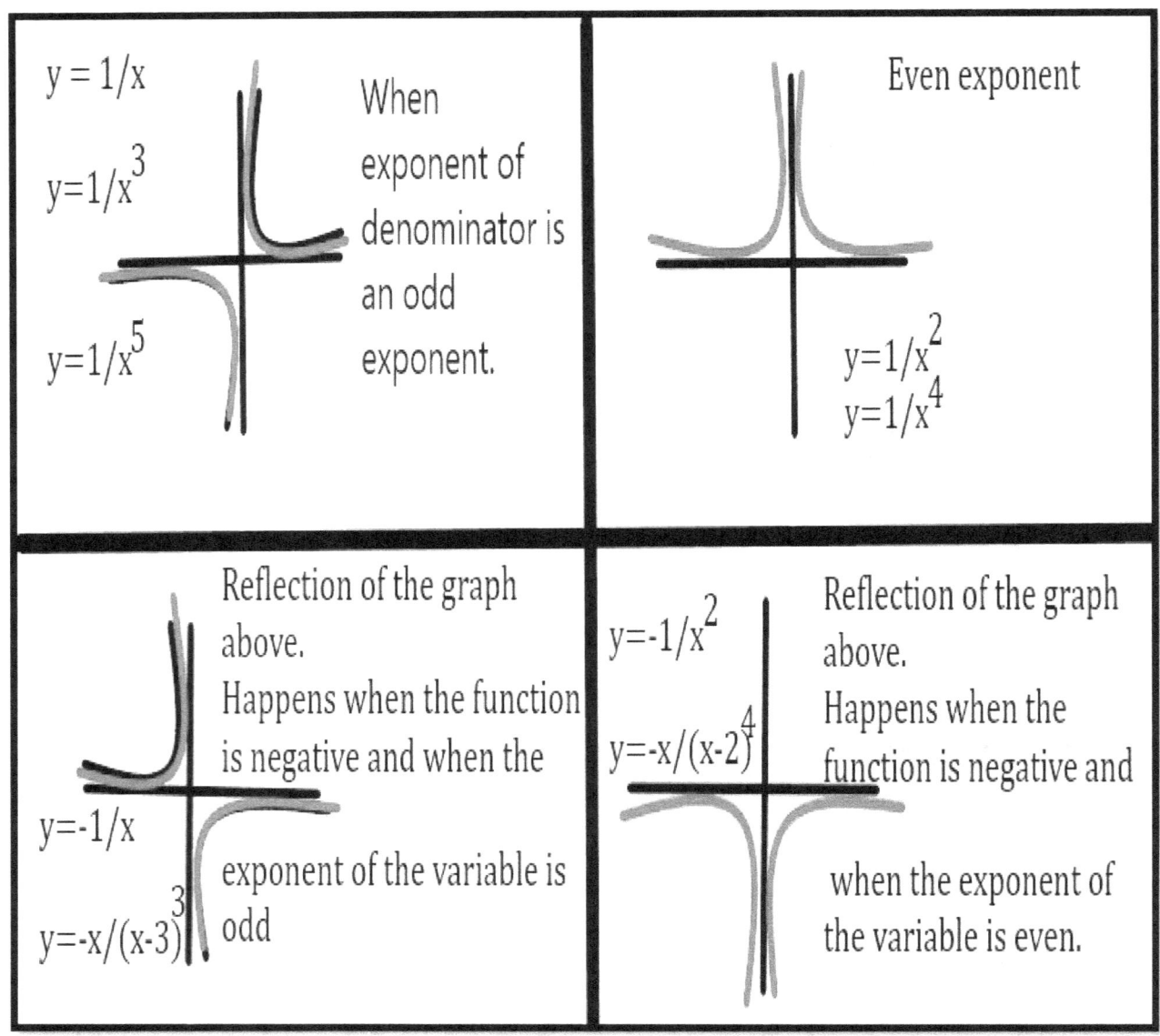

As you can see, it is the exponent of the variable that plays an important role in the shape of this graph, and so does the sign (positive or negative).

Horizontal Asymptotes

In this section, we will establish a general rule for finding horizontal asymptotes.

In the equation $y = \dfrac{5}{x}$ the horizontal asymptote is 0. The reason for this is because there will never be an output of 0 for the function.

What about finding the horizontal asymptote of $y = \dfrac{x+1}{x^2+2}$?

The +1, and the +2 are insignificant, with that being said, we are only concerned with the variables in the numerator and the denominator. Since the exponent of the highest degree term in the denominator is larger than the highest degree term in the numerator, then the horizontal asymptote is $y = 0$. Another important thing to keep in mind is if that you are adding or subtracting outside of the rational expression, such as, $\dfrac{x-1}{x^2+2} + 3$ then you basically say that the horizontal asymptote is $y = 0 + 3$

$y = 3$ is the equation of the horizontal asymptote. So, when the exponent of the highest degree term in the denominator is larger than the highest degree term in the numerator, and you are adding or subtracting something outside of the fraction, then the horizontal asymptote is $y = 0 + d$, where d is the number you are adding or subtracting after the division occurs.

Example 1: $$y = \frac{2x - x}{3x^2 - 4x + 3} - 5$$

The horizontal asymptote is $y = 0 + d$

$d = -5$; therefore, $y = 0 + (-5)$

$y = -5$ is the horizontal asymptote.

When the exponent of the highest degree term in the numerator and the denominator are the same then the horizontal asymptote is the ratio of the coefficients of the leading terms. For example, $$y = \frac{-7x^3 - x}{2x^3 + 3x^2 - 4}$$

$$y = -\frac{7x^3}{2x^3}$$

$y = -\frac{7}{2}$ is the equation of the horizontal asymptote

Example 2: $\dfrac{2x^2 - 9}{x^2 + 3}$

To find the horizontal asymptote, we must find the ratio of the coefficients of the leading term in both the numerator and the denominator.

$$y = \dfrac{2x^2}{x^2}$$

$$y = \dfrac{2}{1} = 2$$

$y = 2$ is the equation of the horizontal asymptote.

Transformations of Rational Functions

Rational functions come in the form $y = \dfrac{a}{x - h} + k$

Where h is the horizontal translation, and k is the vertical translation.

In this form:

- Horizontal Asymptote $y = k$
- Vertical Asymptote $x - h = 0$
- a is the vertical expansion or compression.
- It is also easier to graph rational functions in this form!

You will see that most of the time equations are not in this form. But don't worry, you can re-write equations in this form. Here are a few examples:

Example 1: $y = \dfrac{3x - 7}{x - 4}$

$y = \dfrac{3x - ? + ? - 7}{x - 4}$ We must ignore the -7 for the time being, and we must find a coefficient that we can put beside the $3x$ that will allow us to factor out a $x - 4$. We can find this number by multiplying 3 by the coefficient in the numerator which is -4. So, $3 * -4 = -12$. We can now add and subtract negative twelve inside of the numerator.

$y = \dfrac{(3x - 12) + 12 - 7}{x - 4}$

$$y = \frac{3(x-4)}{x-4} + \frac{12-7}{x-4}$$

$$y = 3 + \frac{5}{x-4}$$

$$y = \frac{5}{x-4} + 3$$

The graph is vertically expanded by a factor of 5, it is translated 4 units right, and it is translated 3 units up.

Horizontal asymptote: $y = k \qquad k = 3$

$y = 3$

Vertical asymptote: $x - h = 0 \qquad h = 4$

$x - 4 = 0$

$x = 4$

This is what the graph looks like:

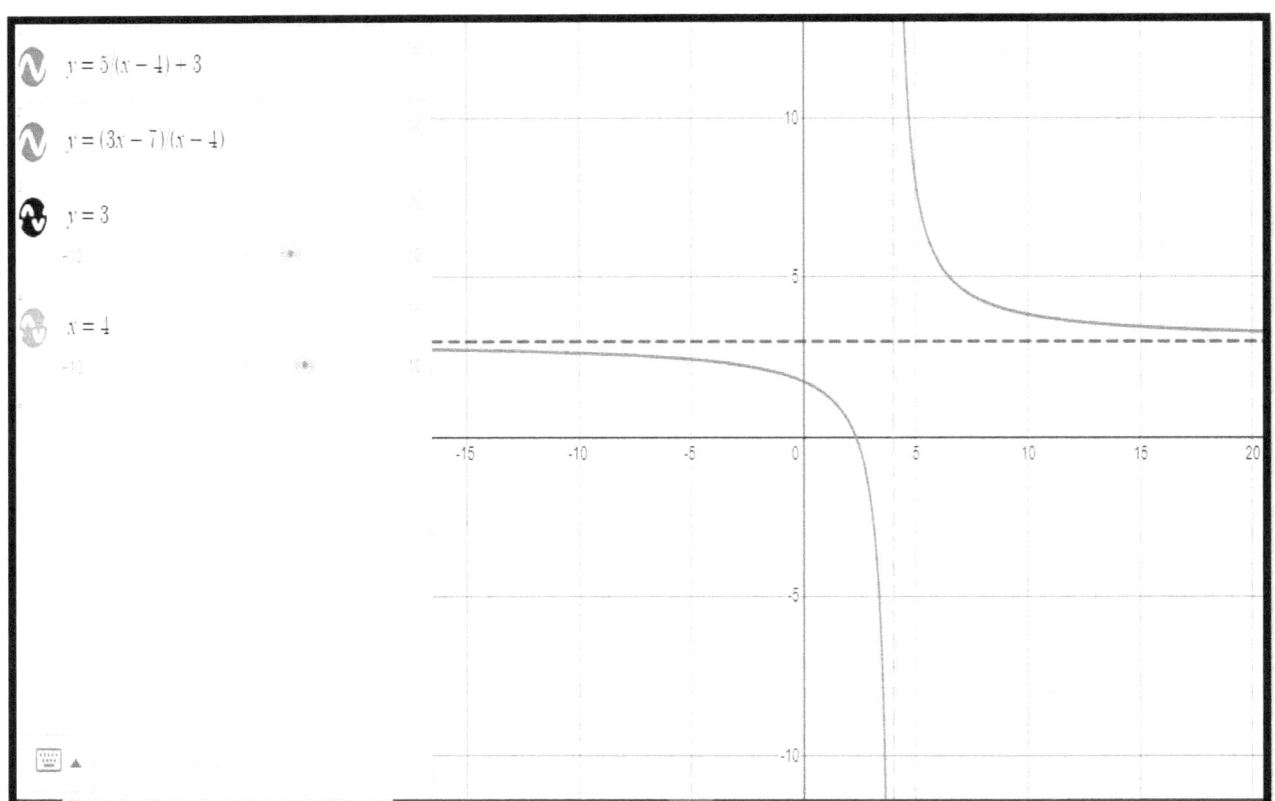

Example 2: $y = \dfrac{2x + 12}{x + 3}$

$y = \dfrac{2x + ? - ? + 12}{x + 3}$

2*3=6

$y = \dfrac{2x + 6 - 6 + 12}{x + 3}$

$$y = \frac{2(x+3)}{x+3} + \frac{12-6}{x+3}$$

$$y = 2 + \frac{6}{x+3}$$

$$y = \frac{6}{x+3} + 2$$

Horizontal asymptote: $y = 2$

Vertical asymptote: $x = -3$

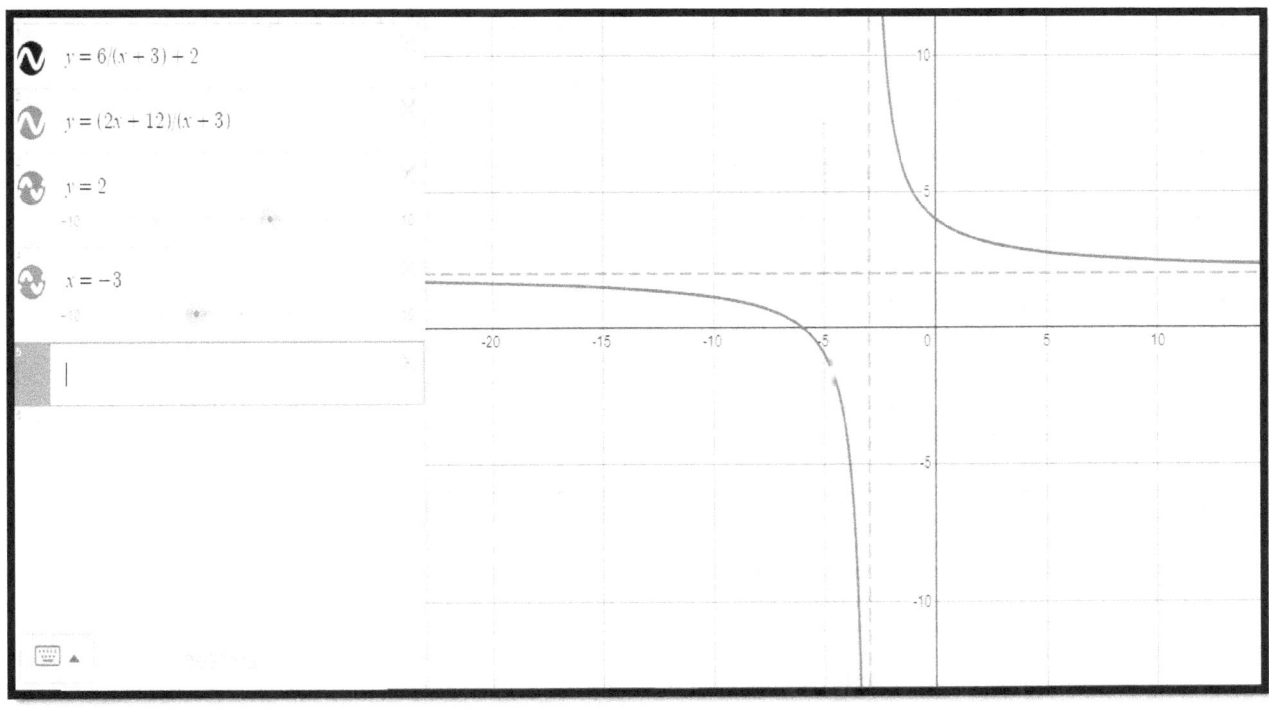

Point of Discontinuity

We will now look at a special case of a rational function where a "hole" occurs. The "hole" occurs when a factor with a variable cancel out of the numerator and the denominator.

For example, $f(x) = \dfrac{x^2 - x - 2}{x - 2}$

$f(x) = \dfrac{(x + 1)(x - 2)}{x - 2}$

$(x - 2)$ cancels out, leaving you with $f(x) = x + 1$ which as you know, is a Linear Function. Don't think because you cancelled out a factor, that all of your problems are solved and that you can just draw a simple linear function with no restrictions. That is not the case, **YOU CAN NEVER FIX A DOMAIN PROBLEM**!!! You can never make a domain better. What happens is you take the equation of the vertical asymptote and you plug it inside of the Linear function and it becomes the "Point of Discontinuity" which means a point that is not included in the domain, and it is not included in the range either.

Equation of Vertical Asymptote: $x = 2$

Restriction: $x \neq 2$

$f(2) = (2) + 1$

$f(2) = 3$

Point (2,3) is the point of discontinuity,

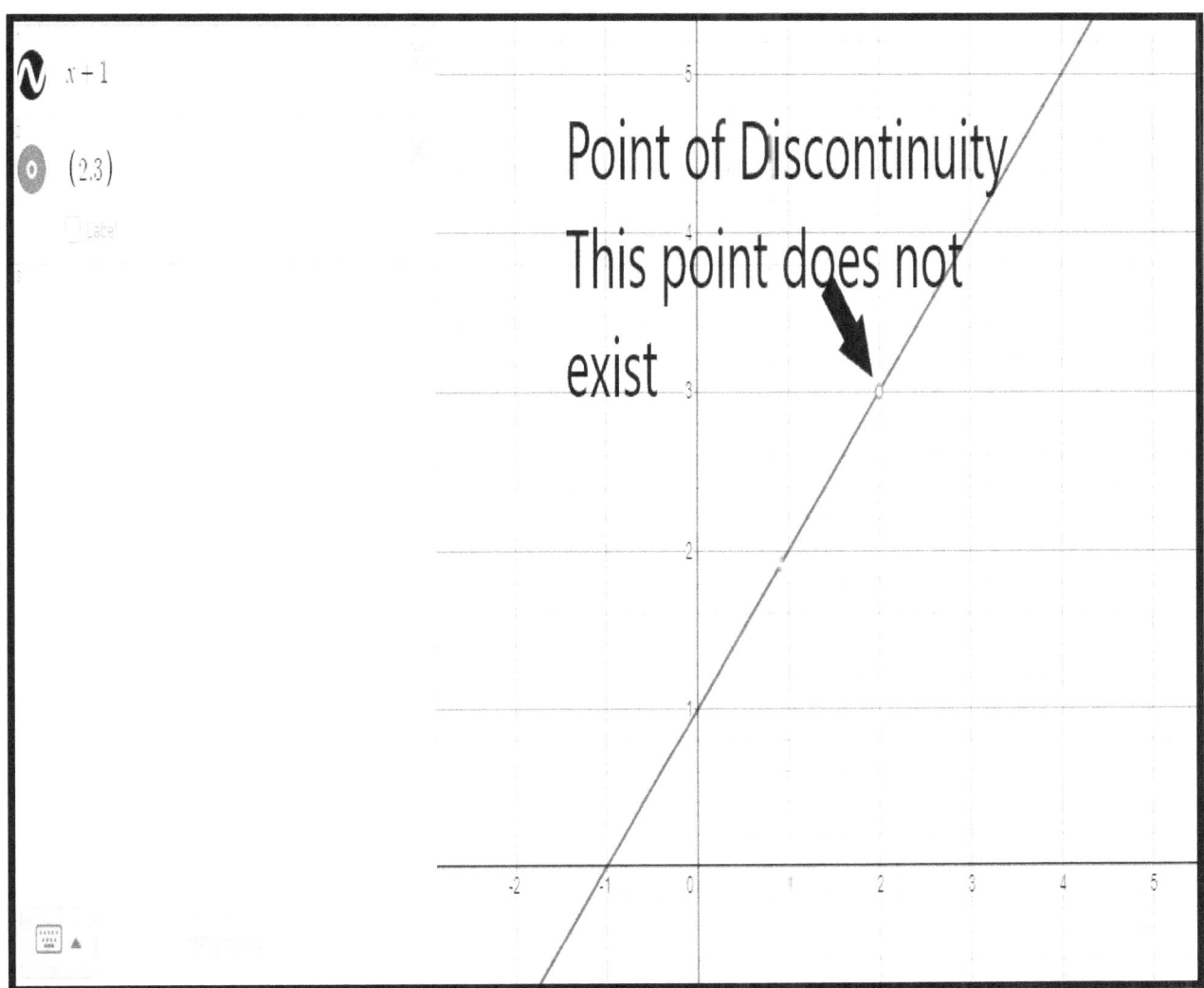

Writing Equations Given Characteristics

Sometimes you will be asked to write an equation based on some given characteristics. X-intercepts of rational functions are from factors found in the numerator.

Example 1: Write the equation of a rational function with a point of discontinuity at $\left(-\frac{1}{2}, -\frac{23}{3}\right)$, a vertical asymptote at $x = -2$ and an x-intercept of 11.

The y-coordinate of the point of discontinuity is very irrelevant when we are writing the equation of a rational function. Since the x-coordinate of the point of discontinuity is from a factor in both the numerator and denominator, then we must work backwards and find a factor when $x = -\frac{1}{2}$

$2 * x = -\frac{1}{2} * 2$

$2x = -1$

$2x + 1 = -1 + 1$

$(2x + 1) = 0$ this factor appears in both the numerator and denominator.

Vertical asymptote: $x = -2$, work backwards to find that this factor is $(x + 2)$. Since the vertical asymptote occurs in the denominator, we will put this in the denominator.

X-intercept is 11.

$x = 11$

$$(x - 11) = 0$$

Since the x-intercept occurs in the numerator, this factor can go in the numerator.

Final equation: $$y = \frac{(2x + 1)(x - 11)}{(2x + 1)(x + 2)}$$

OR

$$y = \frac{2x^2 - 21x - 11}{2x^2 + 5x + 2}$$ when expanded

Practice Questions

1. Solve $\sqrt{x - 4} - 4 = 0$

2. Solve $x = \sqrt{x-4} + 6$, identify any extraneous roots.

3. Solve $\sqrt{x+5} = x - 7$

4. Graph $y = 2\sqrt{-(x-2)} + 1$ using transformations

5. Determine the horizontal asymptote of the following:
   a) $y = \dfrac{2x^3 + 5}{5x^3 - 7}$
   b) $y = \dfrac{x+1}{x-2}$

6. Graph the following, if there is a point of discontinuity, show it with an open circle:
   a) $y = \dfrac{x^2 + x - 2}{x^2 - x - 6}$
   b) $y = \dfrac{2}{x-2} + 1$

Polynomials

Chapter 9

We will start by reviewing long division before moving on to long division with polynomials. If this is confusing to you, do not worry, it takes a bit of practice before anyone can be good at this!

When we do long division, our goal is to have two things at the end. One, the quotient (everything that can be fully divided). Two, the remainder (what is left).

Example 1: $935 \div 7$ Take a very close look at this example.

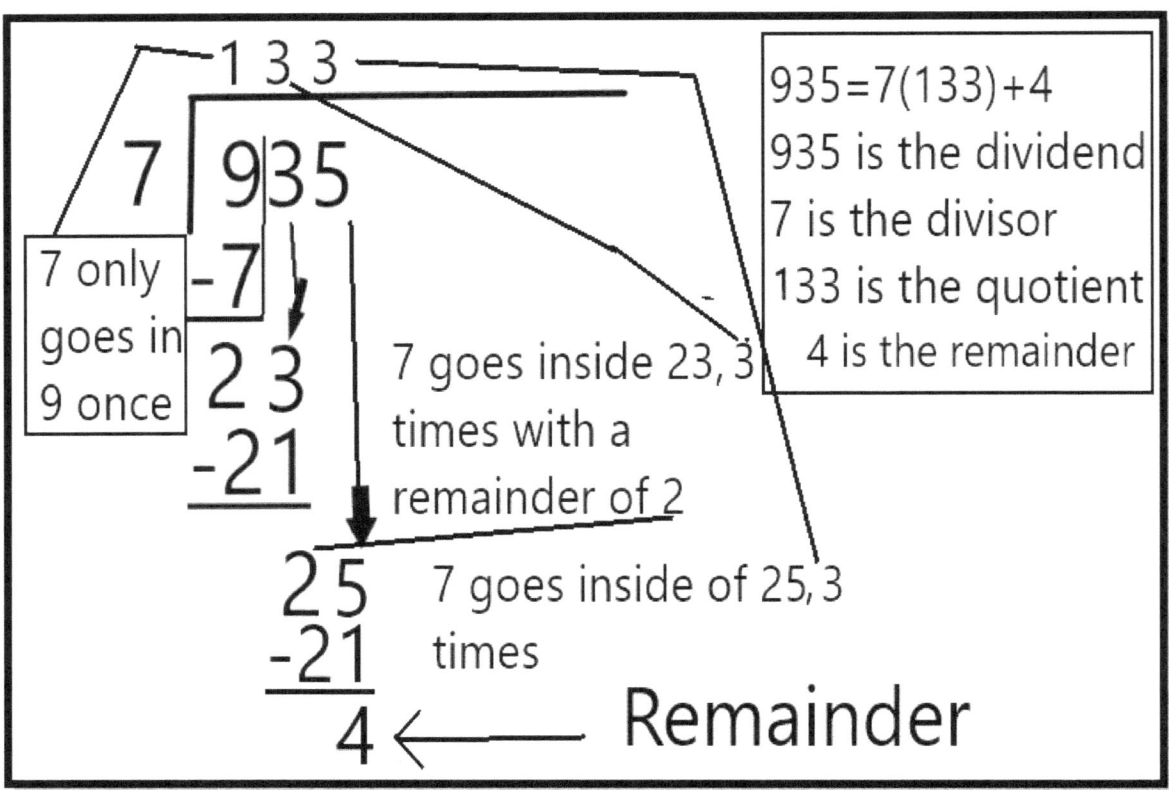

When we do long division with polynomials, we perform similar steps, but with a few twists.

Example 2: $(x^2 - x + 3) \div (x + 1)$

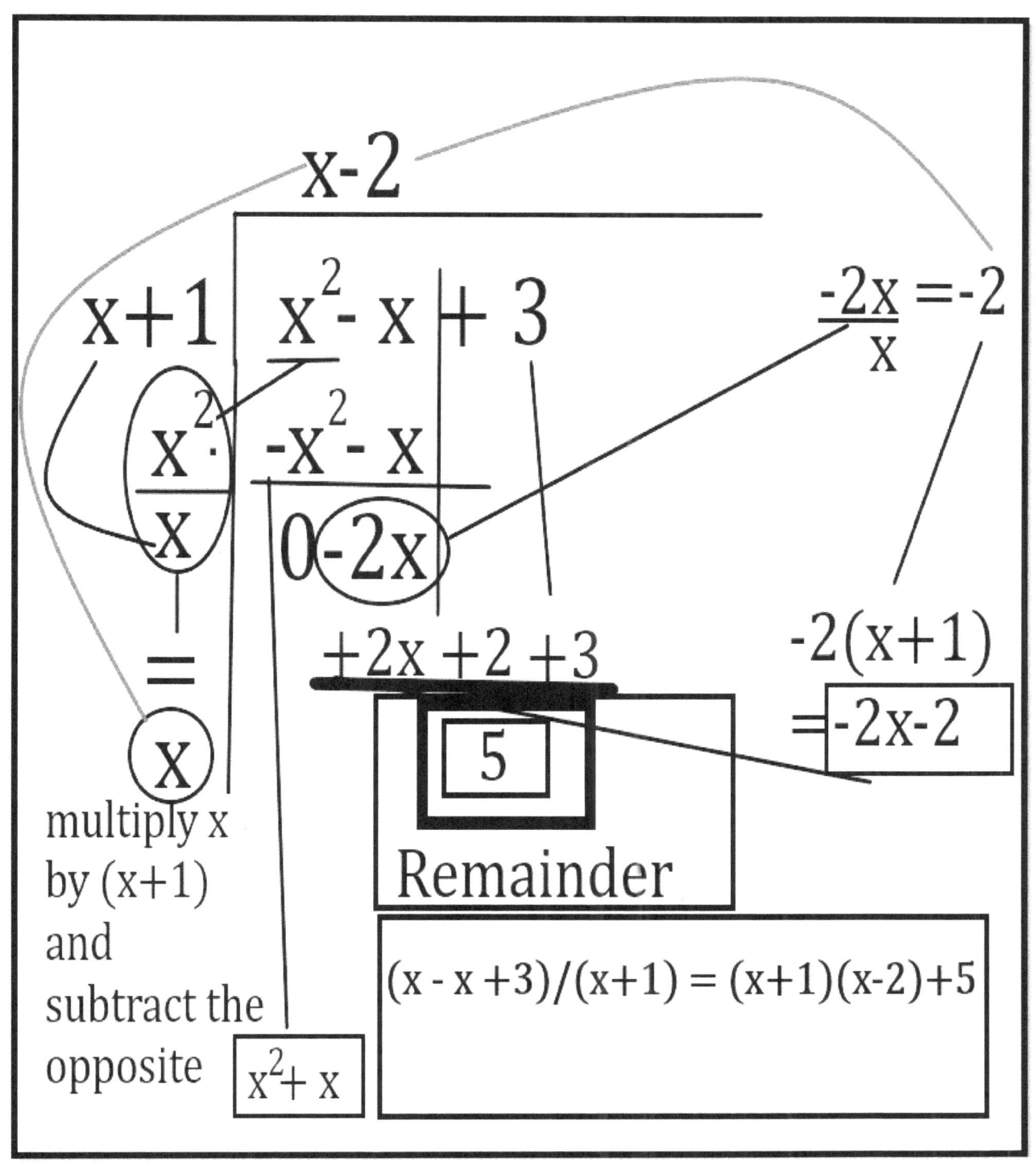

When $(x^2 - x + 3)$ is divided by $(x + 1)$ the quotient is $(x - 2)$ with a remainder of 5.

An easier method as I have mentioned earlier is Synthetic Division. The only drawback is that it only works when the degree of x is 1 and the coefficient in front of x is 1.

When you do synthetic division, you take the coefficients of the terms and you place them in the row inside of that half rectangle. After that you find the

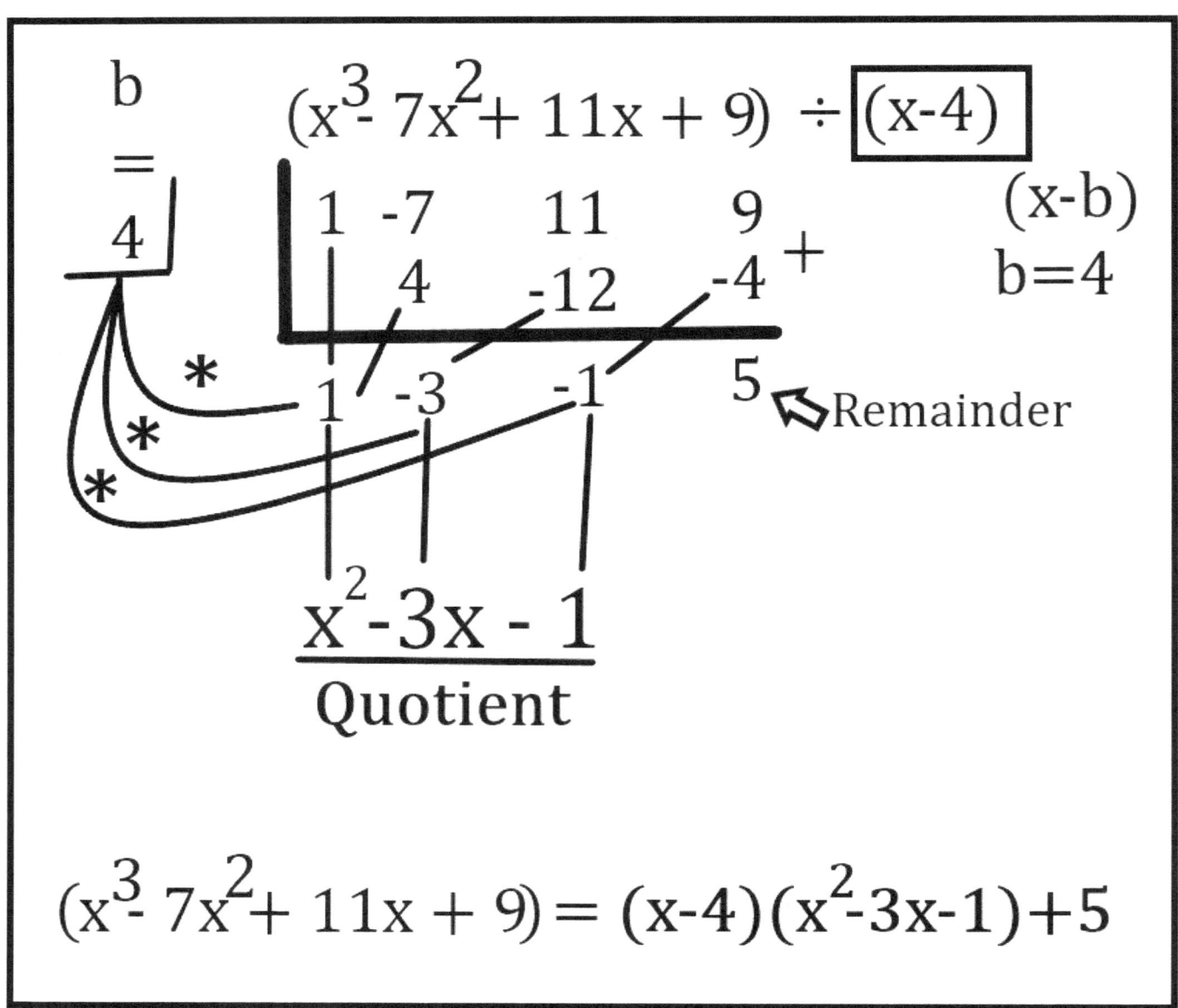

value of b in the expression $(x - b)$. You use this number to multiply the number that goes below the half rectangle and you carry it over to the next column. Re-examine the example in the page before, so you see what I mean by that. Each term in the quotient is always of degree 1 less than the term on top of it. The last number you get will be the remainder. You have to be careful when doing synthetic division, as the terms have to be in descending order from left to right. Here is another example:

The Remainder Theorem

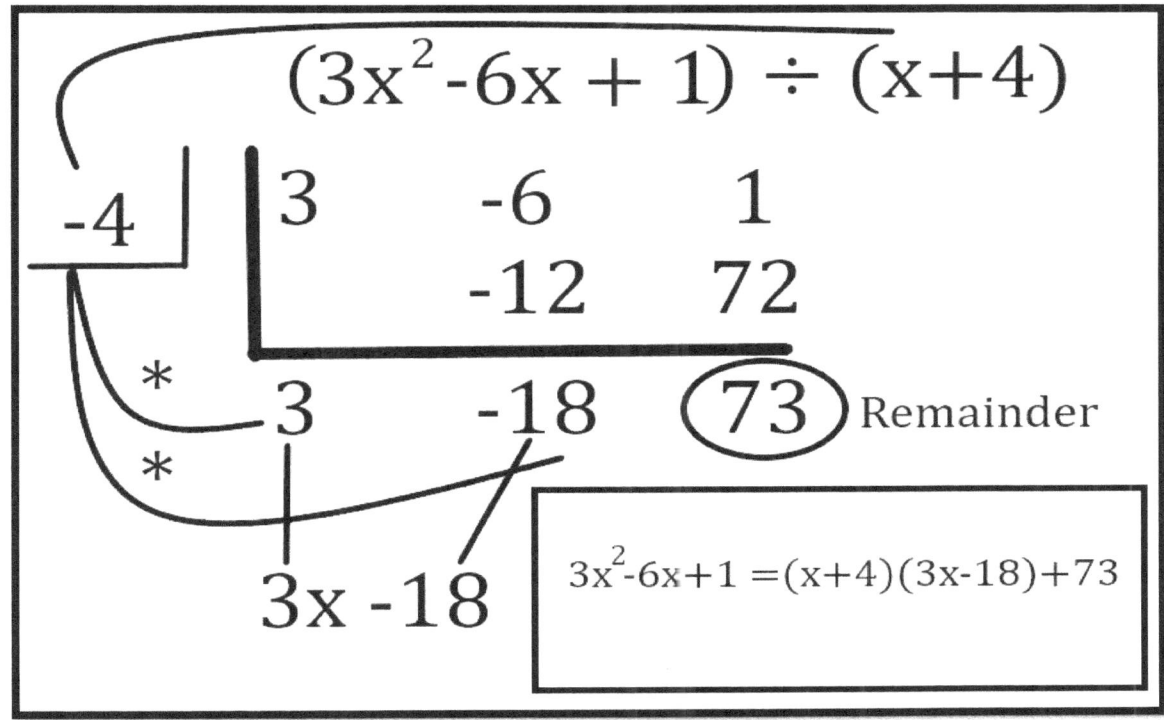

Polynomial $P(x)$	Divisor $(x-b)$	Quotient	Remainder (R)	$P(b)$
$x^2 - 5x + 9$	$x - 2$	$x - 3$	$R = 3$	$P(2) = 3$
$-3x^3 + x^2 - 2x -$	$x - 1$	$-3x^2 - 2x - 4$	$R = -9$	$P(1) = -$

When a polynomial is divided by $x - b$ what is the relationship between the remainder and the values of P(b)?

the remainder can be found when b is substituted into the polynomial P(X).

When a polynomial P(x) is divided by $x - b$, the remainder is P(b). What is nice about this is that if you get a question asking you to find the remainder of a polynomial after it is divided by $x - b$; you can just simply plug in the value of b and the output you get will be the remainder. If the question on the test is asking you to find the quotient as well, you will have no choice but to use long division or synthetic division. In my opinion, synthetic division Is the most time efficient way to find the quotient and the remainder. If you are unsure about your final answer, you can use the Remainder Theorem to verify the remainder.

Example 1: $(x^3 - 4x) \div (x + 4)$

$P(-4) = (-4)^3 - 4(-4)$

$P(-4) = -64 + 16$

$P(-4) = -48$

Remainder $= -48$

Example 2: $(a^4 - 3a^2 + 2a + 1) \div (a - 2)$

$P(2) = (2)^4 - 3(2)^2 + 2(2) + 1$

$P(2) = 16 - 12 + 4 + 1$

$P(2) = 9$

Remainder $= 9$

Using Remainder Theorem to find an unknown coefficient

Sometimes you will be given a polynomial expression, a divisor and the remainder, and an unknown coefficient. To solve for the unknown coefficient, you evaluate $P(b)$, and then you solve for the variable.

Example 1: when $2x^3 - 3x^2 + wx - 1$ is divided by $x - 1$ the remainder is 1. Find w.

$P(1) = 1$

$P(1) = 2(1)^3 - 3(1)^2 + w(1) - 1$

$1 = 2 - 3 + w - 1$

$1 + 1 = -1 + w - 1 + 1$

$2 + 1 = w$

$w = 3$

We will now verify our solution,

$P(1) = 2x^3 - 3x^2 + 3x - 1$

$P(1) = 2(1)^3 - 3(1)^2 + 3(1) - 1$

$P(1) = 2 - 3 + 3 - 1$

$P(1) = 1$

The solution is correct! $w = 3$.

Example 2: $(2x^3 + 5x^2 - cx + 2) \div (x - 2)$

Has a remainder of 3, find c.

$P(2) = 2(2)^3 + 5(2)^2 - c(2) + 2$

$3 = 16 + 20 - 2c + 2$

$3 = 36 - 2c + 2$

$3 = 38 - 2c$

$$3 - 38 = -2c$$

$$-35 = -2c$$

$$-\frac{35}{-2} = -\frac{2c}{-2}$$

$$\frac{35}{2} = c$$

We will now test out our solution to verify that it is correct.

$$P(2) = 2(2)^3 + 5(2)^2 - \frac{35}{2}(2) + 2$$

$$P(2) = 16 + 20 - 35 + 2$$

$$P(2) = 36 - 35 + 2$$

$$P(2) = 1 + 2$$

$$P(2) = 3$$

$$c = \frac{35}{2}$$

The Factor Theorem

If the remainder of a polynomial is 0, after evaluating $P(b)$ then $(x - b)$ is a factor of a polynomial $P(x)$, and b is a root of the polynomial $P(x)$.

Example 1: Solve by factoring,

$x^2 + x - 6 = 0$

$(x - 2)(x + 3) = 0$

$x = 2$

$x = -3$

What is the remainder when $x^2 + x - 6$ is divided by $x - 2$?

$P(x) = x^2 + x - 6$

$P(2) = (2)^2 + 2 - 6$

$P(2) = 4 + 2 - 6$

$P(2) = 0$

Is $x - 2$ a factor of $x^2 + x - 6$? YES!!!

Is 2 a root of the equation? YES!!!

The factor theorem states that, "If $x = b$ is substituted into a polynomial $P(x)$ and the remainder equals zero, then $x - b$ is a factor of $P(x)$.

Example 2: Does $x^4 - 3x^3 + 7x^2 - 6x + 1$ have a factor of $x - 1$?

$P(1) = (1)^4 - 3(1)^3 + 7(1)^2 - 6(1) + 1$

$P(1) = 1 - 3 + 7 - 6 + 1$

$P(1) = 0$

Yes $x - 1$ is a factor of $x^4 - 3x^3 + 7x^2 - 6x + 1$.

Example 3: A polynomial $y = P(x)$ has $P(2) = 0, \ P(-1) = 0, \ and \ P(4) = 3.$ List the factors,

Using the factor theorem, we know that if $P(b) = 0$ then $x - b$ is a factor; therefore, the factors are $(x + 1)$ and $(x - 2)$.

The factor property: If a polynomial has any factor of the form $x - b$ then the number b is a factor of the constant term of the polynomial.

Example 1: $(x^3 - 6x^2 + 3x + 10)$

Potential roots: $\pm 1, \ \pm 2, \ \pm 5, \ \pm 10$

$P(1) = (1)^3 - 6(1)^2 + 3(1) + 10$

$P(1) = 1 - 6 + 3 + 10$

$P(1) = 8$

$P(1)$ is not a root.

$P(-1) = (-1)^3 - 6(-1)^2 + 3(-1) + 10$

$P(-1) = -1 - 6 - 3 + 10$

$P(-1) = 0$

$P(-1)$ is a root; therefore, $(x+1)$ is a factor.

Example 2: $P(x) = x^4 + 3x^3 - 11x^2 + 8x - 12$

Potential roots: $\pm 1, \pm 2, \pm 3, \pm 4, \pm 6, \pm 12$

$P(1) = (1)^4 + 3(1)^3 - 11(1)^2 + 8(1) - 12$

$P(1) = 1 + 3 - 11 + 8 - 12$

$P(1) = -11$ 1 is not a root.

$P(-1) = (-1)^4 + 3(-1)^3 - 11(-1)^2 + 8(-1) - 12$

$P(-1) = 1 - 3 - 11 - 8 - 12$

$P(-1) = -33$ -1 is not a root.

$P(2) = (2)^4 + 3(2)^3 - 11(2)^2 + 8(2) - 12$

$P(2) = 16 + 24 - 44 + 16 - 12$

$P(2) = 0$ 2 is a root and $x - 2$ is a factor.

I will show you how to do this using synthetic division,

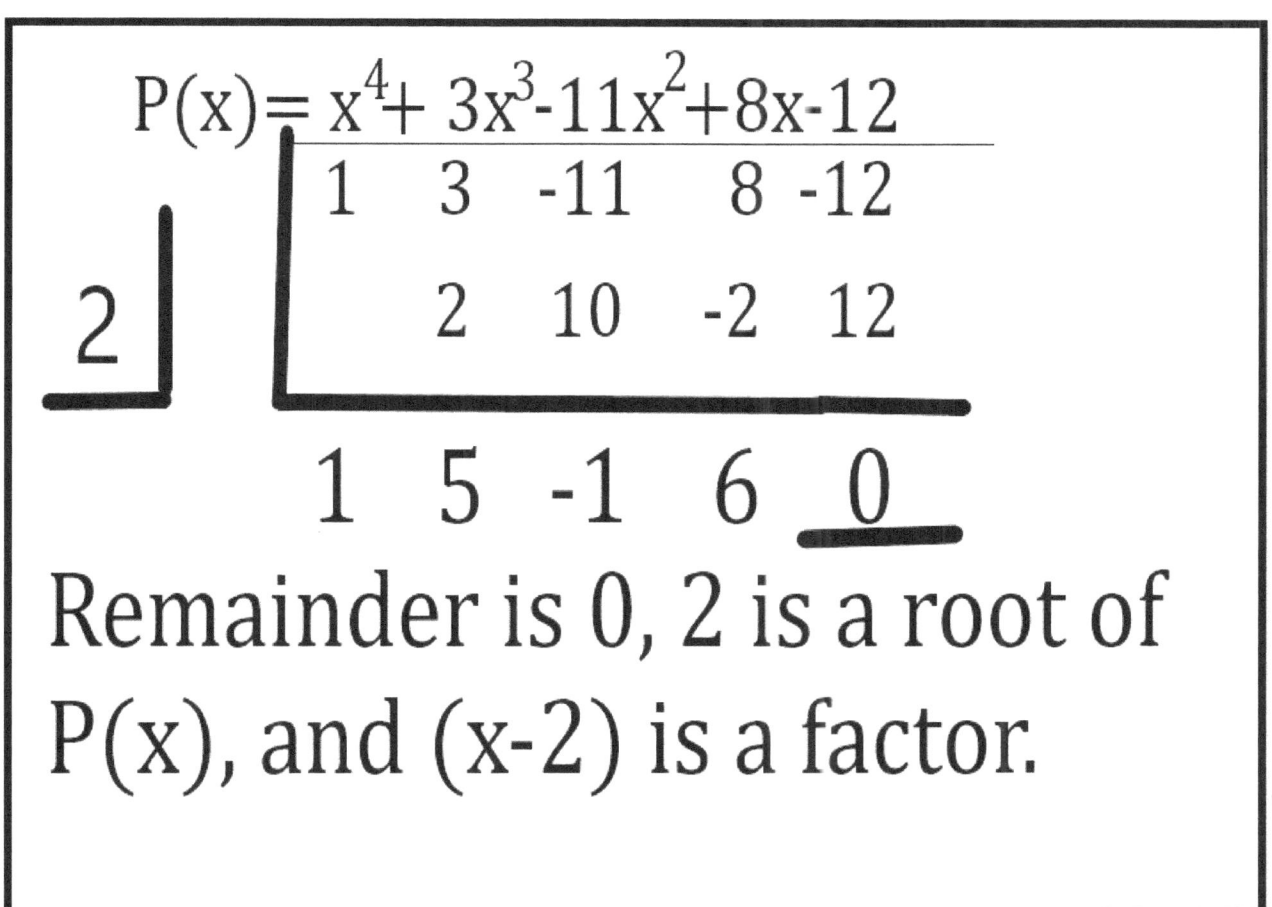

As you can see, synthetic division takes a little bit more time, but it will give you the quotient as well, something that is useful when you are trying to fully factor a polynomial.

Rational Zero Theorem

The rational root theorem is useful when the coefficient of the leading term is not 1. The Rational Zero Theorem states that if a polynomial $P(x)$ has a rational zero $x = \dfrac{b}{a}$ then b is a factor of the constant term of the polynomial and a is a factor of the coefficient of the highest degree term.

For example, $P(x) = 4x^2 - 8x - 3$

Potential roots: $\dfrac{\pm 1, \pm 3}{\pm 1, \pm 2, \pm 4}$

$\pm 1, \pm \dfrac{1}{2}, \pm \dfrac{1}{4}, \pm 3, \pm \dfrac{3}{2}, \pm \dfrac{3}{4}$

This theorem is useful when you are trying to find factors and roots of an equation, and do not know which numbers will work.

Try to find the potential roots of
$2x^3 + 11x^2 + 17x + 5 = 0$

Turn the next page when you have written down the potential roots.

You should have, $\dfrac{\pm 1, \pm 5}{\pm 1, \pm 2}$

$\pm 1, \pm \dfrac{1}{2}, \pm 5, \pm \dfrac{5}{2}$

Factoring Polynomials

In this chapter you will come across polynomials of higher degrees, such as polynomials of third and fourth degree. You will use some of the methods you are already familiar with. The first rule of factoring is to remove a common factor is possible.

Example 1: $5x^2 - 10x - 15$

$5(x^2 - 2x - 3)$ 2 numbers that multiply to -3 and add up to -2. -3 and 1 satisfy this condition.

$5(x + 1)(x - 3)$

Example 2: $x^4 + 3x^2 - 28$ This is still a trinomial, even though the degree of the leading term is 4. Find two numbers that multiply to -28. And add up to 3.

The two numbers that satisfy these conditions are 7 and -4. One more important thing to keep in mind is that

$(x^2)(x^2) = x^4$ this means that instead of having $(x-b)(x-c)$ in the brackets we will have $(x^2-b)(x^2-c)$.

$x^4 + 3x^2 - 28 = 0$

$(x^2 - 4)(x^2 + 7) = 0$

Can we factor this further? Yes, we can, $(x^2 - 4)$ is a difference of squares. ***Remember to factor a difference of squares such as $a^2 - b^2$ we factor it using the form $(a-b)(a+b)$.***

$(x^2 - 4) = (x - 2)(x + 2)$

$(x - 2)(x + 2)(x^2 + 7) = 0$ is the fully factored form of $x^4 + 3x^2 - 28 = 0$.

Example 3: Factor $4x^3 - 3x^2 - x = 0$

Is there a common factor? Yes, x is a common factor.

$x(4x^2 - 3x - 1)$

$x[4x^2 - 3x - 1]$

$x[4x^2 - 4x + x - 1]$

$x[(4x(x - 1) + 1(x - 1)]$

$x(4x+1)(x-1) = 0$

Example 4: Factor $x^4 - 4x^2 - 45 = 0$,

$(x^2+5)(x^2-9)$ (x^2-9) is a difference of squares.

$(x^2+5)(x-3)(x+3) = 0$

Example 5: Factor $30x^4 - 34x^3 - 8x^2 = 0$

Is there a common factor? Yes, $2x^2$ is a common factor.

$2x^2(15x^2 - 17x - 4) = 0$

Find two numbers that multiply to -60 and add up to -17. -20 and 3 satisfy these conditions.

$2x^2[15x^2 - 20x + 3x - 4]$

$2x^2[5x(3x-4) + 1(3x-4)]$

$2x^2(5x+1)(3x-4)$

We will now look at factoring sum and differences of cubes. These are a little bit trickier, but with practice, it will become easy.

$x^3 - y^3$ this is known as a difference of cubes. To factor a difference of cubes, we use the form,

$$x^3 - y^3 = (x-y)(x^2 + xy + y^2)$$

Why does this work? We will expand it and by expanding it we will see what happens.

$(x-y)(x^2 + xy + y^2)$
$= x^3 + x^2y + xy^2 - x^2y - xy^2 - y^3$
$= x^3 + x^2y - x^2y + xy^2 - xy^2 - y^3$
$= x^3 - y^3$

Example 6: $x^3 - 8$

$(x)^3 - (2)^3 = (x-y)(x^2 + xy + y^2)$
$(x)^3 - (2)^3 = (x-2)(x^2 + x(2) + (2)^2)$
$(x^3 - 8) = (x-2)(x^2 + 2x + 4)$

Example 7: $64a^3 - 125b^3$ is a difference of cubes

$(4a)^3 - (5b)^3$
$(4a)^3 - (5b)^3 = (4a - 5b)((4a)^2 + (4a)(5b) + (5b)^2)$
$(64a^3 - 125b^3) = (4a - 5b)(16a^2 + 20ab + 25b^2)$

$x^3 + y^3$ this is known as a sum of cubes. To factor a sum of cubes, we use the form,

$$x^3 + y^3 = (x + y)(x^2 - xy + y^2)$$

Why does this work? We will expand it and by expanding it we will see what happens.

$(x + y)(x^2 - xy + y^2)$
$= x^3 - x^2y + xy^2 + x^2y - xy^2 + y^3$
$= x^3 - x^2y + x^2y + xy^2 - xy^2 + y^3$
$= x^3 + y^3$

Example 8: $8x^3 + 64y^3$

$(2x)^3 + (4y)^3 = (2x + 4y)((2x)^2 - (2x)(4y) + (4y)^2)$
$(8x^3 + 64y^3) = (2x + 4y)(4x^2 - 8xy + 16y^2)$

Factor by grouping when there are 4 terms in a polynomial.
For example, $ax + ay + bx + by$

$= a(x + y) + b(x + y)$
$(a + b)(x + y)$

Example 9: $x^3 + x^2 + 2x + 2$

$[(x^3 + x^2) + (2x + 2)]$

$x^2(x + 1) + 2(x + 1)$

$(x^2 + 2)(x + 1) = 0$

Example 10: $x^5 + 5x^3 - 7x^2 - 35$

$[(x^5 + 5x^3) - (7x^2 + 35)]$

$x^3(x^2 + 5) - 7(x^2 + 5)$

$(x^3 - 7)(x^2 + 5) = 0$

Example 11: $x^3 - 3x^2 - 4x + 12$

$[(x^3 - 3x^2) - (4x - 12)]$

$x^2(x - 3) - 4(x - 3)$

$(x^2 - 4)(x - 3) = 0$ $(x^2 - 4)$ is a difference of squares.

$(x - 2)(x + 2)(x - 3) = 0$

Polynomial Functions

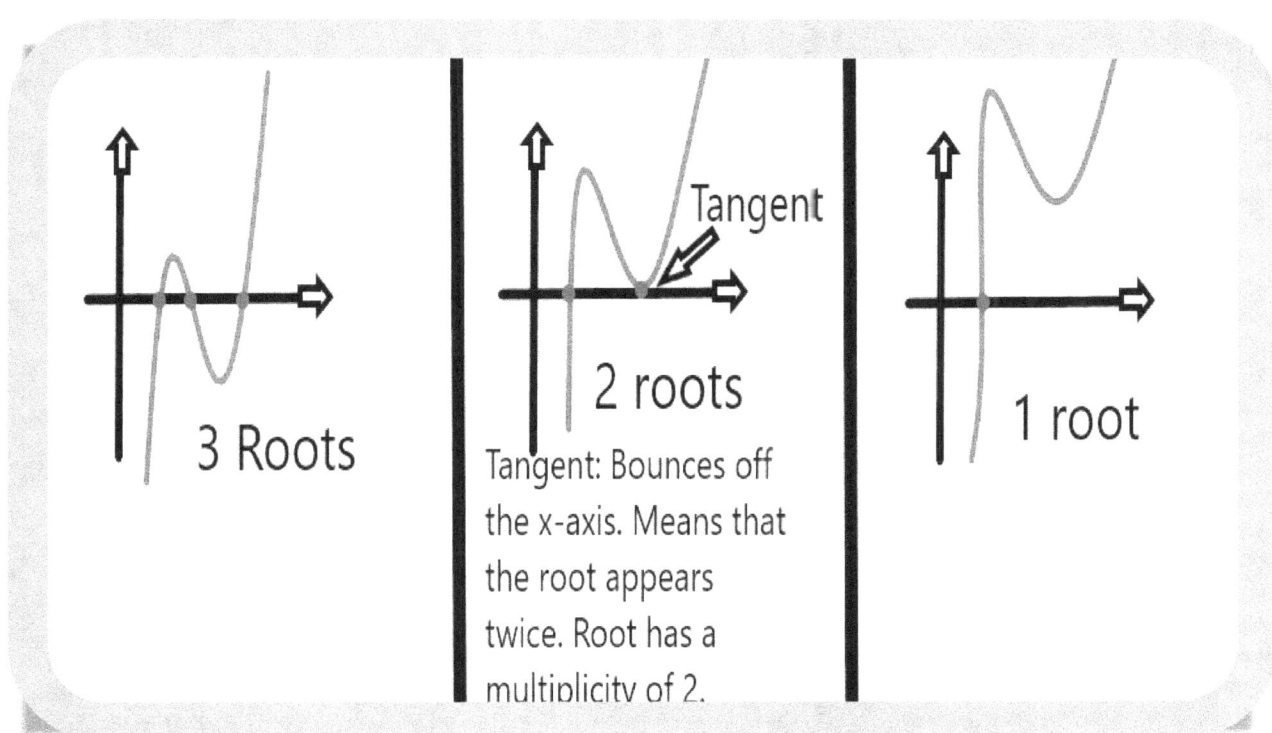

As you can see, the number of roots is determined by the number of times the graph crosses the x-axis. Tangent means that the graph bounces off of the x-axis, since it bounces off of the x-axis it has a multiplicity of two. For example, $(x-1)(x-2)(x-2)$ can also be represented as, $(x-1)(x-2)^2$ since the same root appears twice. It only has two distinct roots. The graph of $P(x) = (x-1)(x-2)^2$ is on the next page,

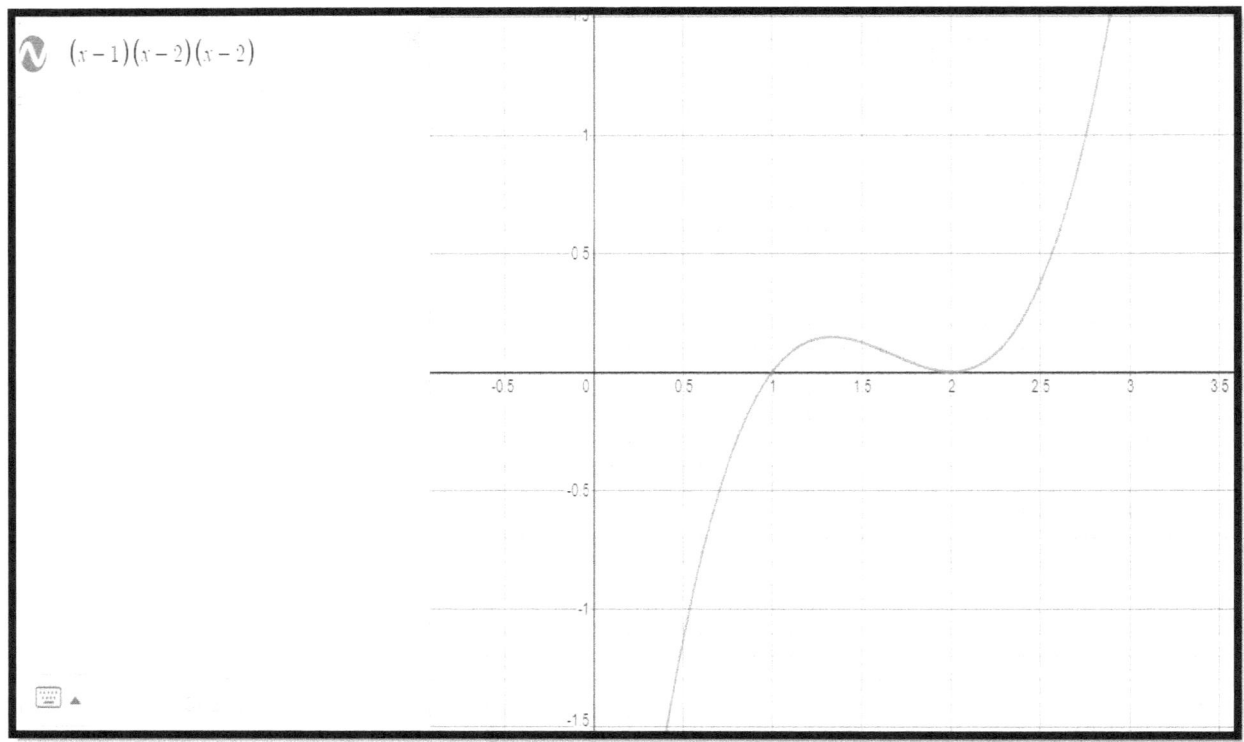

As you can see, the graph bounces on the x-axis at $P(2)$.

Here are two properties of polynomial functions:

- An nth degree polynomial, with n an odd number will have a minimum 1 root, and a max of n roots.
- An nth degree polynomial, with n an even number will have a minimum 0 roots, and a max of n roots.

To find the zeros of polynomial functions, you factor it and find the roots as with any other function you have worked with.

Here are 3 important features that will help you determine the multiplicity of a factor:

- If a factor has a multiplicity of 1, the graph goes directly through the x-axis.

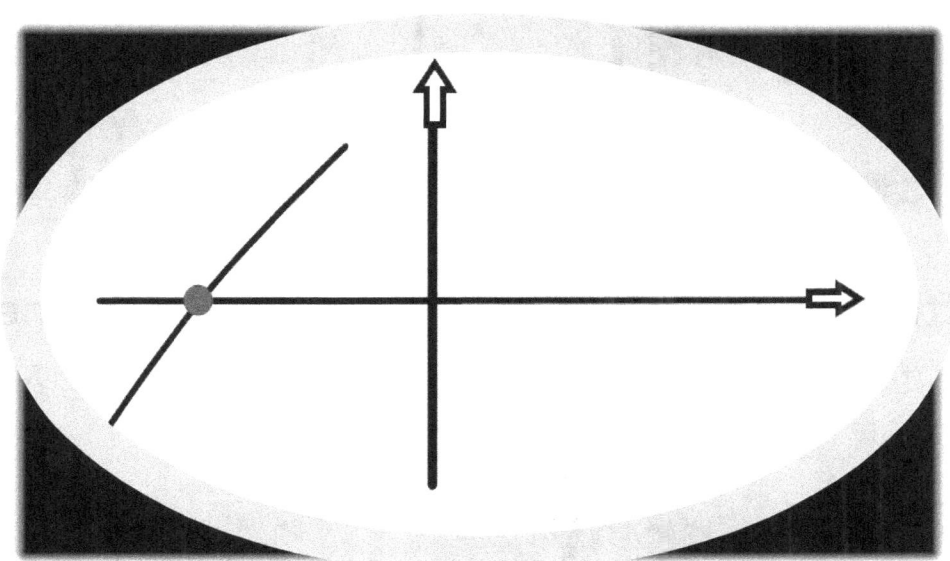

- If a factor has a multiplicity of 2 (even number), the graph has a tangent (bounces off of the x-axis but does not cross).

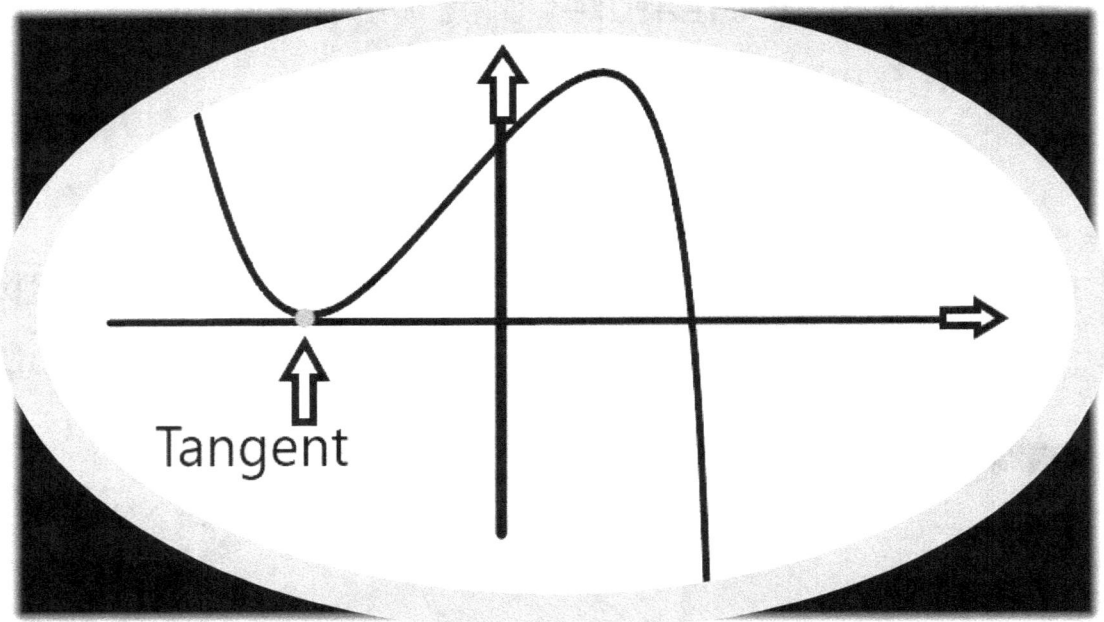

- If a factor has a multiplicity of 3 (odd number), the graph crosses the x-axis at an inflection point.

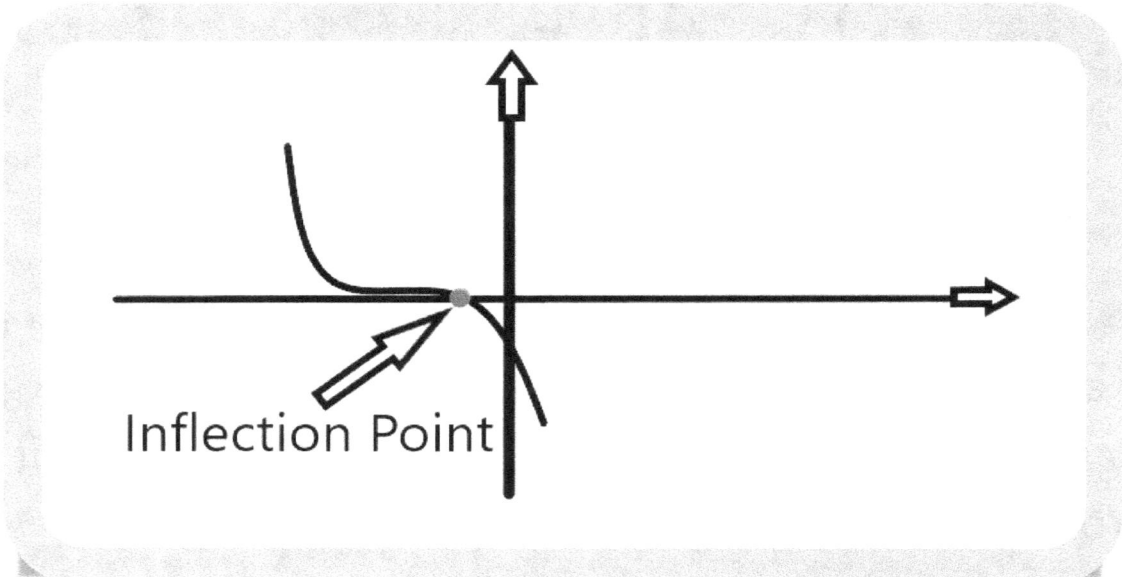

There are two other terms which you should familiarize yourself with, the **Relative Maximum** and the **Relative Minimum**.

- Relative Maximum: A point that is higher than the points directly beside it on both sides.
- Relative Minimum: A point that is lower than the points directly beside it on both sides.

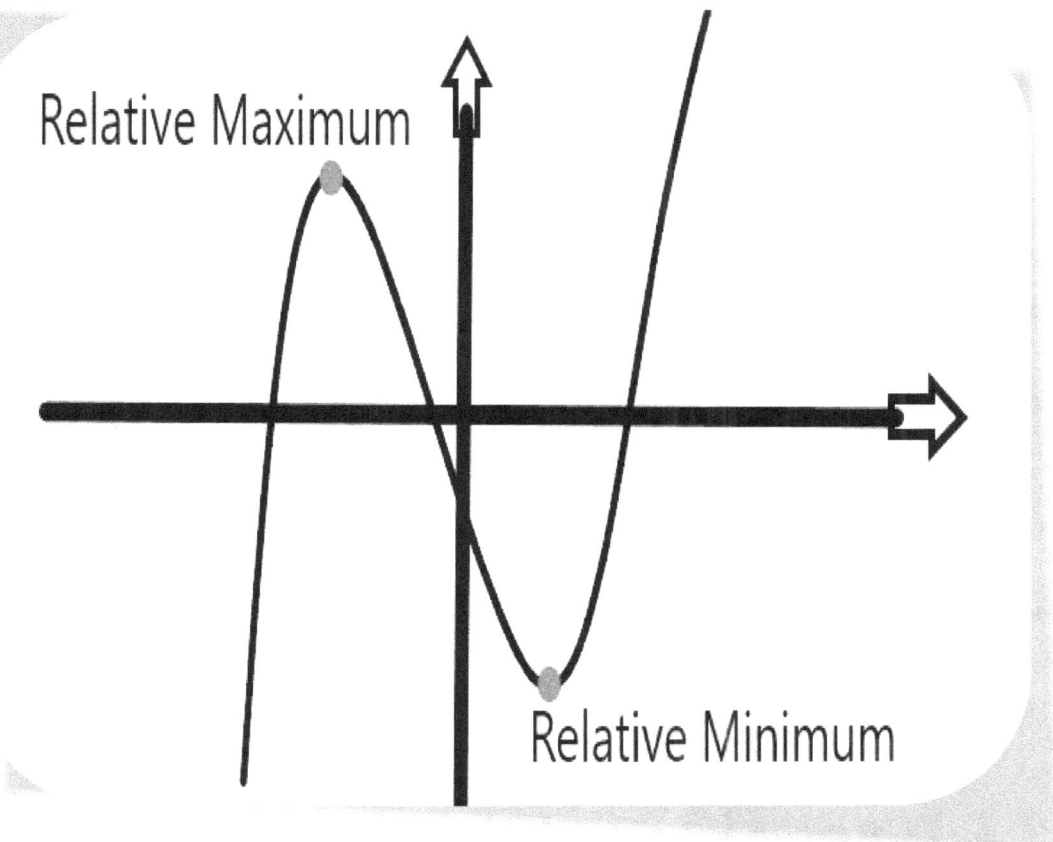

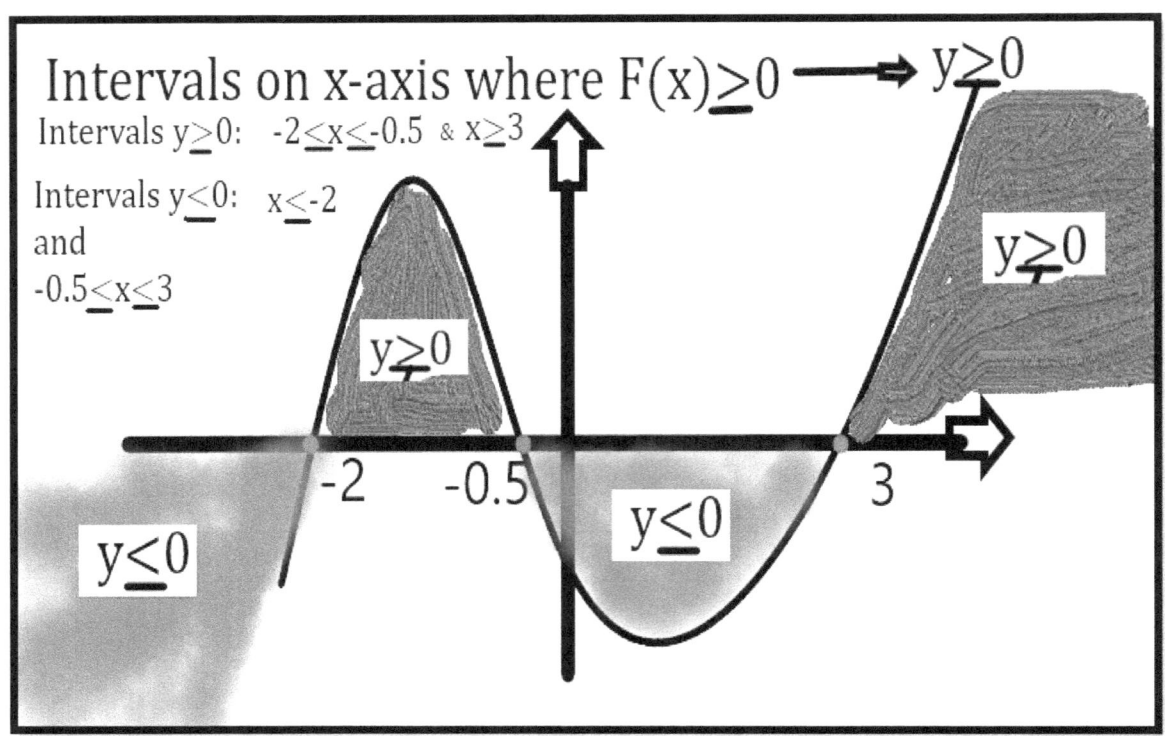

Sometimes you will be given the roots and the multiplicity of the factors, and you will be asked to write an equation based on this information.

Example 1: Write the equation of a polynomial function with the following roots:

$x = -3 \qquad x = 2 \qquad x = -1$

By working backwards, you can find the factors,

$x + 3 = 0$

$x - 2 = 0$

$x + 1 = 0$

Multiply these three factors together,

$(x + 3)(x - 2)(x + 1) = 0$

$(x + 3)[(x - 2)(x + 1)] = 0$

$(x + 3)(x^2 + x - 2x - 2) = 0$

$(x + 3)(x^2 - x - 2) = 0$

$x^3 - x^2 - 2x + 3x^2 - 3x - 6$

Polynomial graph equation: $x^3 + 2x^2 - 5x - 6$

Or $(x + 3)(x - 2)(x + 1) = 0$

If the teacher asks you to expand it then expand it, if it asks you to leave it in factored form then leave it in factored form.

Polynomial graphs are smooth and continuous, polynomials of degree n, have at most $n - 1$ turns.

For example,

- degree 4: 3 turns
- degree 45: 44 turns

The highest degree term is very important:

1. Exponent tells you the basic shape of the graph.
2. Coefficient tells you the direction of the graph.

There are four different shapes of polynomial graphs:

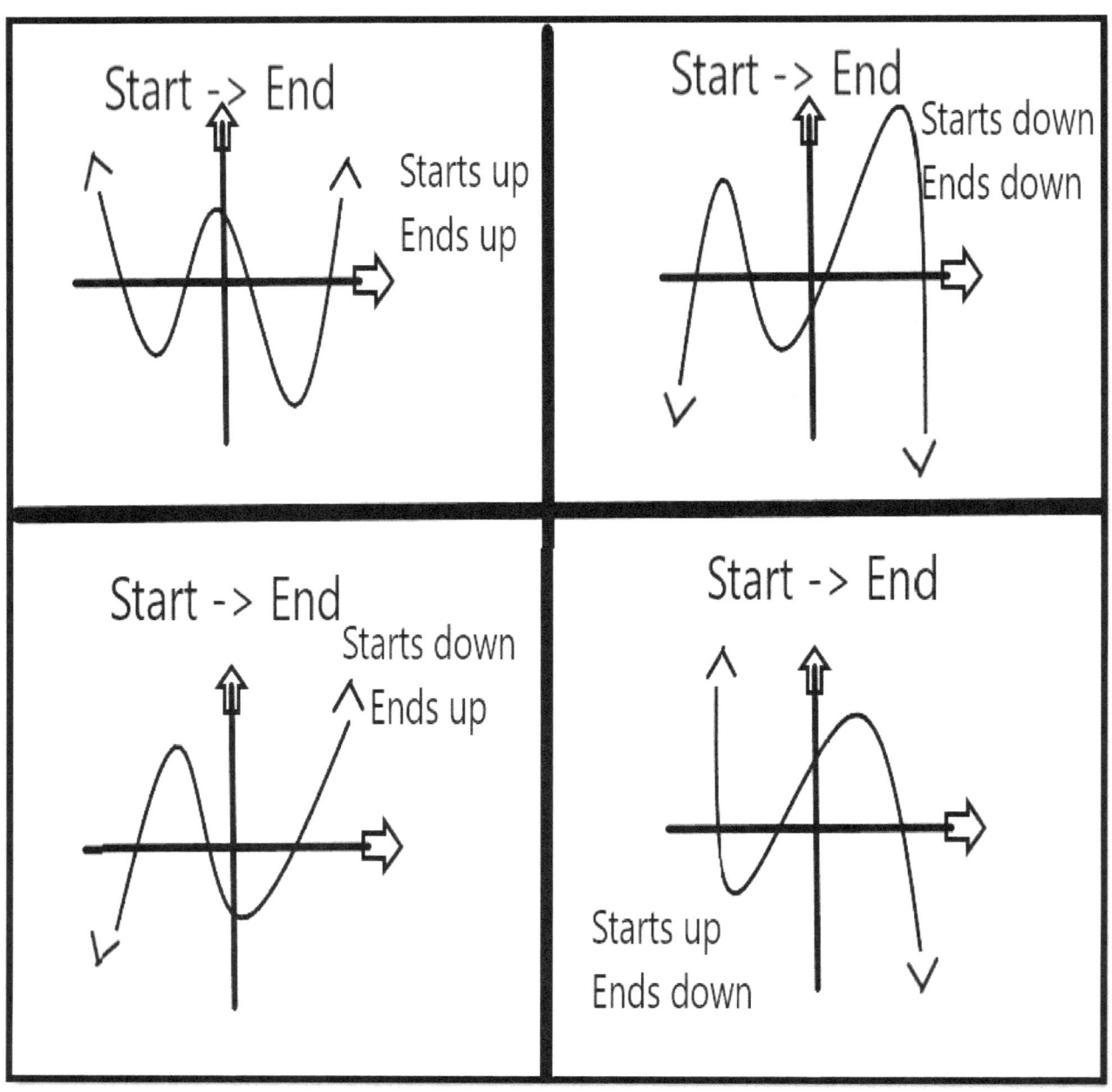

- The graph starts up and ends up if the exponent is even and the coefficient is positive.

- The graph starts down and ends down if the exponent is even and the coefficient is negative.

- The graph starts down and ends up if the exponent is odd and coefficient is positive.

- The graph starts up and ends down if the exponent is odd and the coefficient is negative.

Practice Questions

1. Use either long or synthetic division to determine the division statement for,
$(x^3 + 2x^2 - 7x - 9) \div (x - 1)$.

2. Use the remainder theorem to determine the remainder for the following divisions:
   a) $(x^2 - 2x) \div (x - u)$
   b) $(x^3 - x + 2) \div (x + 4)$

3. Determine the value of c, if $(x^2 + cx - 7) \div (x - 2)$ has a remainder of 3.

4. What are the potential roots of $x^3 + 3x^2 - 4x + 6$?

5. Does $P(x) = x^3 - x^2 b - xb^2 + b^3$ have a factor of $(x - b)$?

6. Does $P(w) = w^2 - 5w + 6$ have a factor of $(w - 3)$?

7. Use any method to factor the following:
   a) $x^3 + 5x^2 + 6x$
   b) $x^3 - 8d^3$
   c) $27v^3 + b^3$
   d) $x^3 - 2x^2 - 4x + 8$

8. Write the equation of a polynomial function with the following information:
   Roots at $x = -4,\ x = 3,\ x = 2,\ the\ root\ x = 2$ has a multiplicity of 2. Leave your answer in factored form.

Exponents & Logarithms

Chapter 10

Before we learn about Logarithms, we will do a brief review on exponents. You probably already know about the basic laws; however, I will go over them just to make sure, as I do not believe in giving people any unexplained surprises. I want to make sure you are as successful as possible, since this is a key course for University entrance. This chapter will be divided into three sections, PART A, PART B, and PART C. There will be questions at the end of each of these three sections.

CHAPTER 10 PART A (EXPONENTS)

An **exponent** is basically a shorthand for a number repeatedly multiplied by itself. The exponent indicates how many times a number or variable should be multiplied by itself. The **base** number tells what number is being multiplied. The *exponent*, the small number written above and to the right of the base number, tells how many times the *base* number is being multiplied.

Here is an example:

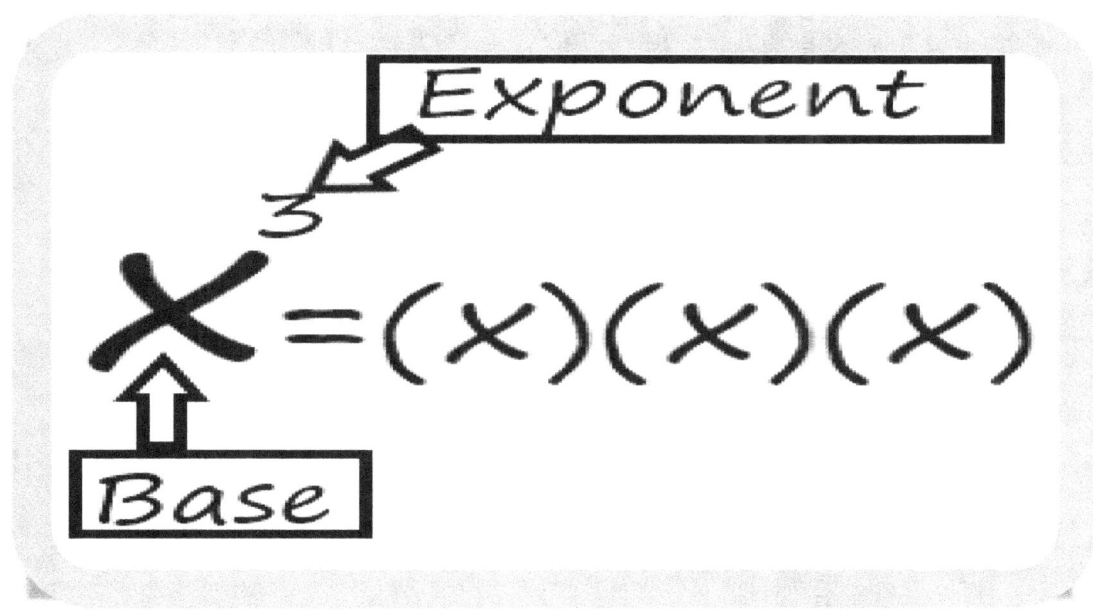

As you can see, the exponent determines how many times the base gets multiplied by itself. Another general rule is that any base to the power of 0 is 1. Ex: $x^0 = 1$. Now that you know the difference between a base and an exponent, we will now look at some "**Laws of Exponents**" that will be useful when simplifying expressions.

Law 1: *Product Law*

When multiplying two or more identical bases, you simply add the two exponents together.

Law 1: $x^a * x^b = x^{a+b}$

Here are some examples,

Example 1: $d * d^5 = d^{1+5} = d^6$

Example 2: $j * k * j^2 * k * d = j^{1+2}k^{1+1}d = dj^3k^2$

Example 3: $2^3 * 2^3 = 2^{3+3} = 2^6 = (2)(2)(2)(2)(2)(2) = 64$

Example 4: $(cv^2)(c^3v) = c^{1+3}v^{2+1} = c^4v^3$

Now that we have looked at the basics of the first law, we will now look at the next law, the Quotient Law

Law 2: Quotient Law,

When dividing two identical bases, you simply subtract the exponent of the base in the denominator from the exponent of the base in the numerator.

Law 2: $\dfrac{x^a}{x^b} = x^{a-b}$

Here are some examples:

Ex 1: $\dfrac{x^3}{x^2} = x^{3-2} = x^1 = x$

Ex 2: $\dfrac{x^a y^c}{x^b y^d} = x^{a-b} y^{c-d}$

Ex 3: $\dfrac{x^5 y^6}{x^2 y^3} = x^{5-2} y^{6-3} = x^3 y^3$

Ex 4: $\dfrac{4^6}{4^3} = 4^{6-3} = 4^3 = (4)(4)(4) = 64$

Law 3: *Power of a Power,*

If you have a product inside parentheses, and a power on the parentheses, such as [$(x^2)^3$], then you simply multiply the inner exponent by the outer exponent.

$(x^a)^b = x^{a*b}$

Word of Warning: This rule does NOT work if you have a sum or difference within parentheses. Exponents unlike multiplication do NOT distribute over subtraction or addition.

Ex: $(x^2 + y^2)^2$ Is **Not** equal to $(x^4 y^4)$, $(x^2 + y^2)^2$ is **Actually** equal to $(x^2 + y^2)(x^2 + y^2) = x^4 + y^4 + 2x^2 y^2$

Here are some examples of when this third law can be applied:

Ex 1: $(x^2)^3 = x^{2*3} = x^6$

Ex 2: $(y^6)^3 = y^{6*3} = y^{18}$

Ex 3: $(2^3)^2 = (2^{3*2}) = 2^6 = (2)(2)(2)(2)(2)(2) = 64$

Ex 4: $(x^3)^y = (x^{3*y}) = x^{3y}$

Law 4: Power of a Product

If you have a product within parenthesis being raised to a power, you multiply both variables inside of the parenthesis by the outer exponent.

$(xy)^a = x^a y^a$

Example 1: $(2m)^3 = (2^3)(m^3) = 8m^3$

Example 2: $(ab)^c = a^c b^c$

Example 3: $(jx)^4 = j^4 x^4$

Law 5: Power of a Quotient:

When you raise a quotient to a power, you raise the numerator and denominator to that power.

$$\left(\frac{x}{y}\right)^a = \frac{x^a}{y^a}$$

Example 1: $\left(\frac{a}{b}\right)^3 = \frac{a^3}{b^3}$

Example 2: $\left(\frac{x}{3y}\right)^2 = \frac{x^2}{(3y)^2} = \frac{x^2}{9y^2}$

Law 6: *Negative exponent rule:*

To simplify a negative exponent in the numerator, you simply move the variable or coefficient with the negative exponent to the denominator and it becomes a positive exponent. To simplify a negative exponent in the denominator, you simply move the variable or coefficient with the negative exponent to the numerator and it becomes a positive exponent.

$$x^{-a} = \frac{1}{x^a}$$

$$\frac{1}{x^{-a}} = x^a$$

Example 1: $x^{-3} = \frac{1}{x^3}$

Example 2: $\dfrac{x^{-2}}{y} = \dfrac{1}{x^2 y}$

Example 3: $\dfrac{a^{-2}x}{y} = \dfrac{x}{a^2 y}$

Example 4: $\dfrac{6a^{-4}}{b^{-2}} = \dfrac{6b^2}{a^4}$

Law 7: *Rational Exponents Rule:*

rational exponents are exponents in the form of $x^{\frac{a}{b}}$, where $\frac{a}{b}$ is the exponent. To write this as a radical expression, you basically take the denominator of the exponent and make it the index, and you raise the radical to the power of the numerator. For example, $x^{\frac{a}{b}} = \sqrt[b]{x^a}$ or $(\sqrt[b]{x})^a$

Here are some more examples:

- $8^{\frac{2}{3}} = \sqrt[3]{8^2} = \sqrt[3]{(8)(8)} = \sqrt[3]{64} = 4$
- Or $8^{\frac{2}{3}} = (\sqrt[3]{8})^2 = (\sqrt[3]{8})^2 = (2)^2 = 4$
- $4^{\frac{1}{2}} = \sqrt{4} = 2$

- $y^{\frac{5}{n}} = \sqrt[n]{y^5}$ or $(\sqrt[n]{y})^5$

Changing Base

Sometimes you will be asked to change the base, to do this, you need to figure out a base and an exponent that works out to the original base.

Example 1: 27^{4a} to base 3

$27 = (3)^3$

$27^{4a} = (3^3)^{4a} = 3^{3 \ast 4a}$

3^{12a}

Example 2: 128^{x-3} to base 2

$128 = (2^7)$

$128^{x-3} = (2^7)^{x-3} = 2^{7(x-3)}$

2^{7x-21}

Example 3: $\dfrac{1}{625^{2x}}$ to base 5.

$$\frac{1}{625} = 5^{-4}$$

$$(5^{-4})^{2x} = 5^{-8x}$$

Two Types of Equations Involving Exponents

There are two different types of equations involving exponents:

1. Equations where the variable is the base. For example,

$$x^{\frac{2}{3}} = 8, \quad 2x^{\frac{3}{4}}, \quad (2x+3)^{\frac{2}{5}} = 4$$

2. Equations where variable is an exponent, this is known as an exponential equation. For example, $5^x = 625$, $3^{x+1} = 81$

We will start by solving equations where the base is a variable.

Example 1: $x^{\frac{3}{2}} = 125$ *Take the reciprocal of the exponent.*

$$\left(x^{\frac{3}{2}}\right)^{\frac{2}{3}} = (125)^{\frac{2}{3}}$$ Use the rational exponent law

$$x = \left(\sqrt[3]{125}\right)^2$$

$$x = (5)^2$$

$$x = 25$$

Example 2: $4x^{\frac{3}{4}} = 108$

$$\frac{4x^{\frac{3}{4}}}{4} = \frac{108}{4}$$

$$x^{\frac{3}{4}} = 27$$

$$\left(x^{\frac{3}{4}}\right)^{\frac{4}{3}} = (27)^{\frac{4}{3}}$$

$$x = \left(\sqrt[3]{27}\right)^4 \qquad x = (3)^4$$

$$x = 81$$

Example 3: $(6x - 28)^{\frac{2}{3}} = 4$

$$\left((6x-28)^{\frac{2}{3}}\right)^{\frac{3}{2}} = (4)^{\frac{3}{2}}$$

$6x - 28 = (\sqrt{4})^3$ *since we took the "square root" of a number, there will be two solutions. 1 solution when the right side is 8, the other solution is when the right side is -8.*

$6x - 28 = 8$	$6x - 28 = -8$
$6x = 8 + 28$	$6x = -8 + 28$
$6x = 36$	$6x = 20$
$\dfrac{6x}{6} = \dfrac{36}{6}$	$\dfrac{6x}{6} = \dfrac{20}{6}$
$x = 6$	$x = \dfrac{10}{3}$

Now we will look at the methods to solve exponential equations (when variable is the exponent).

Example 4: $3^{3x-7} = 9$

- Step 1: Write both sides as the same base (3).
$3^{3x-7} = (3)^2$
$3^{3x-7} = 3^2$

- Step 2: Now that we have both sides as the same base, we do not need to be worrying about the bases on both sides. We can write the exponents as a normal equation ($3x - 7 = 2$).
  The reason for this is because we are basically saying, "Both sides have the same base, therefore, we need to make the exponent of the left side equal to the right side."
  $3x - 7 = 2$

- Step 3: Solve for x.
  $3x - 7 = 2$
  $3x = 2 + 7$
  $3x = 9$
  $$\frac{3x}{3} = \frac{9}{3}$$
  $x = 3$

Example 5: $2^{3x-1} = 16$

$2^{3x-1} = (2)^4$

$2^{3x-1} = 2^4$

$3x - 1 = 4$

$$3x = 4 + 1 \qquad 3x = 5$$

$$\frac{3x}{3} = \frac{5}{3}$$

$$x = \frac{5}{3}$$

Example 6: $\dfrac{1}{343^{x-1}} = 49^{2x-1}$

$$((7)^{-3})^{x-1} = ((7)^2)^{2x-1}$$

$$7^{-3(x-1)} = 7^{2(2x-1)}$$

$$7^{-3x+3} = 7^{4x-2}$$

$$-3x + 3 = 4x - 2$$

$$-3x - 4x = -2 - 3$$

$$-7x = -5$$

$$-\frac{7x}{-7} = -\frac{5}{-7}$$

$$x = \frac{5}{7}$$

CHAPTER 10 PART A: PRACTICE QUESTIONS

1. Use the Exponent Laws to simplify the following:

A) $(3x^{-2}y^4)^2$

B) $\left(\dfrac{x^2}{b^{-1}}\right)^{-2}$

C) $v\left(\dfrac{a}{c}\right)\left(\dfrac{a^{-1}}{c^{-1}}\right)\left(\dfrac{b}{d}\right)\left(\dfrac{d}{b}\right)$

2. Solve the following Exponential equations:

A) $2^{x-2} = 8^x$

b) $3^{c-6} = 81$

3. Solve for x, $\quad x^{\frac{3}{2}} = 64$

CHAPTER 10 PART B (LOGARITHMS)

We will begin this chapter by determining what the Log Key does.

Log 1= 0   Log 0 = error   Log (-1) = error
Log 10= 1   Log 0.1= -1   Log (-2) = error   Log 100 = 2   Log 0.01 = -2   Log (-3) = error
Log 1000= 3   Log 0.001= -3

Logarithmic Form	Exponential Form
$\log 1 = 0$	$10^0 = 1$
$\log 10 = 1$	$10^1 = 10$
$\log 100 = 2$	$10^2 = 100$
$\log 1000 = 3$	$10^3 = 1000$

What is $\log 10,000$?

$\log 10,000 = 4$ can be written as $10,000 = 10^4$

What is another name for a logarithm?
A logarithm is just an "EXPONENT"
That is what a log KEY does on your calculator, it gives you an exponent with a base of 10. Called a "Common Logarithm"

Logarithm: An exponent

Common Logarithm: Base 10

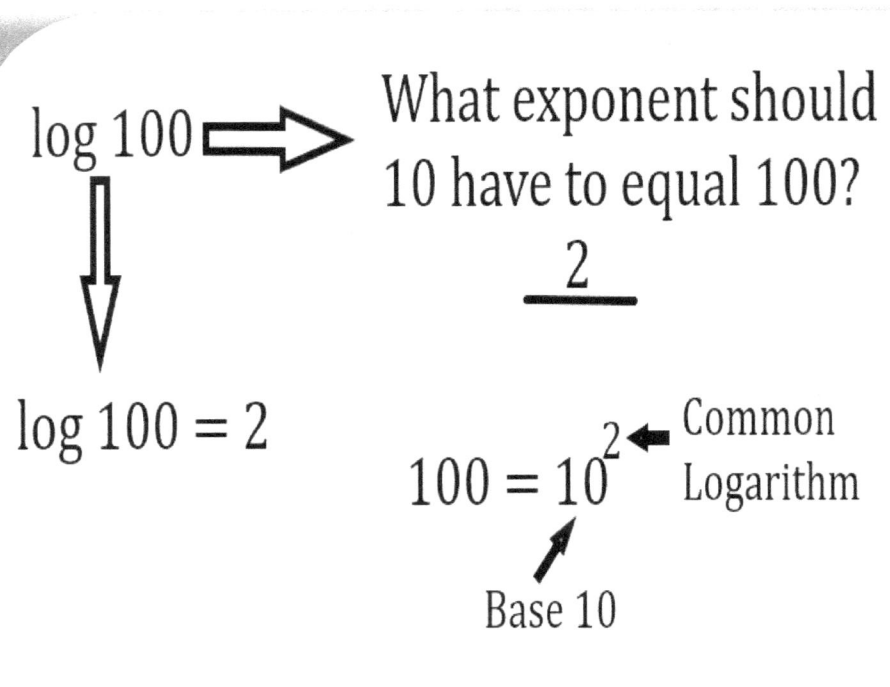

$\log x = y$ means $x = 10^y$

$x > 0$

To convert logarithmic form to exponential form you do something called "Boot The Log",

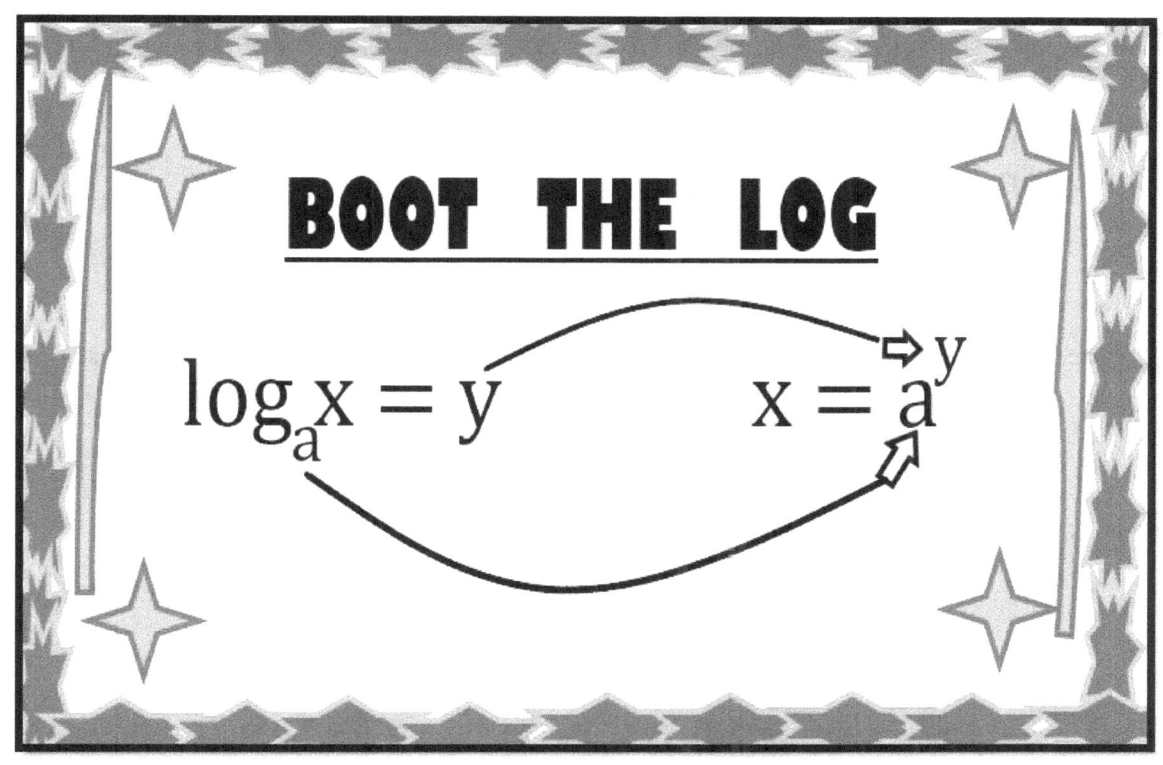

If a question is asking you to find a logarithm, do you know what it is asking?

It is asking you to find the exponent!

Determine the logarithm of: $\log 50$

What exponent should 10 have to equal 50?

Exponential form: $50 = 10^{1.69897004}$

Logarithmic form: log50 = 1.69897004

Earlier I said that the x, in $\log x = y$ must be greater than 0, $x > 0$, is because it does not make sense to raise an exponent to a power to get 0 as a result or a negative number. Try it on your calculator, it is not possible when dealing with real numbers.

We will now look at logarithms with a base other than 10.

For example, $\log_3 9$ what exponent should 3 have to equal 9? The answer is 2.

$9 = 3^2$ Exponential form

$\log_3 9 = 2$ Logarithmic form

Definition of a logarithm of base a ($\log_a x = y$) where, $x > 0$, $a > 0$, $a \neq 1$

Example 1: Convert $a^b = c$ to logarithmic form:

$a^b = c$ is written as $\log_a c = b$ in logarithmic form.

Example 2: Convert $\log_6 \sqrt{6} = \dfrac{1}{2}$ to exponential form:

$\log_6 \sqrt{6} = \dfrac{1}{2}$ is written as $\sqrt{6} = 6^{\frac{1}{2}}$ in exponential form.

Restrictions of Logarithms

IMPORTANT

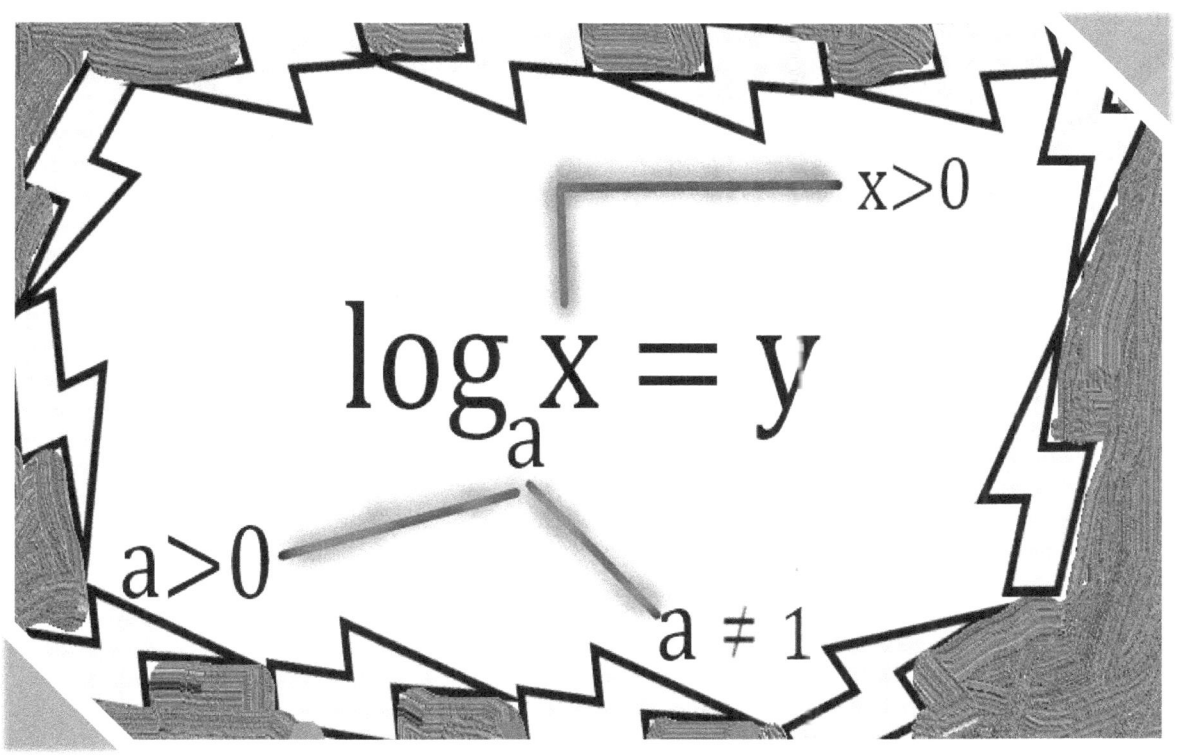

$$a > 0, \quad a \neq 1, \quad x > 0$$

Domain and Range

When we determine the domain of a Logarithmic function, we are concerned with the **Argument**.

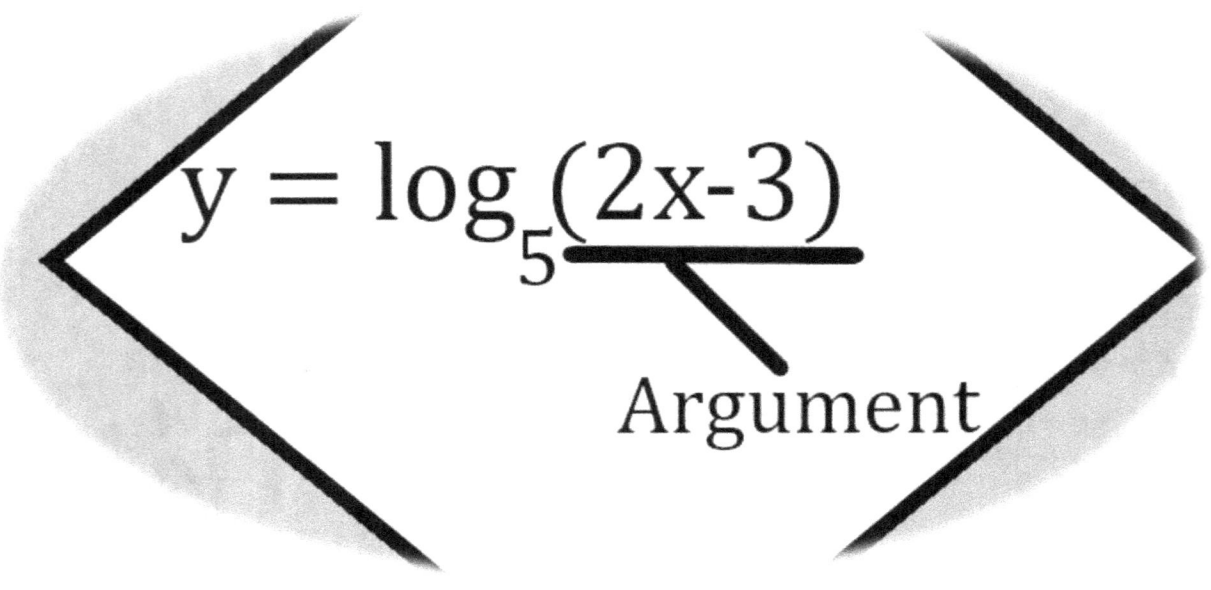

Domain: $\quad 2x - 3 > 0$

$2x > 3$

$\dfrac{2x}{2} > \dfrac{3}{2}$

$x > \dfrac{3}{2}$

Range: the range will depend on what x-values are allowed in the function

Domain: $x > \dfrac{3}{2}$

What y-values are produced when x gets close to 1.5?

$x = 1.5001$

$y = \log_5 (2x - 3)$

$y = \log_5 (2(1.5001) - 3)$

$y = \log_5 (0.0002)$

When $x = 1.5001$ $y = -5.292029674$

When $x = 1.50000001$ $y = -9.584059348$

When $x = 1000$ $y = 4.72177313$

When $x = 1000000$ $y = 9.014734975$

Range: $y \in R$ y will yield any value.

The graph of this function is on the next page,

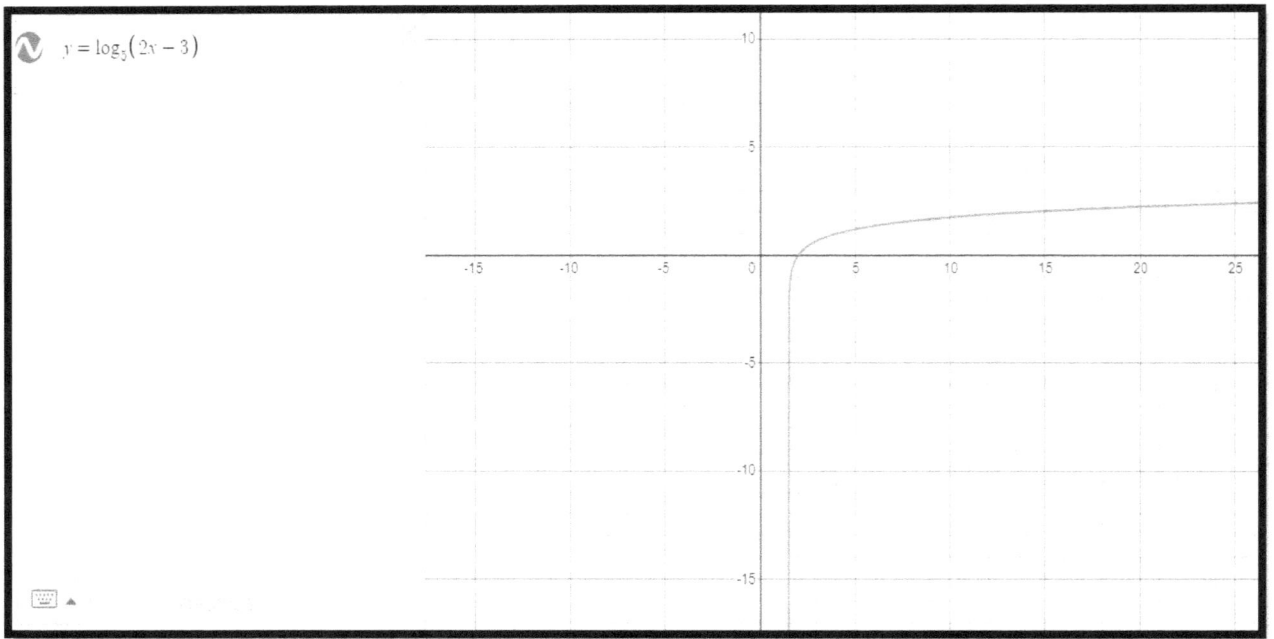

The graph above, is the graph of $y = \log_5(2x-3)$

Here is another example of solving for the domain of a logarithmic function:

$y = \log_7(4x-3)$

$4x - 3 > 0 \quad 4x > 3$

$\dfrac{4x}{4} > \dfrac{3}{4}$

Domain: $x > \dfrac{3}{4}$

Range: $y \epsilon R$

Laws of Logarithms

POWER RULE : $\log_a x^n$ is the same as $n\log_a x$

We will start by looking at the power rule of Logarithms.

$\log x^n$ and $n\log x$ are both the same thing. We will look at some examples that will help establish this rule:

$\log 5^2 = 1.397940009$ is the same as
$2\log 5 = 1.397940009$

$\log \pi^3 = 1.49$ is the same as $3\log \pi = 1.49$

$\log \dfrac{1}{4} = -0.60$ is the same as $-\log 4 = -0.60$

$\dfrac{1}{4} = 4^{-1}$

Example 1: Solve $3^x = 14$ take log of both sides

$$\log 3^x = \log 14$$

$$x\log 3 = \log 14$$

$$\frac{x\log 3}{\log 3} = \frac{\log 14}{\log 3}$$

$$x = \frac{\log 14}{\log 3}$$

$$x = 2.40217$$

Example 2: $2^{x-3} = 21$

$$(x-3)\log 2 = \log 21$$

$$\frac{(x-3)\log 2}{\log 2} = \frac{\log 21}{\log 2}$$

$$x - 3 = 4.39232$$

$$x = 4.39232 + 3$$

$$x = 7.39232$$

Example 3: Express 19 as a power of 4.

$$4^x = 19$$

$$\log 4^x = \log 19$$

$$x\log 4 = \log 19$$

$$\frac{x\log 4}{\log 4} = \frac{\log 19}{\log 4}$$

$$x = \frac{\log 19}{\log 4} \approx 2.12396$$

CHANGE THE BASE RULE: $\log_b a = \dfrac{\log_c a}{\log_c b}$

When you change the base the bases of both logarithms become 10. c is usually 10.

Example 1: $\log_3 7 = \dfrac{\log 7}{\log 3} \approx 1.7712$

Example 2: $\log_8 6 = \dfrac{\log 6}{\log 8} \approx 0.8616$

LAW OF MULTIPLICATION: $\log_a xy = \log_a x + \log_a y$

Example 1: $\log 7 + \log 4 = \log(7 * 4) = \log 28$

Example 2: $\log 5 + \log v = \log(5 * v) = \log 5v$

Example 3: $\log 4 + \log 2 = \log(4 * 2) = \log 8$

Example 4: Write $\log 8$ as a sum of two logarithms,

$\log 12 = \log? + \log?$ $\log 12 = \log 2 + \log 6$

LAW OF DIVISION: $\log_a\left(\dfrac{x}{y}\right) = \log_a x - \log_a y$

Example 1: $\log\left(\dfrac{5}{16}\right) = \log 5 - \log 16 \approx -0.505149978$

Example 2: $\log 7 - \log \pi = \log\left(\dfrac{7}{\pi}\right) \approx 0.347948167$

Example 3: $\log 3 - \log 4 = \log\left(\dfrac{3}{4}\right)$

Example 4: $2\log 7 - 3\log 6 = \log 7^2 - \log 6^3 = \log\left(\dfrac{7^2}{6^3}\right)$

$= \log\left(\dfrac{49}{216}\right)$

Example 5: Write $\log 8$ as a difference of two logarithms,

$\log 8 = \log\left(\dfrac{?}{?}\right) = \log\left(\dfrac{16}{2}\right)$

$\log 8 = \log 16 - \log 2$

RECIPROCAL RULE: $$log_a b = \frac{1}{log_b a}$$

Why does this rule work, here is why:

$$\frac{1}{log_b a} = \frac{1}{\frac{\log a}{\log b}} = 1 \div \frac{\log a}{\log b}$$

$$= 1 * \frac{\log b}{\log a}$$

$$= \frac{\log b}{\log a}$$

$$= log_a b$$

It can even be proven by using the **Extended Power Rule**

$$log_{a^n} b^n = \frac{\log b^n}{\log a^n}$$

$$= \frac{n \log b}{n \log a} = \frac{\log b}{\log a}$$

$$= log_a b$$

Example 1: $(\log_x y)(\log_y x)$ *as you can see they both have different bases, we can use the reciprocal rule so they have the same base.*

$$\left(\frac{1}{\log_y x}\right)\left(\frac{\log_y x}{1}\right) = 1$$

$\log_a a^x = x$

For example, $\log_7 7^3 = 3$

$3\log_7 7 = 3\left(\dfrac{\log 7}{\log 7}\right)$

$= 3$

This makes sense because if you think about it, if you have the same base and the same argument, your final answer will always be the exponent of the argument.

Another important thing to take note of is that in the form $a^{\log_a x} = x$ the answer will always be the x value.

Example 2: $2^{\log_2 3} = 3$

what exponent should 2 have to equal 3?

≈ 1.58496

$2^{1.58496} = 3$

Example 3: $3^{\log_3 150} = 150$

Example 4: $4^{\log_4 520} = 520$

Example 5: $7^{\log_7 6} = 6$

Example 6: $a^{(\log_a 14 - \log_a 2)}$

$a^{\log_a \left(\frac{14}{2}\right)} = a^{\log_a 7}$

$= 7$

Example 7: Simplify $\log_2 \sqrt[6]{32}$

$$= \log_2 32^{\frac{1}{6}} = \frac{1}{6}\log_2 32$$

$$= \frac{1}{6}\left(\frac{\log 32}{\log 2}\right) = \frac{\log 32}{6\log 2}$$

$$= \frac{5}{6}$$

Example 8: $4^{2\log_4 3x}$

$$= 4^{(\log_4 (3x)^2)}$$

$$= 4^{\log_4 9x^2}$$

$$= 9x^2$$

POWER RULE: $\log_a x^n$ is the same as $n\log_a x$

CHANGE THE BASE RULE: $\log_b a = \dfrac{\log_c a}{\log_c b}$

LAW OF MULTIPLICATION: $\log_a xy = \log_a x + \log_a y$

LAW OF DIVISION: $\log_a\left(\dfrac{x}{y}\right) = \log_a x - \log_a y$

RECIPROCAL RULE: $\log_a b = \dfrac{1}{\log_b a}$

EXTENDED POWER RULE: $\log_{a^n} b^n = \dfrac{\log b^n}{\log a^n}$

OTHER: $a^{\log_a x} = x$, $\log_a a^x = x$

Solving Exponential and Logarithmic Equations

You use logarithms when you cannot make both sides of the equation have the same base.

Example 1: $5^x = 20$ you cannot make the bases the same; therefore, use logarithms.

$\log 5^x = \log 20$

$x \log 5 = \log 20$

$\dfrac{x \log 5}{\log 5} = \dfrac{\log 20}{\log 5}$

$x = \dfrac{\log 20}{\log 5}$ Exact value

Or

$x \approx 1.86135$ Decimal

Example 2: $\left(\dfrac{1}{36}\right)^{2x-1} = 216^x$

$(6^{-2})^{2x-1} = (6^3)^x$

$6^{-4x+2} = 6^{3x}$

$-4x + 2 = 3x$

$-4x - 3x = -2$

$-7x = -2$

$\dfrac{-7x}{-7} = \dfrac{-2}{-7}$

$x = \dfrac{2}{7}$

Example 3: $4^{3x+2} = 3^{4x}$

$\log 4^{3x+2} = \log 3^{4x}$

$(3x+2)\log 4 = 4x \log 3$

$3x\log 4 + 2\log 4 = 4x\log 3$ Let's bring the logs with the x's on both sides

$3x\log 4 - 4x\log 3 = -2\log 4$

$$x(3\log 4 - 4\log 3) = -2\log 4$$

$$\frac{x(3\log 4 - 4\log 3)}{3\log 4 - 4\log 3} = -\frac{2\log 4}{3\log 4 - 4\log 3}$$

$$x = -\frac{2\log 4}{3\log 4 - 4\log 3} \quad \text{exact value}$$

$$x = 11.76980 \quad \text{decimal value}$$

Example 4: $\log x = 2$ boot the log

$$x = 10^2$$

$$x = 100$$

Example 5: $2\log x = \log 81$

$\log x^2 = \log 81$ since both sides have a log and an argument, we can CUT the log on both sides. Yes, let's grab our chainsaws and cut that big log down. Sorry, I just had to throw that in there. I can't be serious the whole way through this book.

$$\log x^2 = \log 81$$

$$x^2 = 81$$

$$\sqrt{x^2} = \pm\sqrt{81}$$

$$x = \pm 9$$

Example 6: $\log(3x+2) + \log(x-1) = 2$

$\log(3x+2)(x-1) = 2$

$\log(3x^2 - 3x + 2x - 2) = 2$ we can boot the log and factor the equation, kick the log as hard as you can (without breaking your legs) and get rid of that thing giving kids nightmares.

$3x^2 - x - 2 = 10^2$

$3x^2 - x - 2 = 100$

$3x^2 - x - 2 - 100 = 0$

$3x^2 - x - 102 = 0$ two numbers that multiply to -306 and add up to -1. -18 and 17 satisfy these conditions

$3x^2 - 18x + 17x - 102 = 0$

$3x(x-6) + 17(x-6)$

$(3x+17)(x-6) = 0$ find the zeros, keep restrictions in mind. Test out the solution to see which one works.

$3x + 17 = 0 \quad 3x = -17 \qquad\qquad x = -\dfrac{17}{3}$

$x - 6 = 0 \qquad\qquad x = 6$

test: $x = -\dfrac{17}{3}$

$\log\left(3\left(-\dfrac{17}{3}\right)+2\right) + \log\left(\left(-\dfrac{17}{3}\right)-1\right) = 2$

$\log(-17+2) + \log\left(-\dfrac{17}{3}-1\right) = 2$

$x = -\dfrac{17}{3}$ is not a solution, you cannot take the log of a negative number.

test: $x = 6$

$\log(3(6)+2) + \log((6)-1) = 2$

$\log(18+2) + \log 5 = 2 \qquad \log 20 + \log 5 = 2$

$\log(20 * 5) = 2 \qquad\qquad \log 100 = 2$

$2 = 2 \qquad\qquad\qquad x = 6$ is the only solution.

Example 7: $\quad 2\log_2 x = 3 + \log_2(x-2)$

$2\log_2 x - \log_2(x-2) = 3$

$\log_2 x^2 - \log_2(x-2) = 3$

$\log_2\left(\dfrac{x^2}{x-2}\right) = 3$

boot the log,

$$\frac{x^2}{x-2} = 2^3$$

$$\frac{x^2}{x-2} = 8$$

$$\frac{(x-2)x^2}{x-2} = 8(x-2)$$

$x^2 = 8x - 16$

$x^2 - 8x + 16 = 0$ two numbers that multiply to 16 and add up to -8. -4 and -4

$(x-4)(x-4) = 0$

$x = 4$

$x = 4$ test

$2\log_2(4) = 3 + \log_2(4-2)$

$2(2) = 3 + \log_2 2$

$4 = 3 + 1$

$4 = 4$

$x = 4$ **is the solution**

CHAPTER 10 PART B: PRACTICE QUESTIONS

1. Write the following logarithms as a single logarithm:
   a) $\log 6 + \log 5$
   b) $\log 7 - \log 2$
   c) $2\log x - 3\log y$
   d) $3^{\log_3 7}$

2. Solve the following Logarithmic equations:
   a) $\log(x-2) = 1$
   b) $\log 15 - \log 5 = \log x$
   c) $\log x + \log x = \log 36$

3. Solve the following exponential equations by either using logarithms, or converting to common bases, and represent as exact value or approximate:
   a) $3^x = 11$
   b) $7^{3x} + 3 = 21$
   c) $3^x = 27^{x-1}$

Chapter 17: Logarithms and Exponents PART C

Applications of Exponential equations

&

Exponential and Logarithmic Functions

In this section, you will learn about how Exponential equations can be applied in real life. After that, you will learn about the graphs of Exponential and Logarithmic Functions.

We will start by looking at the Compound Interest formula (interest that accumulates over time on a principal amount).

$$A = P(1+r)^t$$

- A — Final Amount
- P — Initial
- r — Rate of yearly interest
- t — years

The reason there is a 1 inside of the brackets, is because we are calculating the final amount, and this means 100% of the initial amount plus the interest for a certain number of years.

Example 1: You would like to invest in the stock market where you know historically the average return compounded annually is 8% you must invest $1200 for t years. Write an equation modelling this situation.

- Rate of yearly interest is 8%
- Initial amount is $1200

What this means is that there is an interest amount of 8% that is added to this $1200 every year.

$r = 0.08$ 8%

$P = 1200$ $1200

$t = number\ of\ years$

$A = final\ amount$

$A = P(1 + r)^t$

Putting all of this information together we get the

Final equation: $A = 1200(1 + 0.08)^t$

Example 2: Write an exponential equation and determine the accumulated amount if $4000 invested at 10.5% compounded annually for 10 years.

$P = 4000$

$r = 0.105$

$t = 10$

$A = P(1+r)^t$

$A = 4000(1+0.105)^{10}$

$A = 4000(1.105)^{10}$

$A = 4000(2.714080847)$

$A = \$10{,}856.32$ *if you invest $4000 at 10.5% compounded annually for 10 years, the final amount will be* $\$10{,}856.32$

Remember to always start with what is inside of the brackets first, raise it to the power of the exponent, and then multiply it by the initial amount. Do not enter the entire equation into your calculator as some do not have a BEDMAS system programmed and will calculate your input from left to right, instead of performing the operations in the correct order. If your calculator does perform operations in the appropriate order then you can go ahead and enter the entire equation; however, I think your teacher will want you to show your work.

Example 3 a) : In fog, for every metre d, you move away from a light source, the intensity of light is reduced by 4%. Write an equation to model this situation. What percent of light is remaining if you are standing a distance of 12 metres from the light source?

We can use the same equation form; however, we will use different letters that model the situation more appropriately.

$I = \%$ of intensity of light remaining

$P =$ initial intensity of light remaining $= 100\%$

$r =$ rate is $4\% = 0.04$

$t =$ distance d, from light source $= 12$

$I = P(1-r)^d$

$I = 100(1-0.04)^{12}$

$I = 100(0.96)^{12}$

$I = 100(0.612709757)$

$I \approx 61.27\%$

Example 3 b) : How far is the light source if 30% of light is remaining?

$I = 30\%$

$I = 100(1 - 0.04)^d$

$30 = 100(0.96)^d$ *divide both sides by 100*

$$\frac{30}{100} = \frac{100(0.96)^d}{100}$$

$\log 0.3 = d\log(0.96)$ *take the log of both sides, and use the power rule, bring the d exponent in front of the logarithm.*

$$\frac{\log 0.3}{\log 0.96} = \frac{d\log 0.96}{\log 0.96}$$

$$d = \frac{\log 0.3}{\log 0.96}$$

$d = 29.49 \ metres$

Example 4: A Sports Car is priced at $65,000 and depreciates at a rate of 9% per year. V(value), t(years).

Let's model this problem with an equation,

$V = value$

$t = years$

$P = Original\ value = \$65,000$

$r = $ rate of depreciation$(9\%) = 0.09$

$V = P(1 - r)^t$

$V = 65000(1 - 0.09)^t$

$V = 65000(0.91)^t$

Example 4.b) How much is this Sports Car worth after 5 years?

$t = 5$

$V = 65000(0.91)^5$

$V = 65000(0.624032145)$

$V = \$40{,}562.09$

This Sports Car is worth $\$40{,}562.09$ after 5 years.

Example 4.c) After how many years will this Sports Car be worth $10,000.

$V = \$10{,}000 \quad t = ? \quad P = \$65{,}000 \quad r = 0.09$

$10\,000 = 65000(1 - 0.09)^t$

$10\,000 = 65\,000(0.91)^t$

$\dfrac{10000}{65000} = \dfrac{65000(0.91)^t}{65000} \quad$ *Divide both sides by 65000*

$0.153846154 = (0.91)^t$ *use power rule of logarithms*

$\log 0.153846154 = t \log 0.91$

$\dfrac{\log 0.153846154}{\log 0.91} = t \left(\dfrac{\log 0.91}{\log 0.91} \right)$

$t = 19.85 \; years \approx 20 \; years$

After 19.85 years (approximately 20), the Sports Car will be worth $10,000.

We will now look at the equation for Compound Interest (other than annual one we have looked at).

$A = P\left(1 + \dfrac{r}{n}\right)^{nt}$

As you can see, you are dividing r by n, and you are multiplying n by t. The reason for this is because the interest gets compounded more than once a year. Earlier, each example we looked at was examples of interest accumulating every year.

Compounded semi-annually (every 6 months): $n = 2$

Compounded Quarterly (every 3 months): $n = 4$

Compounded Monthly (every month): $n = 12$

Compounded Weekly: $n = 52$

Compounded Daily: $n = 365$ In this course, we assume there is no Leap Year for the calendar year.

Example 5: $2000 to double at an interest rate of 4.5% compounded quarterly.

$A = Final\ amount$ Since we are finding the doubling time

$A = \$4000$

$r = 4.5\% = 0.045$

$n = 4$ compunded quarterly (every 3 months)

$P = \$2000$ \qquad $t = years$

$$A = 2000\left(1 - \frac{0.045}{n}\right)^{nt}$$

$$A = 2000\left(1 - \frac{0.045}{4}\right)^{4t}$$

Before we move on, I want to talk about this modified formula. As you can see t represents the number of years, now it says 4t. Why? This is because you are saying, "Interest is accumulating 4 times a year." Why are we dividing r by 4 then? This is because, since we already have the 4t, we are saying, "This is happening 4 times a year, whatever is inside the brackets, so we need to divide r by 4."

$$4000 = 2000\left(1 - \frac{0.045}{4}\right)^{4t}$$

$4000 = 2000(1 - 0.01125)^{4t}$

$4000 = 2000(0.98875)^{4t}$

$\dfrac{4000}{2000} = \dfrac{2000(0.98875)^{4t}}{2000}$

$2 = (0.98875)^{4t}$ As you can see, we were looking for the doubling time of this investment, as a result, we got 2 on the left side.

Take the log of both sides,

$\log 2 = 4t \log 0.98875$ We can divide both sides by $4\log 0.98875$.

$\dfrac{\log 2}{4\log 0.98875} = \dfrac{4t \log 0.98875}{4\log 0.98875}$

$t = \dfrac{\log 2}{4\log 0.98875}$

$t \approx 15.5 \; years$

Example 6: What initial investment do you need to make to become millionaire in 25 years if you receive 12% per annum, compounded monthly?

$A = 1,000,000 \qquad P = ? \qquad t = 20 \qquad r = 0.12$

$n = 12$

$$A = P\left(1 + \frac{r}{n}\right)^{nt}$$

$$1{,}000{,}000 = P\left(1 + \frac{0.12}{12}\right)^{(12)(25)}$$

$$1{,}000{,}000 = P\left(1 + \frac{0.12}{12}\right)^{300}$$

300 means that it will gain interest 300 times in 25 years.

$$1{,}000{,}000 = P(1 + 0.01)^{300}$$

$$1{,}000{,}000 = P(1.01)^{300}$$

$$1{,}000{,}000 = P(19.78846626\ldots)$$

$$\frac{1{,}000{,}000}{19.78846626\ldots} = \frac{P(19.78846626\ldots)}{19.78846626\ldots}$$

$$P = \$50{,}534.49$$

Example 7: At what rate of interest would an investor need to be paid for an initial investment of $18,000 to grow into $67,000 in 20 years if the investment was compounded semi-annually?

$A = 67{,}000 \qquad P = 18{,}000 \qquad t = 20 \qquad n = 2$

$r = ?$

$$A = P\left(1 + \frac{r}{n}\right)^{nt}$$

$$67{,}000 = 18{,}000\left(1 + \frac{r}{2}\right)^{(2)(20)}$$

$$67{,}000 = 18{,}000\left(1 + \frac{r}{2}\right)^{40}$$

$$\frac{67{,}000}{18{,}000} = \frac{18{,}000\left(1 + \frac{r}{2}\right)^{40}}{18{,}000}$$

$$\frac{67{,}000}{18{,}000} = \left(1 + \frac{r}{2}\right)^{40}$$

$$\frac{67}{18} = \left(1 + \frac{r}{2}\right)^{40}$$

Do you remember how to remove an exponent from the other side? You take the reciprocal of that exponent.

$$\left(\frac{67}{18}\right)^{\frac{1}{40}} = \left(\left(1 + \frac{r}{2}\right)^{40}\right)^{\frac{1}{40}}$$

$$\left(\frac{67}{18}\right)^{\frac{1}{40}} = 1 + \frac{r}{2}$$

$$\sqrt[40]{\frac{67}{18}} = 1 + \frac{r}{2}$$

$$2(\sqrt[40]{\frac{67}{18}} - 1) = \frac{r}{2}(2)$$

$r = 0.0668$

Multiply by 100 *to get rate of interest* $= 6.68\%$

Doubling Time

Doubling time is the amount of time it takes before an investment, population, etc., doubles.

Example 1: An investment of $3000 is compounded annually at 12%. How long will this investment take to double?

$A = 2(3000) = 6000 \quad P = 3000 \quad r = 0.12$

$6000 = 3000(1 + 0.12)^t$

$$\frac{6000}{3000} = \frac{3000(1.12)^t}{3000}$$

$2 = 1.12^t$

$\log 2 = t\log 1.12$

$$\frac{\log 2}{\log 1.12} = t$$

$t = 6.12 \; years$ 6.12 is called the "doubling time"

Sometimes you will be asked to write the equation in terms of its doubling time, to do this you write the equation like this:

$6000 = 3000 * 2^{\frac{t}{6.12}}$

You are multiplying 3000 by 2 because you are doubling it, and you divide t by 6.12 because that way, after 6.12 years, 6.12 divided by 6.12 is 1; therefore, the exponent becomes 1 and you are simply multiplying 3000 by 2. Look,

$6000 = 3000 * 2^{\frac{6.12}{6.12}}$

$6000 = 3000 * 2^1$ your doubling time equation must make the 2 have an exponent of 1 when you plug in the doubling time.

$6000 = 6000$

Example 2: In 1985, a certain city in British Columbia had a population of 40,000 and was growing at a rate of 6.1% per year. Write the equation and determine the doubling time. Write population as a function of time(years).

$P(t) = 40000 * 2 = 80,000$

$P(t) = 40000(1.061)^t$

$80{,}000 = 40{,}000(1.061)^t$

$\dfrac{80{,}000}{40{,}000} = \dfrac{40{,}000(1.061)^t}{40{,}000}$

$2 = 1.061^t$

$\log 2 = t \log 1.061$

$\dfrac{\log 2}{\log 1.061} = t$

$t = 11.71 \; years$ *11.71 is the doubling time*

Doubling time equation: $80{,}000 = 40{,}000 * 2^{\frac{t}{11.71}}$

Exponential Decay and Half-Life

We will now look at exponential decay and its half life.

Example 1: The battery in a laptop computer loses 15% of its charge every hour. Assume the battery is 100% charged. Write a function to represent percentage of battery left (P) as a *function of h(hours) since the laptop was fully charged. Determine its half-life.*

$P(h) = 100(0.85)^h$ *0.85 is in the equation because 100%-15%=85%.*

$50 = 100(0.85)^h$ *there is a 50 on the left side because half of 100% is 50%.*

$$\frac{50}{100} = \frac{100(0.85)^h}{100}$$

$$\frac{1}{2} = 0.85^h$$

$\log 0.5 = h\log 0.85$

$$\frac{\log 0.5}{\log 0.85} = h\frac{\log 0.85}{\log 0.85}$$

$$h = \frac{\log 0.5}{\log 0.85}$$

$h = 4.27\ hours$

To write the equation with its half-life, write the equation with the base $\frac{1}{2}$.

$$50 = 100 * \left(\frac{1}{2}\right)^{\frac{h}{4.27}}$$

Example 2: Radioactive Iodine was released in the atmosphere at the Chernobyl Nuclear Accident in Russia, April 26, 1986. Its half-life is 8.1 days.

a) Write an equation which represents the percent, A, of Iodine left after t, days.

$$A = 100\left(\frac{1}{2}\right)^{\frac{t}{half\ life}}$$

$$A = 100\left(\frac{1}{2}\right)^{\frac{t}{8.1}}$$

Example 3: 20% of a certain isotope decays in 7 days. What is the half-life of the isotope?

$$80 = 100\left(\frac{1}{2}\right)^{\frac{7}{n}}$$

$$\frac{80}{100} = \frac{100\left(\frac{1}{2}\right)^{\left(\frac{7}{n}\right)}}{100}$$

$$0.8 = \left(\frac{1}{2}\right)^{\frac{7}{n}}$$

take the log of both sides and use the power rule of logarithms.

$$\log 0.8 = \frac{7}{n}\log\frac{1}{2}$$

$$\frac{\log 0.8}{\log 0.5} = \frac{7}{n}$$

$$\frac{7}{n} = 0.321928095$$

$$7 = 0.321928095n$$

$$\frac{7}{0.321928095} = n$$

$$n = 21.74 \text{ days.}$$

Half life equation: $50 = 100\left(\frac{1}{2}\right)^{\frac{t}{21.74}}$

How Many Times as Great?

$$A = P(5)^{\frac{20}{35}}$$

2.51 times as great

To figure out how many times as great, we are only concerned with this section

Example 1: The population of a certain bacteria can increase tenfold in 5 days. The initial number of bacteria is 1000. How many are there after 15 days?

$A = P(C)^{\frac{t}{n}}$ Tenfold, C=10   n= how often does it grow tenfold, every 5 days.

$A = 1000(10)^{\frac{15}{5}}$
$A = 1000(10)^3$
$A = 1,000,000$ *there are 1,000,000 bacteria after 15 days.*

Divide to find how many times as great one quantity is of another.

$$\frac{1000000}{1000} = 1000$$ times as great after 15 days.

Example 2: A certain bacteria can multiply fivefold every 35 minutes.

a) Write the equation

$$A = P(5)^{\frac{t}{35}}$$

b) In 20 minutes how many times as great?

Graphs of Exponential Functions

An exponential function is a function whose equation is in the form:

$$y = Ab^x$$

where A is the coefficient, x is the variable, and b is the base

Example 1: $y = 2^x$

$y - intercept:$ $y = 2^0$ $y = 1$ **1 is the y-intercept.**

$x - intercept:$ $0 = 2^x$ this does not make any sense, therefore, there is no x-intercept.

Domain: $\{x \mid xER\}$ all real numbers since x can have any value.

Range: $\{y \mid y > 0, YER\}$ y has to be greater than 0 because as you saw earlier there is no x-value that will result in 0.

Asymptote: $y = 0$

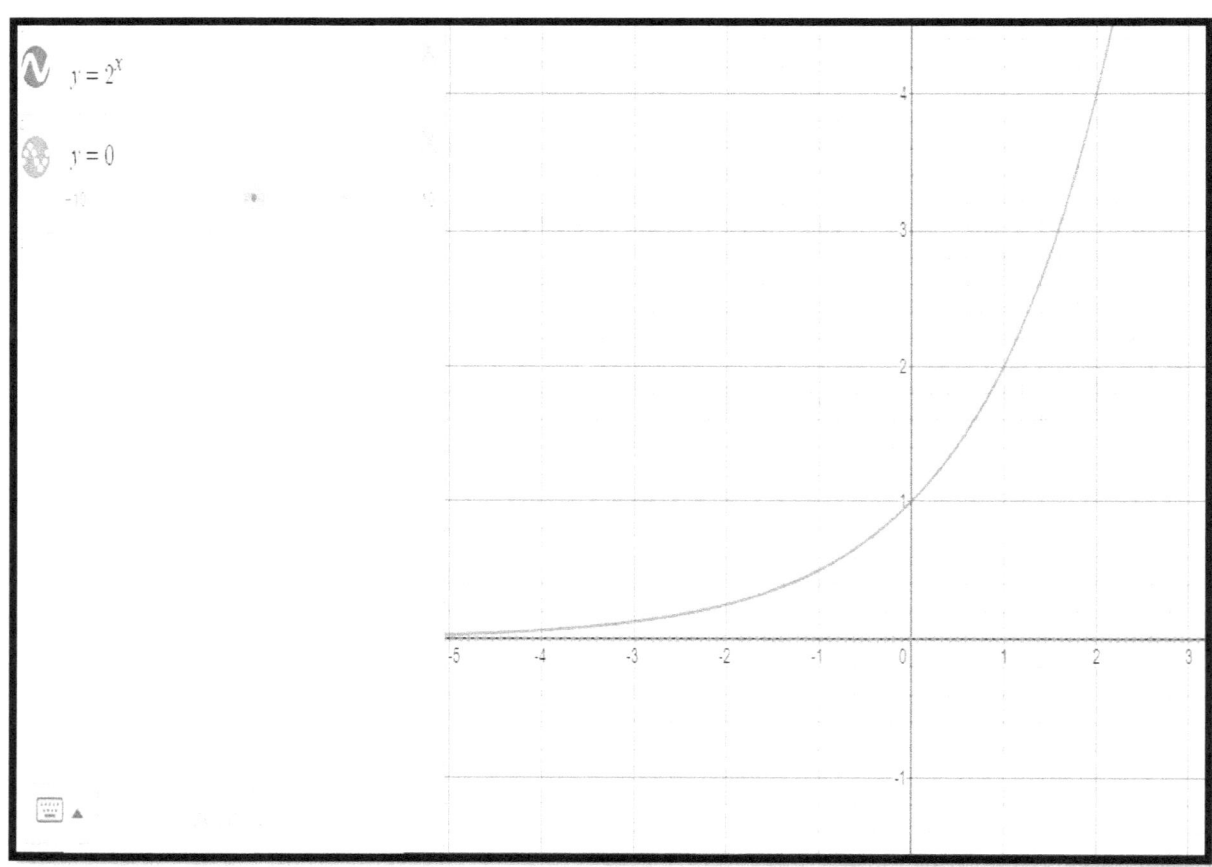

Example 2: $y = \left(\dfrac{1}{2}\right)^x$

y – intercept: $y = \left(\dfrac{1}{2}\right)^0$ $y = 1$ **y-intercept is 1**

x – intercept: $0 = \left(\dfrac{1}{2}\right)^x$ this does not make any sense, therefore, there is no x-intercept.

Domain: $\{x|X \in R\}$

Range: $\{y|y > 0, Y \in R\}$

Asymptote: $y = 0$

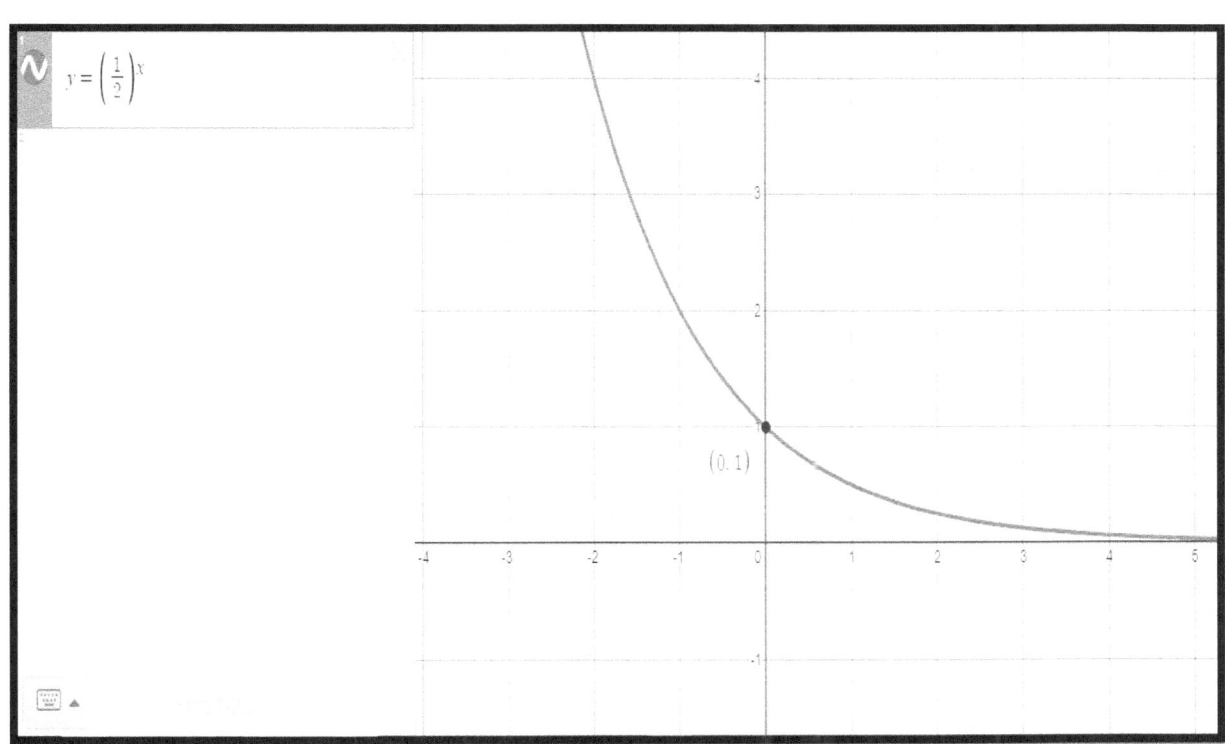

As you can see, it has been reflected in the y-axis, why?

Read the next page to find out why.

$y = \left(\frac{1}{2}\right)^x$ came from the graph $y = 2^{-x}$,

Remember to simplify a negative exponent you take its reciprocal.

$y = 2^{-x}$ becomes $y = \left(\frac{1}{2}\right)^x$

To graph this, you use the equation $y = 2^{-x}$ and you graph the function $y = 2^x$ then you reflect it in the y-axis.

Example 3: $y = 2^{x-3} - 5$

Translated 3 units right and 5 units down

$y - $ intercept: $y = 2^{0-3} - 5 \qquad y = 2^{-3} - 5$

$y = \frac{1}{8} - 5 \qquad y = \frac{1}{8} - \frac{40}{8} \qquad y = -\frac{39}{8} \ or -4\frac{7}{8}$

$x - $ intercept: $0 = 2^{x-3} - 5$

$5 = 2^{x-3}$

$\log 5 = \log 2^{x-3}$

$\log 5 = (x-3)\log 2$

$$\frac{\log 5}{\log 2} = \frac{(x-3)\log 2}{\log 2}$$

$$\frac{\log 5}{\log 2} = x - 3$$

$$\frac{\log 5}{\log 2} + 3 = x$$

$\log_2 5 + 3 = x$ exact value of x-intercept, or $x = 5.32$ as a decimal approximation.

Asymptote: $y = 2^{x-3} - 5$

$y = -5$ is the equation of the asymptote because 2^{x-3} cannot equal 0; therefore, the output will never be $y = -5$. It will always be greater than -5.

Range: $\{y | y > -5, Y \in R\}$ Domain: $\{x | X \in R\}$

This is what the graph looks like:

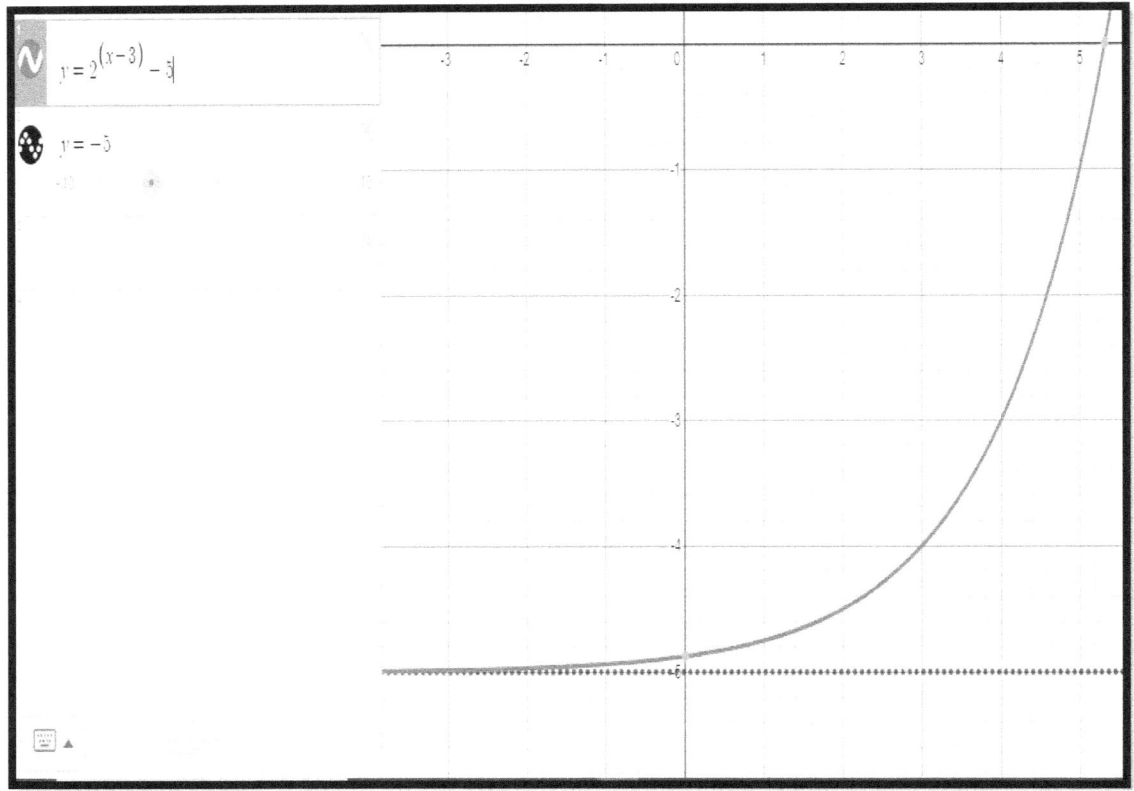

Logarithmic Functions

Logarithmic functions have a vertical asymptote instead of a horizontal one. To graph a logarithmic function, you must convert it to exponential form as it will be much easier. You can do this by booting the log.

Example 1: Graph $y = \log_2 x$

$x = 2^y$

So, instead of inputting x-values, you will be inputting y-values and the output will be the x-values.

This is what the graph looks like,

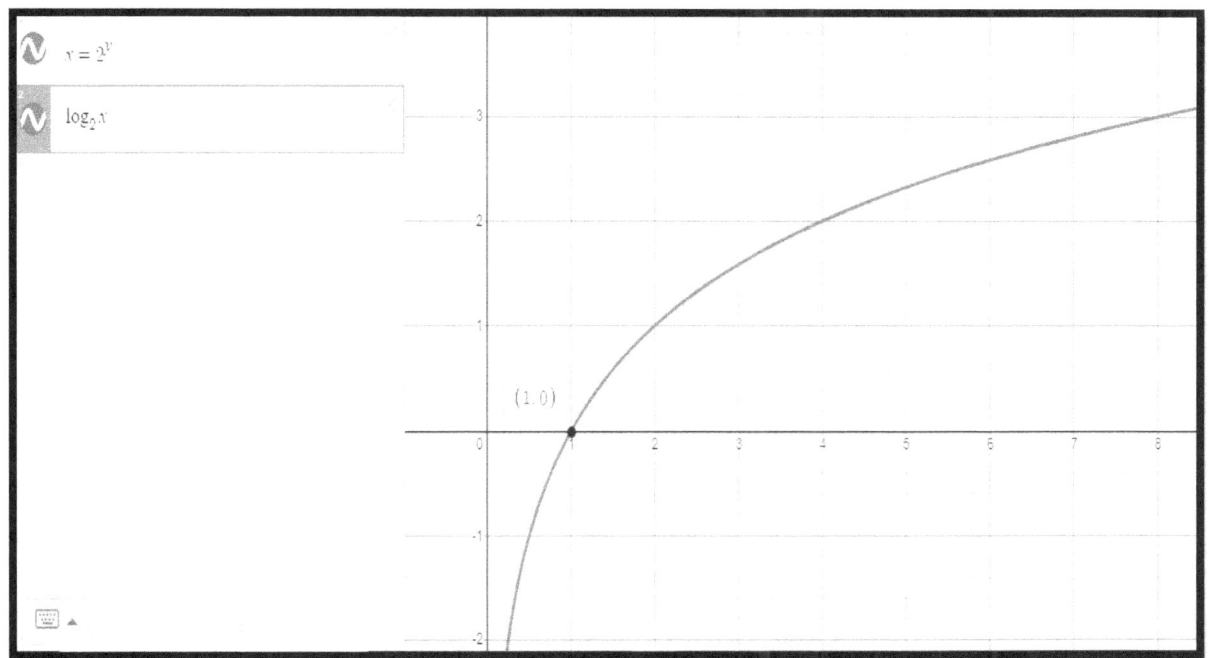

Asymptote: $x = 0$ Range: $\{y | y ER\}$

Domain: $\{x | x > 0, xER\}$ x-intercept: 1

Example 2: $\log_2(x+3) + 2$ In a case like this, it is best to just plug in several values and figure out the graph as when you boot the log, your translations will be confusing for you to read. Here is why

Boot the log, $y = \log_2(x+3) + 2$

$y - 2 = \log_2(x+3)$

$2^{y-2} = x + 3$

$2^{y-2} - 3 = x$

$x = 2^{y-2} - 3$ 2 units up and 3 units left,

y-intercept: $y = \log_2(0+3) + 2$

$y = \log_2 3 + 2$

$y = \dfrac{\log 3}{\log 2} + 2$ $y \approx 3.58$

x-intercept: $0 = \log_2(x+3) + 2$

$-2 = \log_2(x+3)$ boot the log

$2^{-2} = x + 3$

$\dfrac{1}{2^2} - 3 = x$

$\dfrac{1}{4} - 3 = x$ $x = -2.75$

Domain: $x + 3 > 0$

Domain: $\{x | x > -3, X\mathbb{ER}\}$

Asymptote: $x = -3$

Range: $Y\mathbb{ER}$

The graph is below,

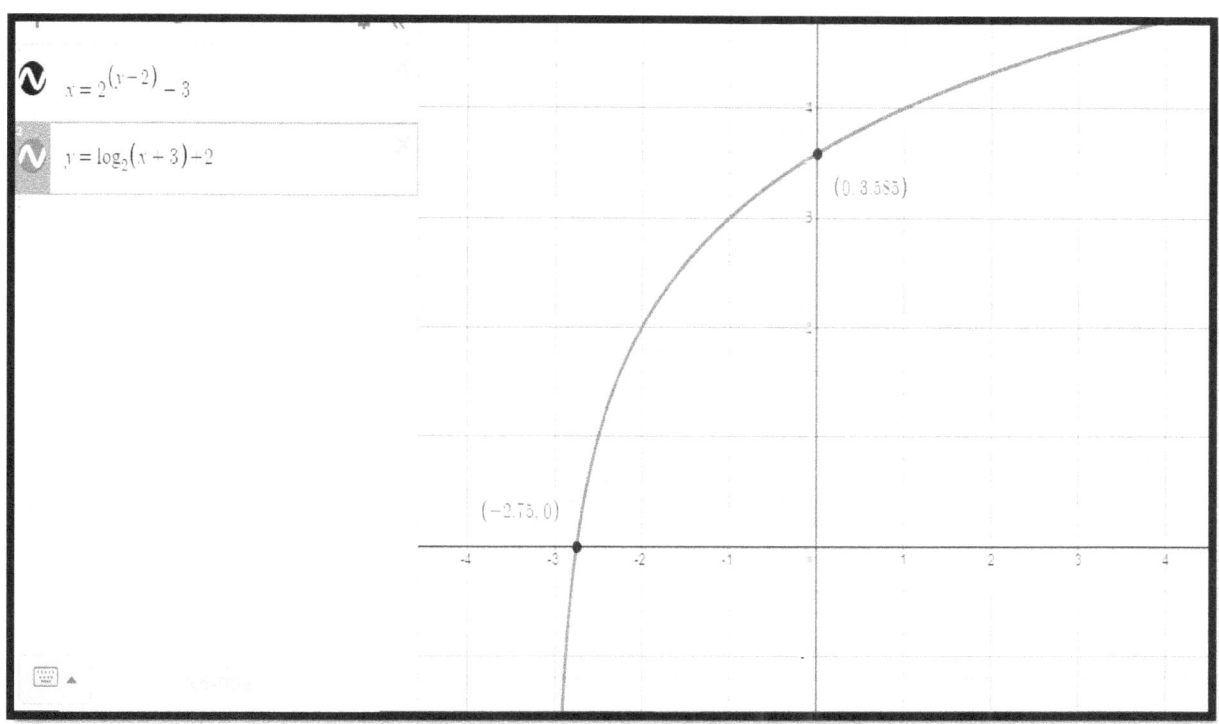

$y = \log_a x$ is the same as $x = a^y$

If $a > 1$ then the graph goes up to the right

If $0 < a < 1$ then the graph goes down to the right

Inverse Exponential Equations

In this section, we will look at the quick method that is used to determine inverse exponential equations in terms of "y=".

Example 1: Determine the inverse of $y = 2^x$.

Swap x and y,

$x = 2^y$ use logarithms to solve for y,

$\log x = \log 2^y$ use the power rule to continue simplifying,

$\log x = y \log 2$

$\dfrac{\log x}{\log 2} = y$ use the change of base law to continue simplifying,

$\log_2 x = y$ $\qquad y = \log_2 x$

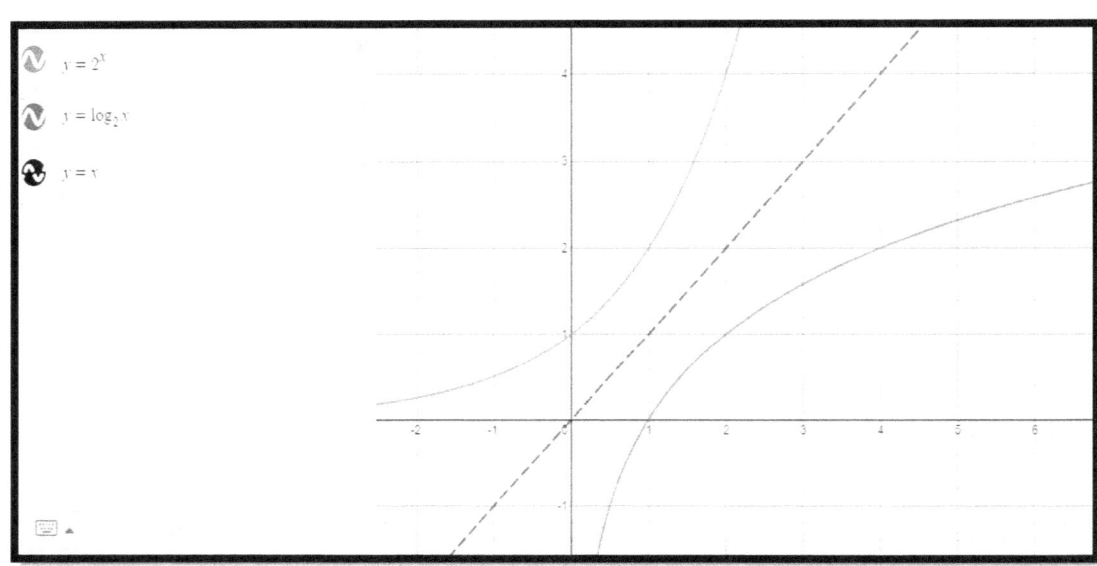

As you can see, the graph is a perfect reflection in the line $y = x$.

Example 2: Write the inverse of $y = 5^x$

$x = 5^y$

$$\log x = \log 5^y$$

$$\log x = y\log 5$$

$$y = \frac{\log x}{\log 5}$$

$$y = \log_5 x$$

Example 3: Write the inverse of $y = \left(\frac{1}{2}\right)^x$

$$x = \left(\frac{1}{2}\right)^y$$

$$\log x = \log \left(\frac{1}{2}\right)^y$$

$$\log x = y\log \frac{1}{2}$$

$$\frac{\log x}{\log \left(\frac{1}{2}\right)} = y$$

$$y = \log_{\frac{1}{2}} x$$

CHAPTER 10 PART C: PRACTICE QUESTIONS

1. Jeremy invests $1200 at a rate of 5% compounded monthly.

   a) How much money would Jeremy have after 8 years?

   b) How long does it take for the investment to grow to $4000?

2. At what rate of interest would an investment have to be paid for an initial $900 to grow into $1400 in 12 years if the investment was compounded quarterly?

3. The intensity of light decreases by 4% for each meter that it descends below the surface of the water. At what depth is the intensity only 30% of that at the surface?

4. Sketch the graph of $y = 2^{x-3} - 4$, state the equation of the asymptote, x-intercept, the y-intercept, and the domain and range.

5. Sketch the graph of $y = \log_2(x+2)$, state the equation of the asymptote, x-intercept, the y-intercept, and the domain and range.

6. Sketch the graph of $y = \log_3 x$, state the equation of the asymptote, x-intercept, the y-intercept, and the domain and range.

7. What is the inverse of $y = 8^x$?

8. What is the inverse of $y = 4^{x-2}$?

9. What is the inverse of $y = \log_3 x$?

Circular Functions

Chapter 11

This chapter, due to its size, will be divided into two sections (Radian measure and exact values, and the other section will be Trigonometric Functions).

Circular Functions Section A:

In this section, we will look at a new kind of measure, Radian measure. What is a Radian? A radian is the measure of the angle at the centre of a circle opened up, by an arc equal to the length of the radius of the circle. The radius of the circle fits inside of the circle 6.28 times approximately, or 2π times exactly. Remember the formula for the circumference of a circle? $C = 2\pi r$, as you can see, the circumference is 2π times the radius, which means all the way around.

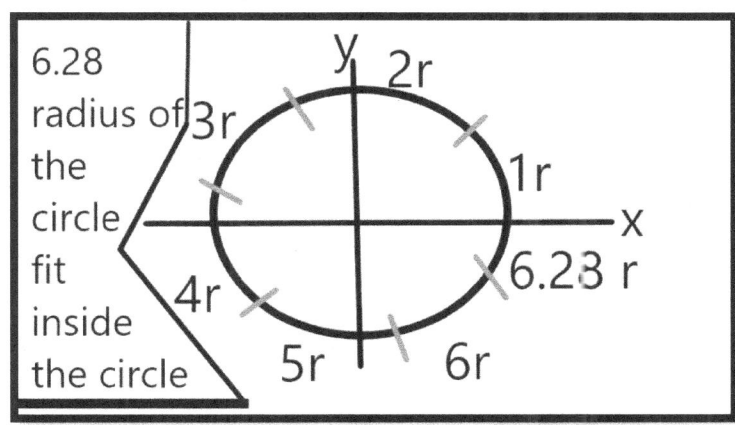

What is the measure of the angle subtended (opened up) by an arc whose arc length is the circumference of the circle?

$2\pi \; radians$ $6.28 \; radians$

Exact angle   Decimal approximation

$2\pi \; radians = 360°$

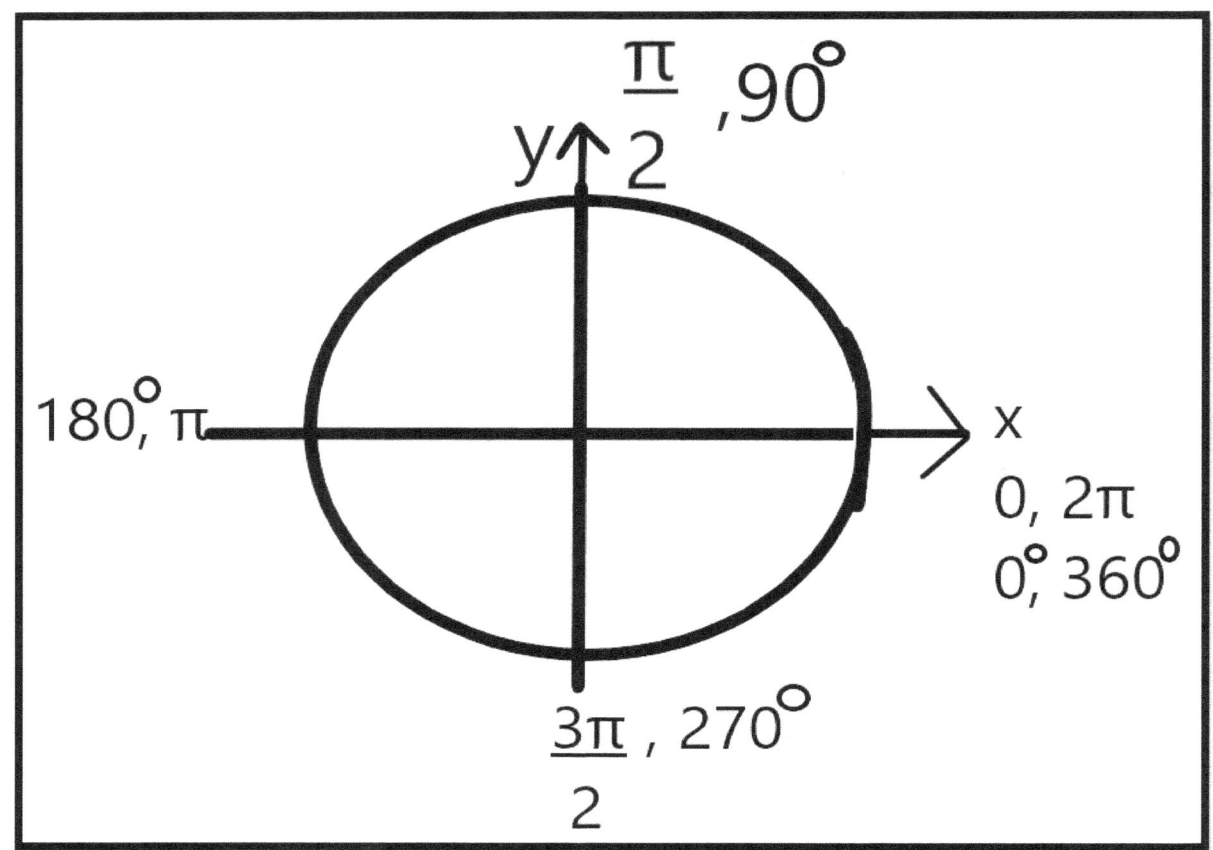

As you can see:

- $\frac{\pi}{2}$ radians is equal to $90°$. (1/4 of circle)
- π radians is equal to $180°$.
- $\frac{3\pi}{2}$ radians is equal to $270°$.
- 2π radians is equal to $360°$.

When we talk about radians it is always in terms of π.

π radians is half-way across the circle, $180°$.

$\pi\ radians = 180°$

To convert radians to degrees, multiply radians by: $\dfrac{180°}{\pi}$.

Example 1: Convert 5 radians to degrees, round it to two decimal places.

$\pi\ radians = 180°$

$\dfrac{\pi\ radians}{\pi} = \dfrac{180°}{\pi}$

$1\ radian = \dfrac{180°}{\pi}$

$5 * (1\ radian) = \left(\dfrac{180°}{\pi}\right) * 5$

$5\ radians = \dfrac{900°}{\pi}$

$5\ radians = 286.48°$

Example 2: Convert 3 radians to degrees,

$1\ radian = \dfrac{180°}{\pi}$

$$3 * (1\ radian) = \left(\frac{180°}{\pi}\right) * 3$$

$$3\ radians = \frac{540°}{\pi}$$

$$3\ radians = 171.89°$$

Example 3: Convert $\frac{5\pi}{6}$ *to degrees,*

$$\pi\ radians = 180°$$

$$\left(\frac{5}{6}\right)\pi\ radians = 180°\left(\frac{5}{6}\right)$$

$$\frac{5\pi}{6}radians = \frac{900°}{6}$$

$$\frac{5\pi}{6}radians = 150°$$

To convert degrees to radians, multiply degrees by $\frac{\pi}{180°}$.

Example 4: $135°$ *to radian measure,*

$$\pi\ radians = 180°$$

$$\frac{\pi \text{ radians}}{180} = \frac{180}{180}$$

$$\frac{\pi \text{ radians}}{180} = 1°$$

$$135° \left(\frac{\pi \text{ radians}}{180°}\right) = 1° * 135°$$

$$\frac{135\pi}{180} = 135°$$

$$\frac{3\pi}{4} \text{radians} = 135°$$

Example 5: Convert $60°$ to radians in terms of π,

$$1° = \frac{\pi}{180}$$

$$60°(1°) = \frac{\pi}{180}(60°)$$

$$60° = \frac{60\pi}{180}$$

$$60° = \frac{\pi}{3} \text{ radians}$$

We will now learn how radians can be used to determine arc length.

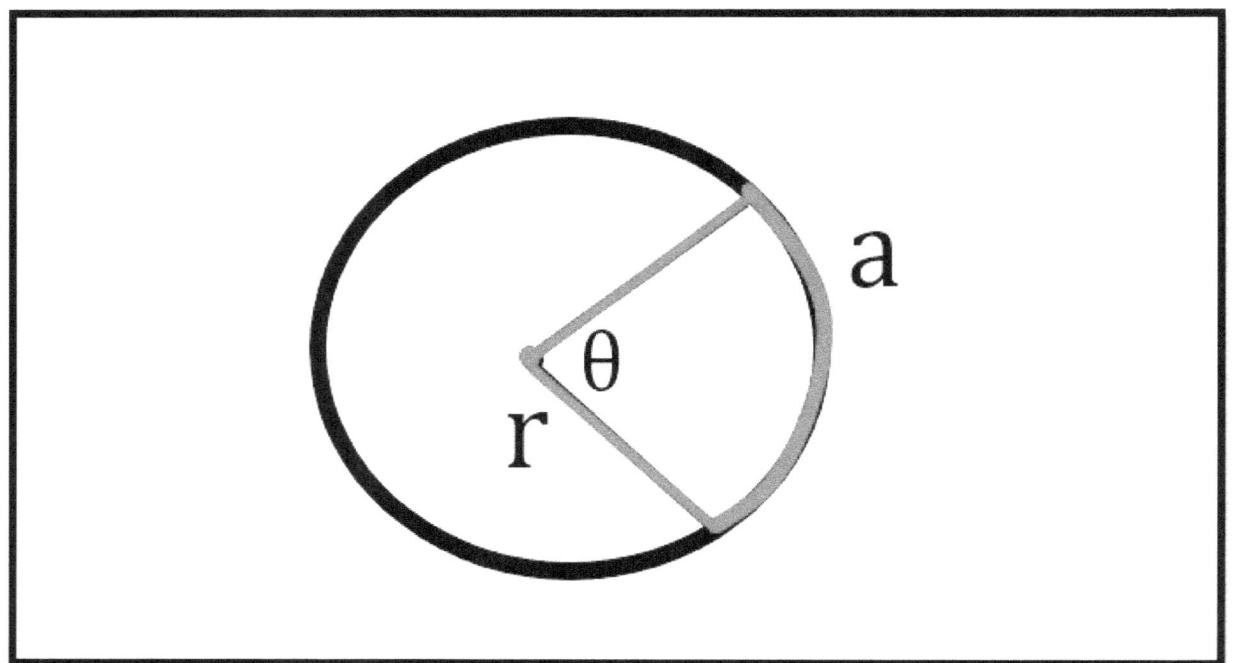

Here is how the formula for arc length is obtained,

$$\frac{arc\ length}{circumference} = \frac{angle\ at\ centre}{complete\ rotation\ 2\pi}$$

$$\frac{a}{2\pi r} = \frac{\theta}{2\pi}$$

$$\frac{a}{2\pi r}(2\pi) = \frac{\theta}{2\pi}(2\pi)$$

$$\frac{a2\pi}{2\pi r} = \theta$$

$$\frac{a}{r} = \theta$$

$a = r\theta$ *formula for arc length*

Example 1: Determine the length of the arc that subtends an angle of 5.2 radians at the centre of a circle with a radius of 12.7cm.

$a = r\theta$ $r = 12.7 cm$ $\theta = 5.2\ radians$

$a = (12.7)(5.2)$

$a = 66.04\ cm$

Example 2: Determine the length of an arc that subtends an angle of $280°$ at the centre of a circle with a radius of 7cm.

$a = r\theta$ $r = 7cm$ $\theta = radians$

First, we need to convert $280°$ to radians,

$$280° = \frac{\pi}{180}(280)$$

$$280° = \frac{280\pi}{180}$$

$$280° = \frac{14\pi}{9} \qquad \theta = \frac{14\pi}{9} = 4.89$$

$$a = r\theta$$

$$a = (7)(4.89)$$

$$a = 34.23 \ cm$$

Example 3: A circle has a radius of 15 cm. Determine the measure of the angle subtended by an arc of length of 22cm.

$$a = r\theta \qquad a = 22cm \qquad r = 15cm \qquad \theta = ?$$

$$22 = (15)\theta$$

$$\frac{22}{15} = \frac{15\theta}{15}$$

$$\theta = \frac{22}{15} = 1.47 \ radians$$

Angles in Standard Position

Remember in Chapter 1, we sketched angles in Standard position? In this section, we will sketch angles in radian measure.

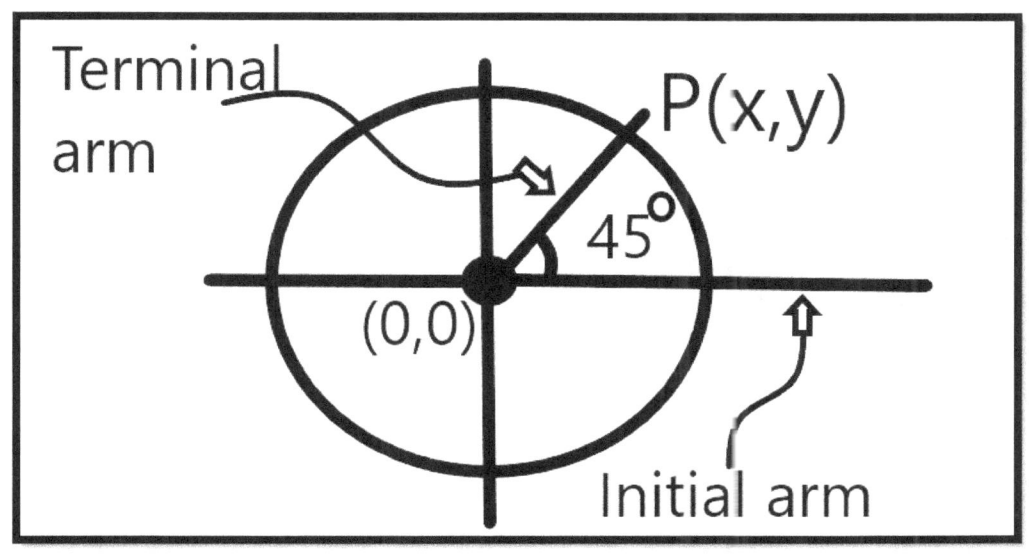

Example 1: Draw $\dfrac{5\pi}{6}$ in standard position,

- *Step 1: Divide π radians into 6 parts*
- *Step 2: The terminal arm lies on the 5th part*

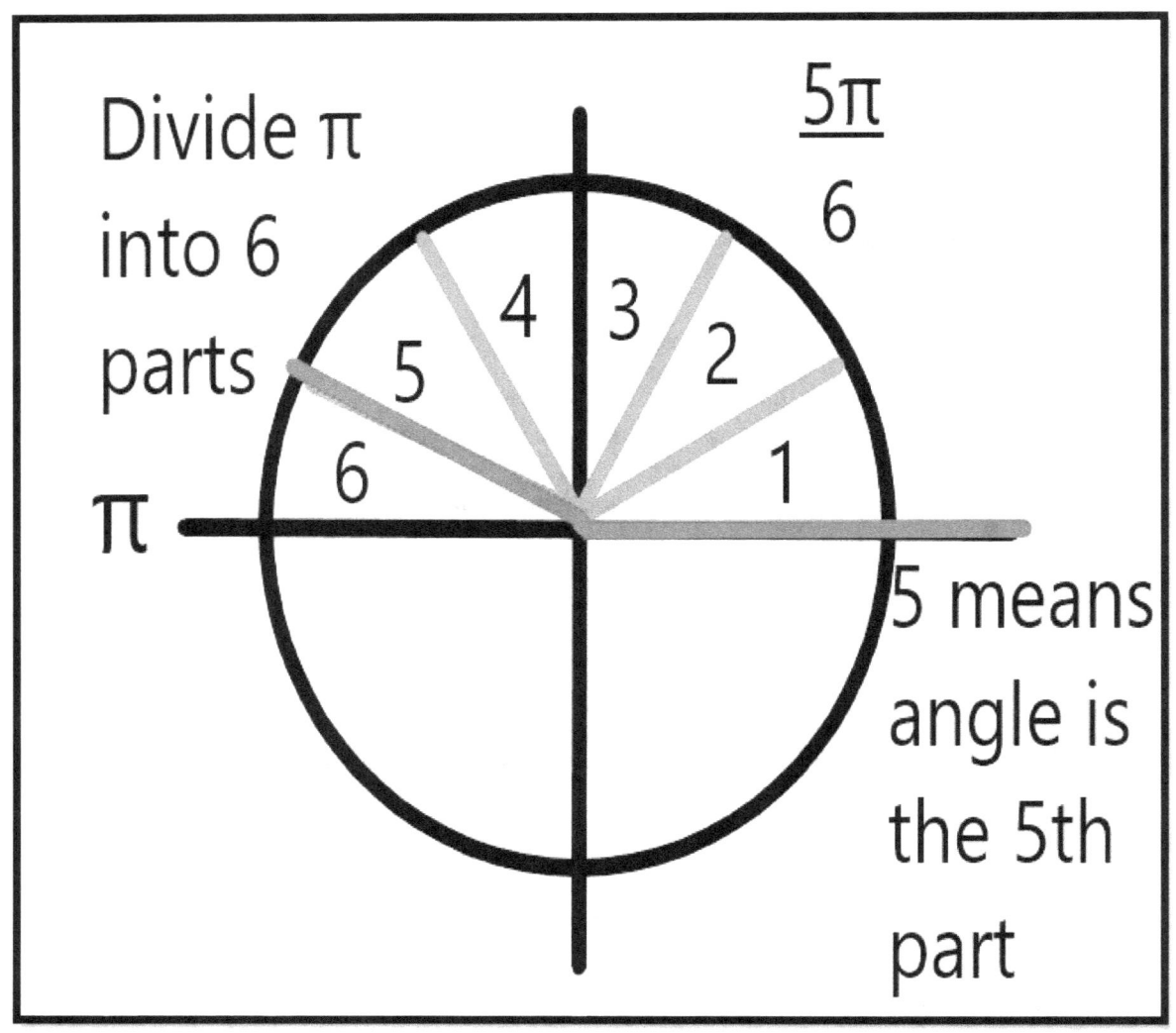

If $\theta > 0$ $(positive)$, the rotation is counter clockwise.

If $\theta < 0$ $(negative)$, the rotation is clockwise.

To find a coterminal angle (angle that lies at the same place as another angle but has a different measure):

- Add multiples of $360°$ to θ if θ is in degrees.

$$Coterminal\ angle = \theta + n(360°)$$

- Add multiples of 2π to θ if θ is in radians.
$$Coterminal\ angle = \theta + n(2\pi)$$

Example 1: What are some coterminal angles of $30°$?
$30 + 360 = 390°$
$30 + 2(360) = 750°$

Why are they coterminal angles? They have the same initial arm and terminal arm, and the position of Point P is the same for each angle.

To find a negative coterminal angle, subtract $360°$ from θ if working with degrees, or subtract 2π from θ if working with radians.

Example 2: Sketch the angle $-\dfrac{\pi}{3}$ in standard position and give two coterminal angles.

Since you are working with a negative angle, you will rotate the terminal arm clockwise. Divide π into 3 sections,

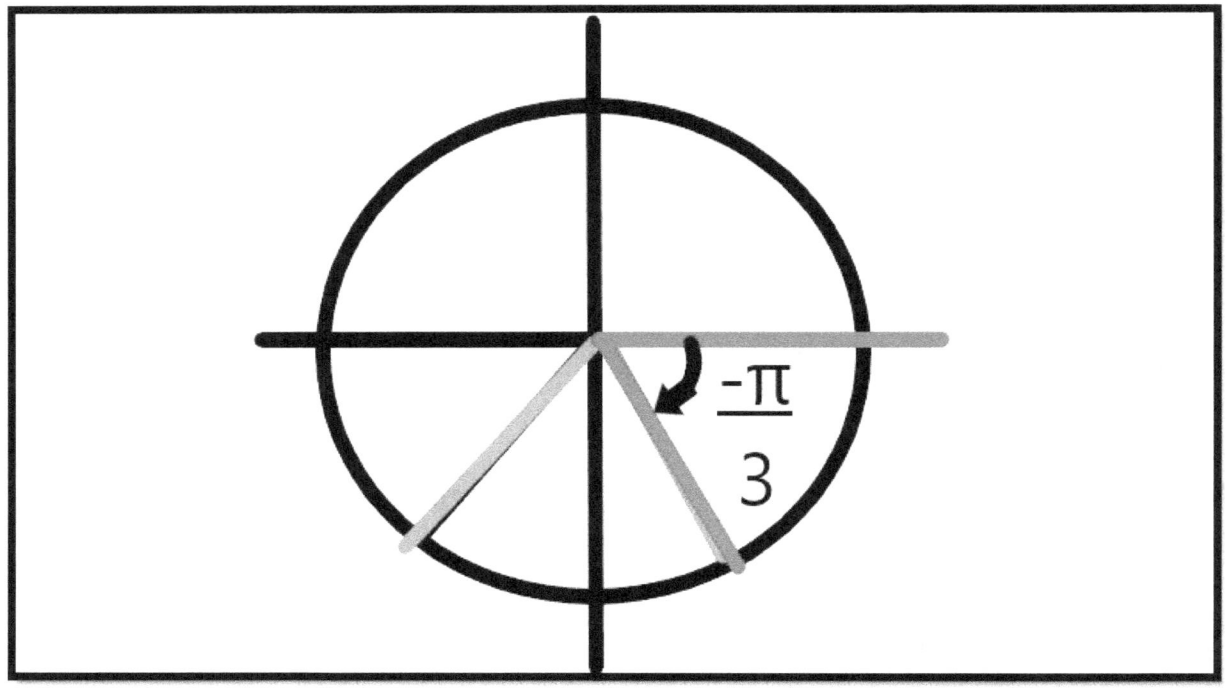

$Coterminal\ 1 = -\dfrac{\pi}{3} + n(2\pi)$

$= -\dfrac{\pi}{3} + 2\pi$

$= -\dfrac{\pi}{3} + \dfrac{6\pi}{3}$

$= \dfrac{5\pi}{3}$

$Coterminal\ 2 = -\dfrac{\pi}{3} + n(2\pi)$

$= -\dfrac{\pi}{3} + 2(2\pi)$

$= -\dfrac{\pi}{3} + 4\pi$

$= -\dfrac{\pi}{3} + \dfrac{12\pi}{3}$

$= \dfrac{11\pi}{3}$

What is the general solution for all coterminal angles for $-\dfrac{\pi}{3}$?

$-\dfrac{\pi}{3} + n(2\pi)$

There are an infinite number of angles coterminal to any angle.

Example 3: Sketch $-\dfrac{4\pi}{3}$ in standard position, in this example I will show you how I sketch these angles by hand,

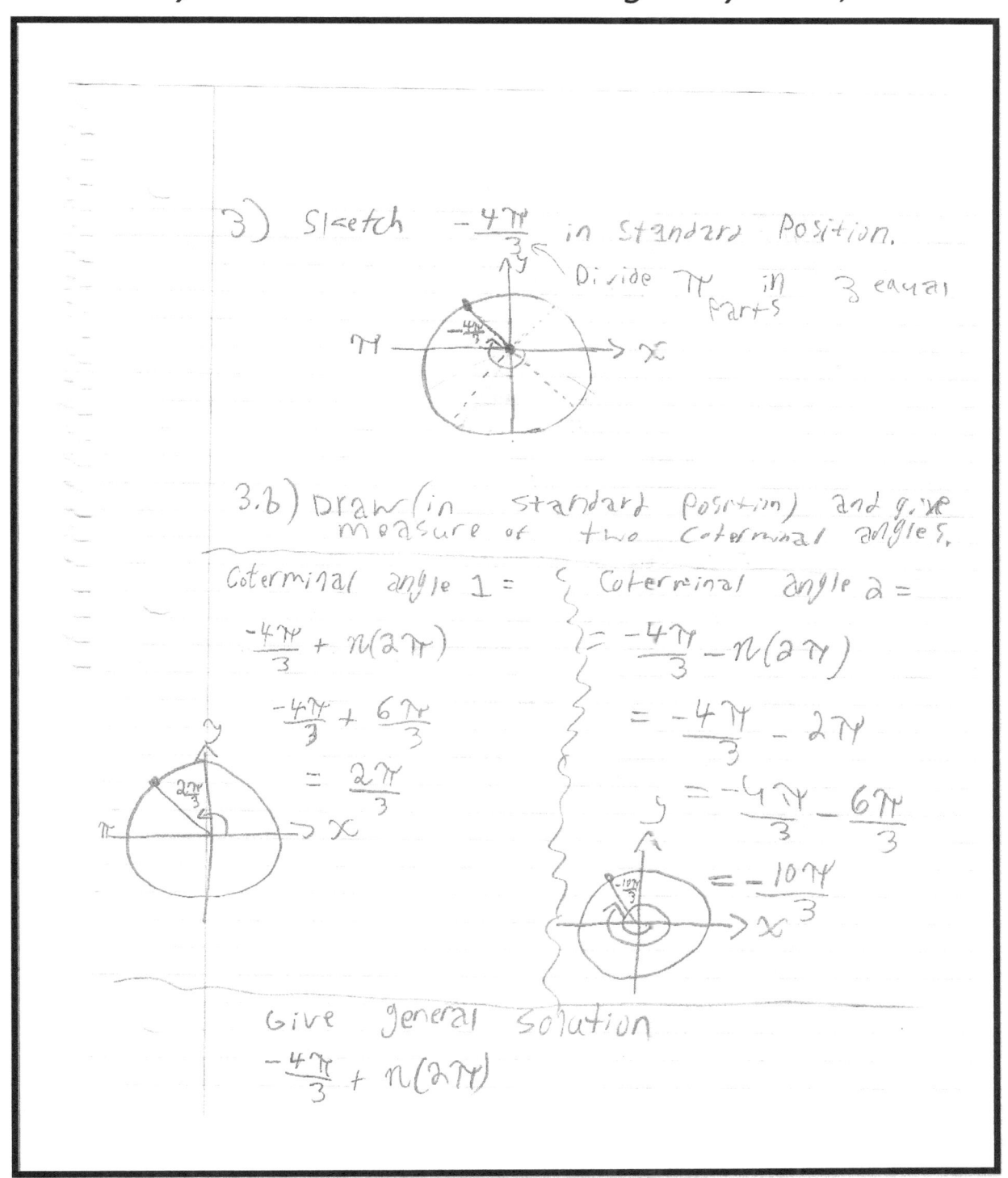

Sine, Cosine and Tangent Functions of angles

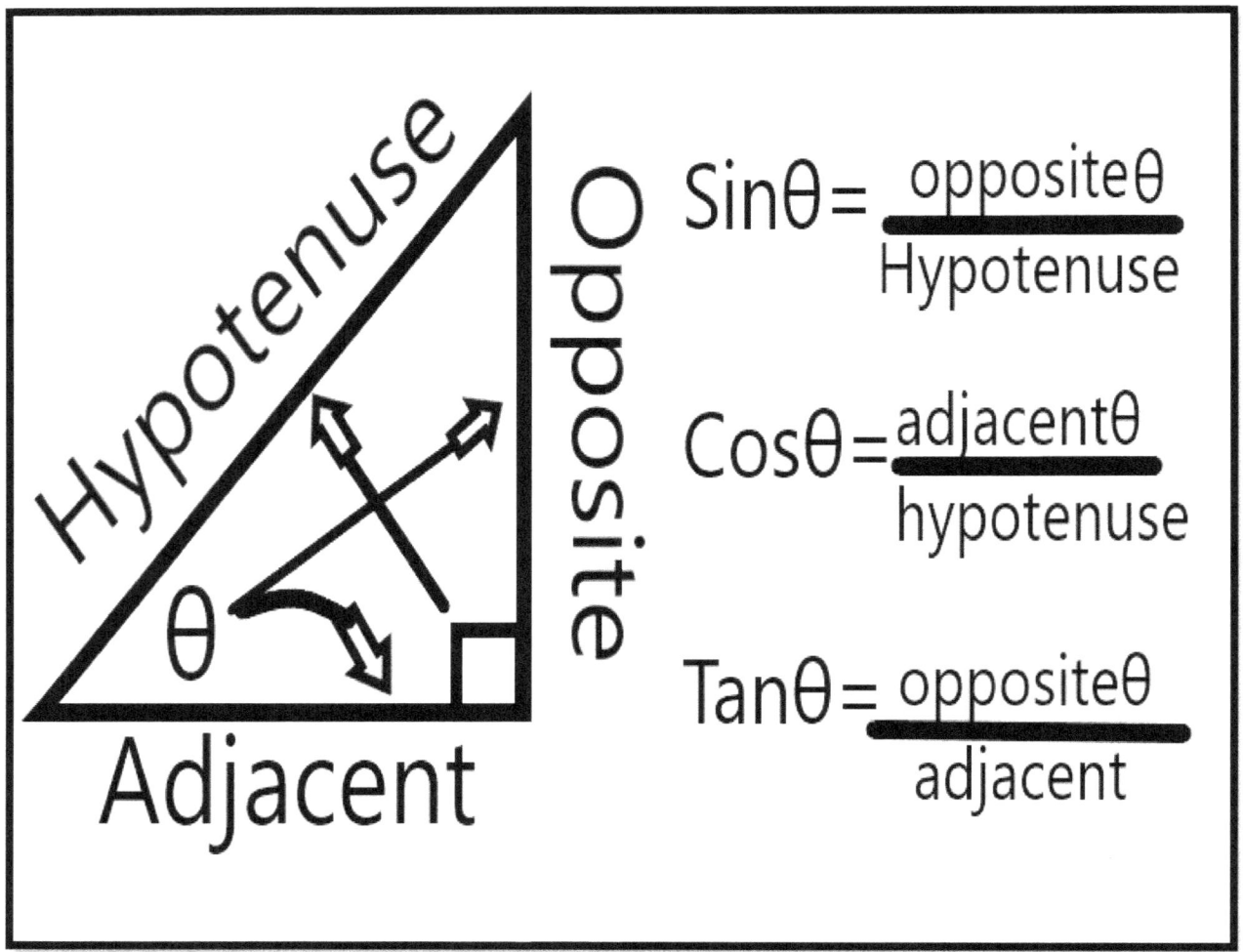

On the next page, we will define the trigonometric functions using the unit circle.

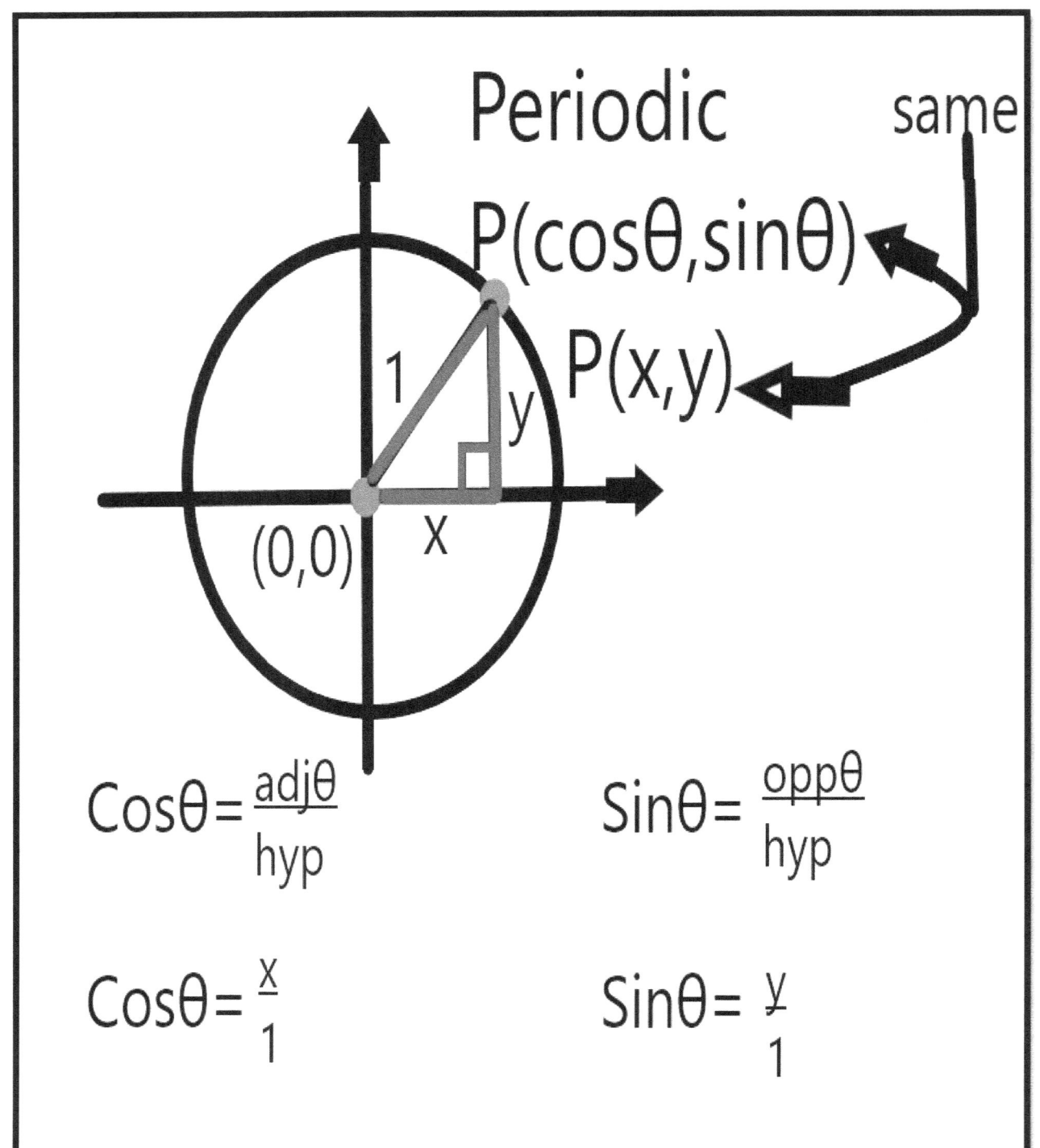

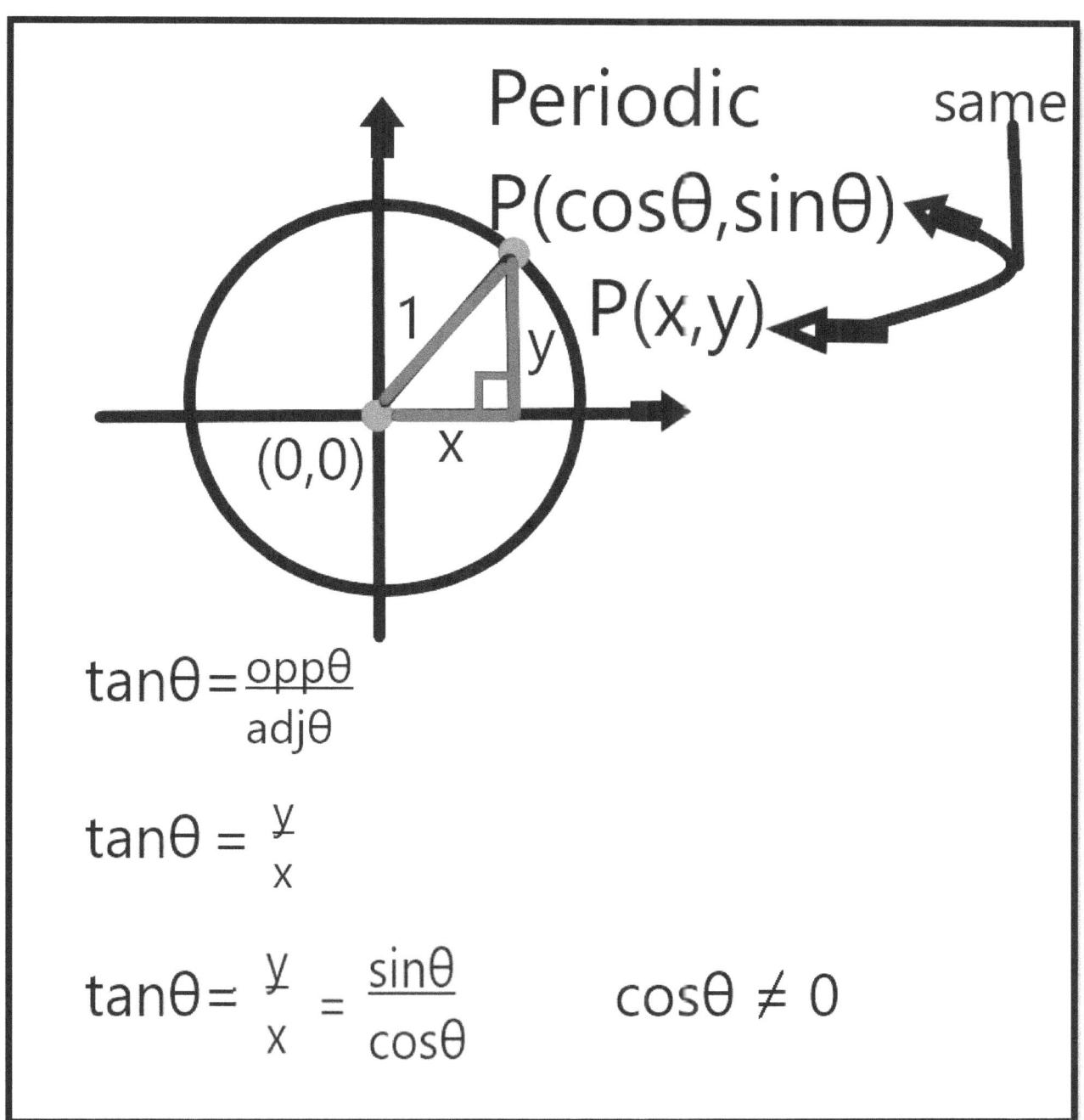

As you can see, $tan\theta = \frac{sin\theta}{cos\theta}$ where $cos\theta \neq 0$

Example 1:

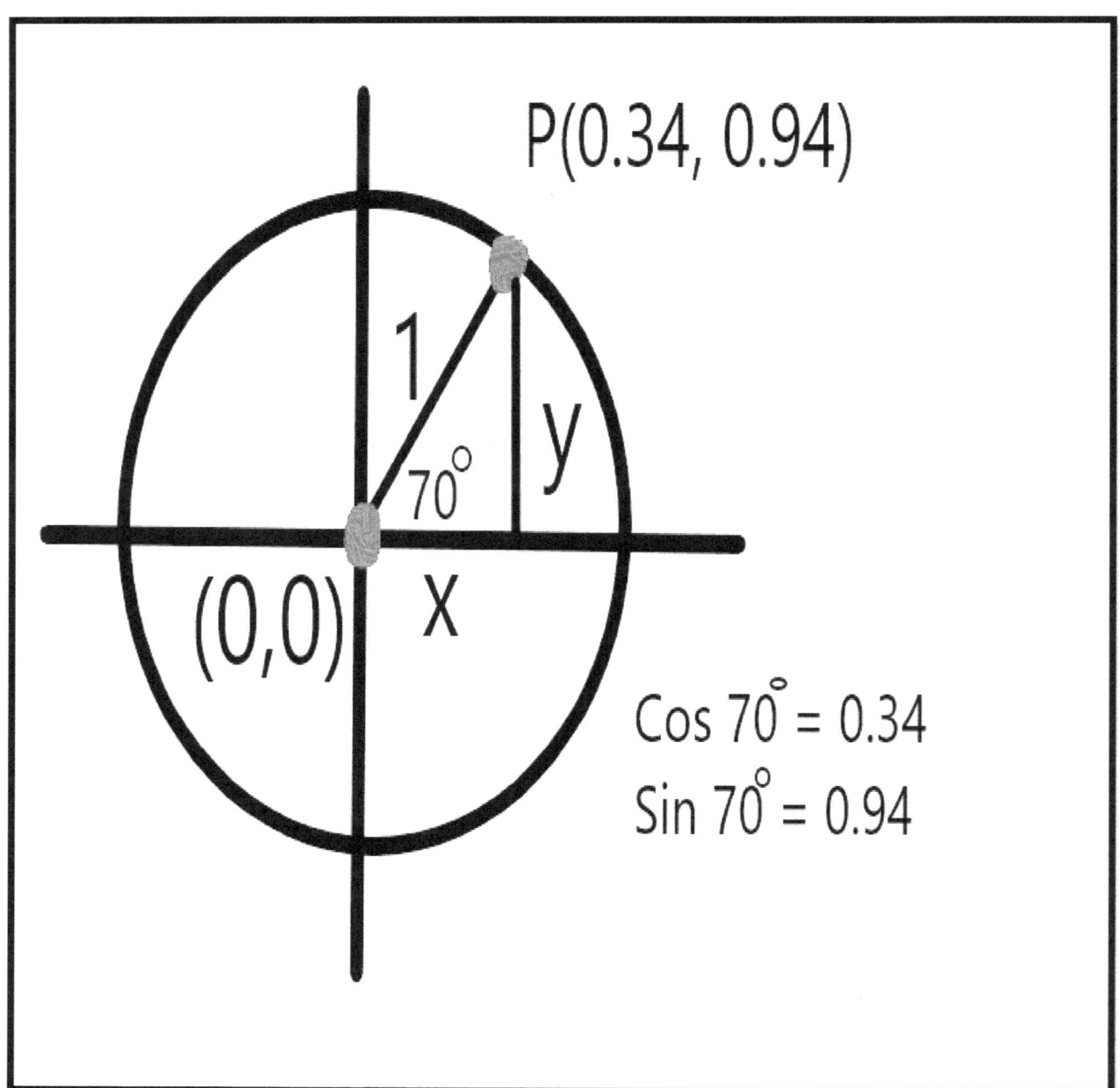

As you can see, the cosine of the angle was the x-coordinate of the point P, and the sine of the angle was the y-coordinate of the point P.

Example 2: Point A(-4,5) is in Quadrant 2 and lies on the terminal arm of angle θ in standard position. Point P is the point of intersection of terminal arm of θ and unit circle centred at (0,0). Determine the x-coordinate of P.

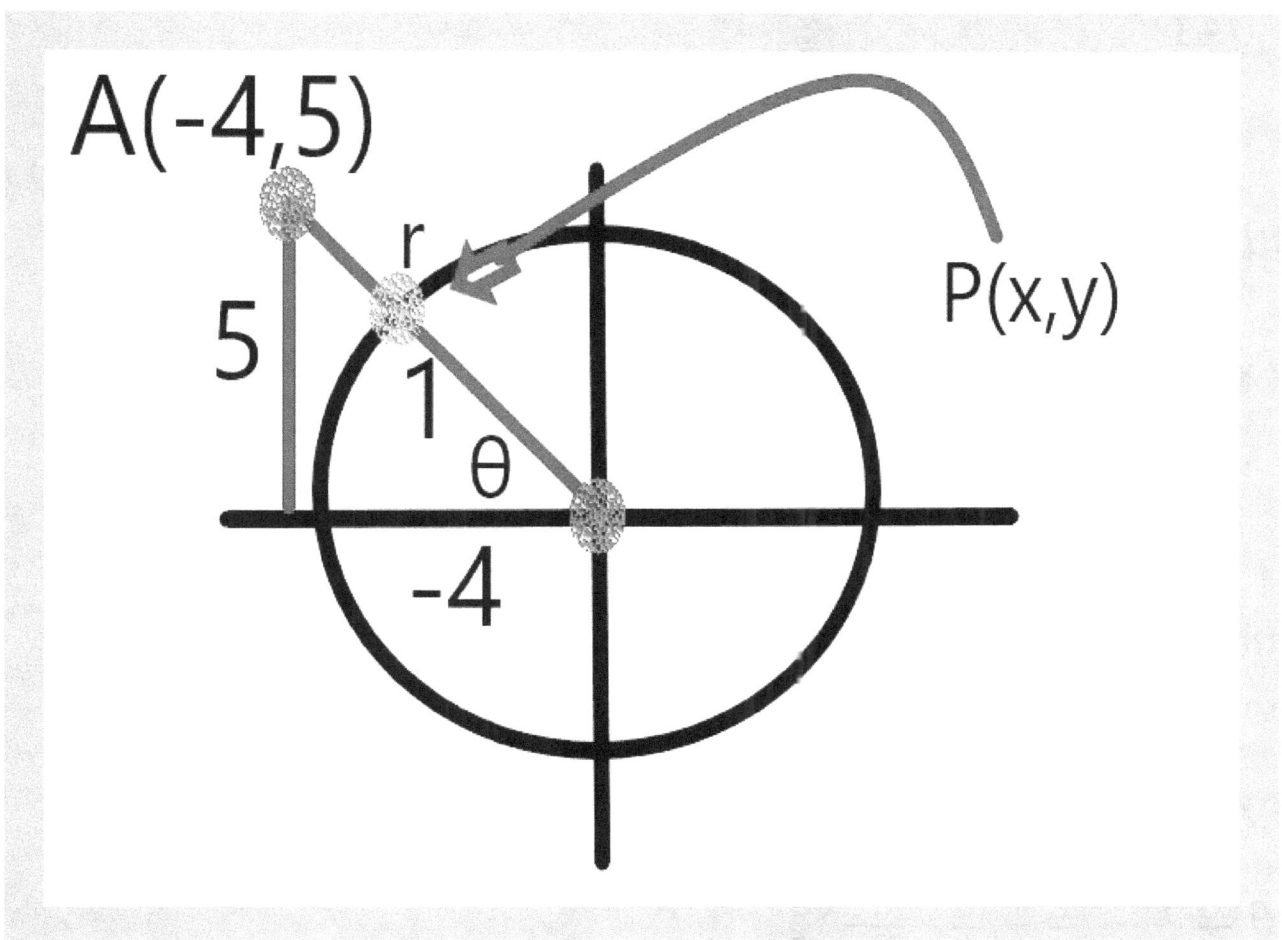

Remember that cosine is the ratio of the adjacent side over the hypotenuse.

$$\cos\theta = \frac{adj}{hyp} = \frac{x}{r} \qquad x = -4 \qquad y = 5$$

$$r^2 = x^2 + y^2$$

$$r^2 = (-4)^2 + (5)^2$$

$$r^2 = 16 + 25$$

$$r^2 = 41$$

$$r = \sqrt{41}$$

$$\cos\theta = \frac{x}{r} = -\frac{4}{\sqrt{41}} = -0.62$$

The x-coordinate of Point P is -0.62.

Remember that Sine is the ratio of the opposite side over the hypotenuse.

The y-coordinate of Point P is:

$$\sin\theta = \frac{opp}{hyp} = \frac{y}{r} = \frac{5}{\sqrt{41}}$$

Exact value: Point P $\left(-\dfrac{4}{\sqrt{41}}, \dfrac{5}{\sqrt{41}}\right)$

Decimal Approximation: Point P(-0.62, 0.78)

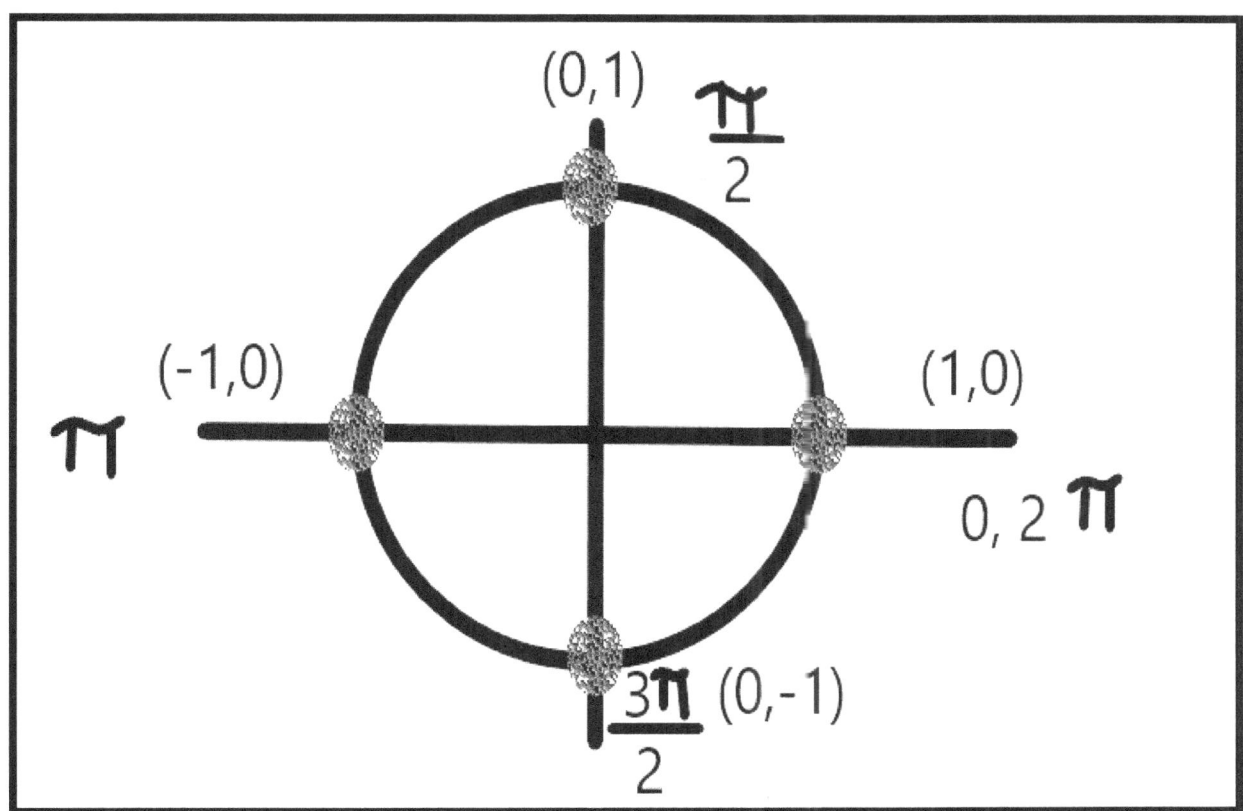

Point(1,0):

$$tan\theta = \frac{y}{x} = \frac{0}{1} = 0$$

Point(0,1):

$$tan\theta = \frac{y}{x} = \frac{1}{0} = undefined$$

Point(-1,0):

$$tan\theta = \frac{y}{x} = \frac{0}{-1} = 0$$

Point(0,-1):

$$tan\theta = \frac{y}{x} = -\frac{1}{0} = undefined$$

Tangent is undefined whenever cosine$^\theta$ equals 0.

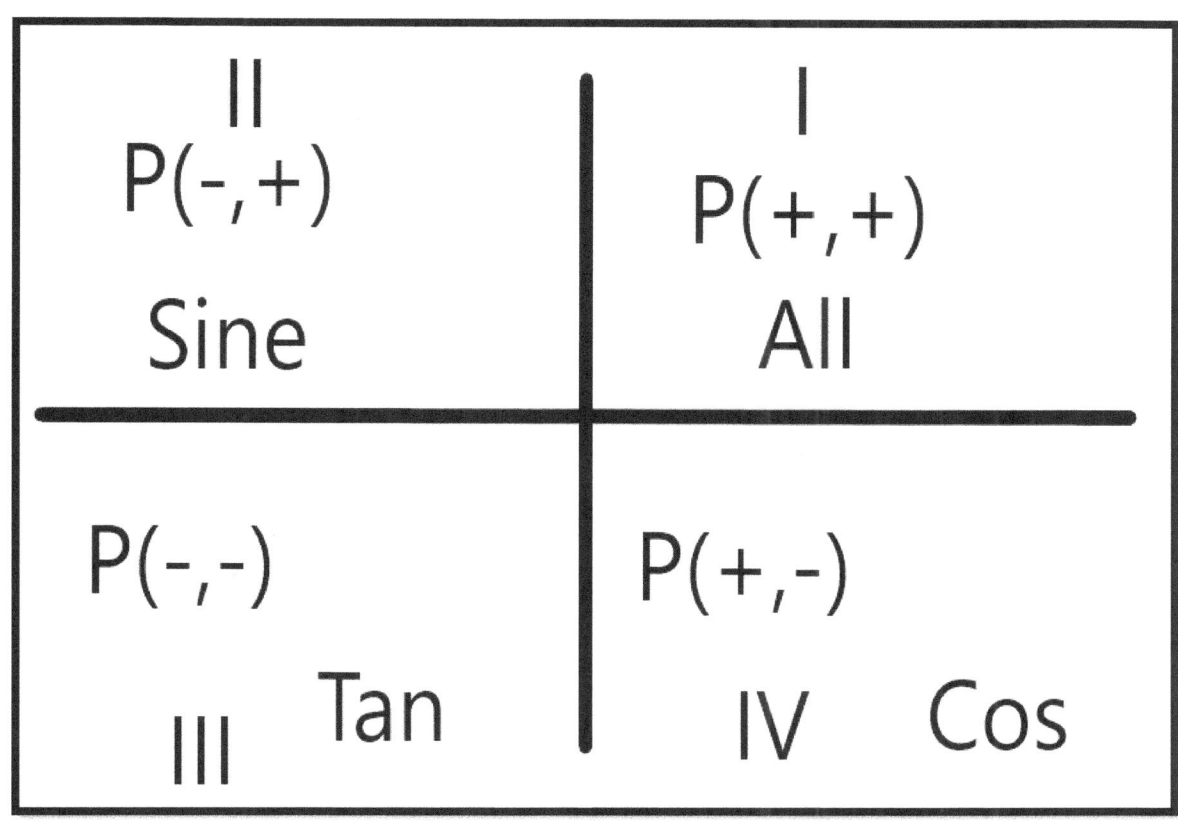

Here is an acronym to remember which trig function is positive in which quadrant; ASTC, A means all(Quad 1), S means Sine(Quad 2), T means Tangent(Quad 3), and C means Cosine(Quad 4).

ASTC, you can say "All Students Take Calculus", "Actually, Someone Took Calculus?", whatever helps you remember.

Quadrant:

1. Sine, Cosine, Tangent are positive
2. Sine is positive
3. Tangent is positive
4. Cosine is positive

Special Triangles

There are some special triangles for which we can find the Exact values for Trig Ratios.

These are for triangles with the ratios: $30° - 60° - 90°$ or $\frac{\pi}{6} - \frac{\pi}{3} - \frac{\pi}{2}$ in radians.

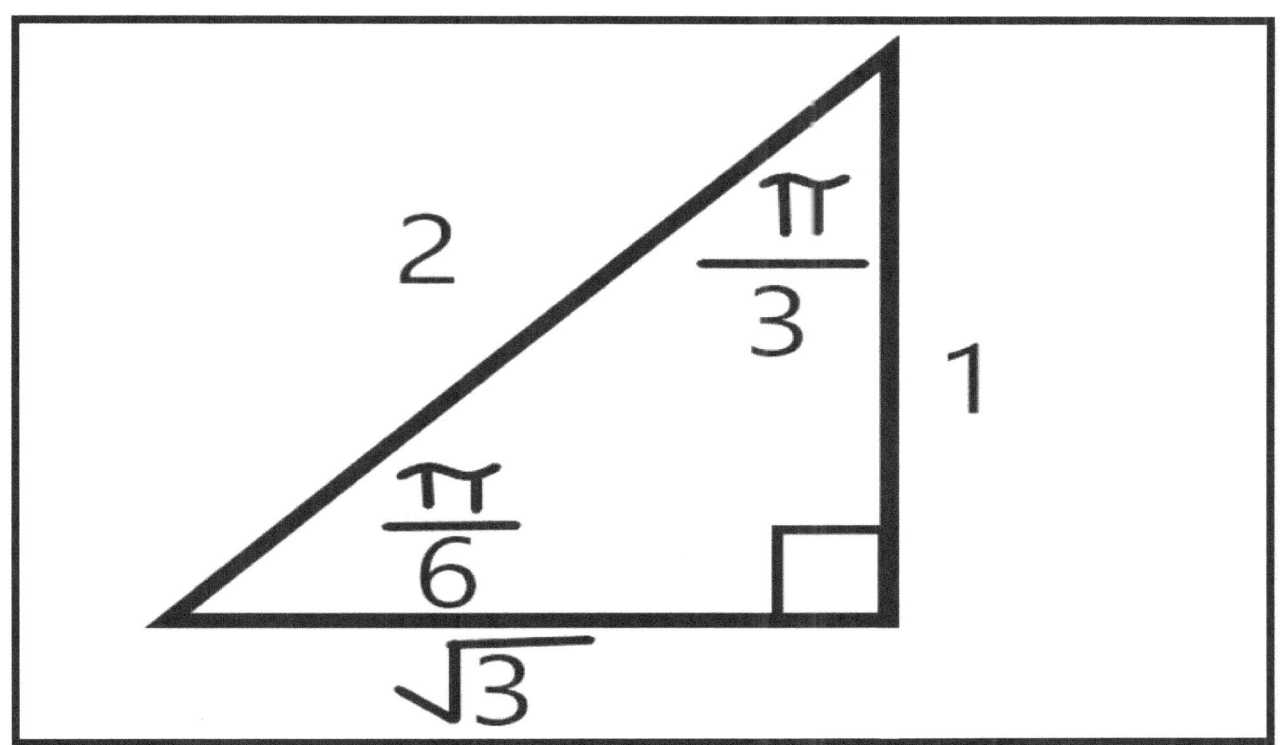

These are the exact values of these angles:

$\sin\left(\dfrac{\pi}{3}\right) = \dfrac{\sqrt{3}}{2}$ $\qquad$ $\sin\left(\dfrac{\pi}{6}\right) = \dfrac{1}{2}$

$\cos\left(\dfrac{\pi}{3}\right) = \dfrac{1}{2}$ $\qquad$ $\cos\left(\dfrac{\pi}{6}\right) = \dfrac{\sqrt{3}}{2}$

$\tan\left(\dfrac{\pi}{3}\right) = \dfrac{\sqrt{3}}{1} = \sqrt{3}$ $\qquad$ $\tan\left(\dfrac{\pi}{6}\right) = \dfrac{1}{\sqrt{3}}$

And the other special triangle is for triangles with the following ratio: $45° - 45° - 90°$ or $\dfrac{\pi}{4} - \dfrac{\pi}{4} - \dfrac{\pi}{2}$ in radians.

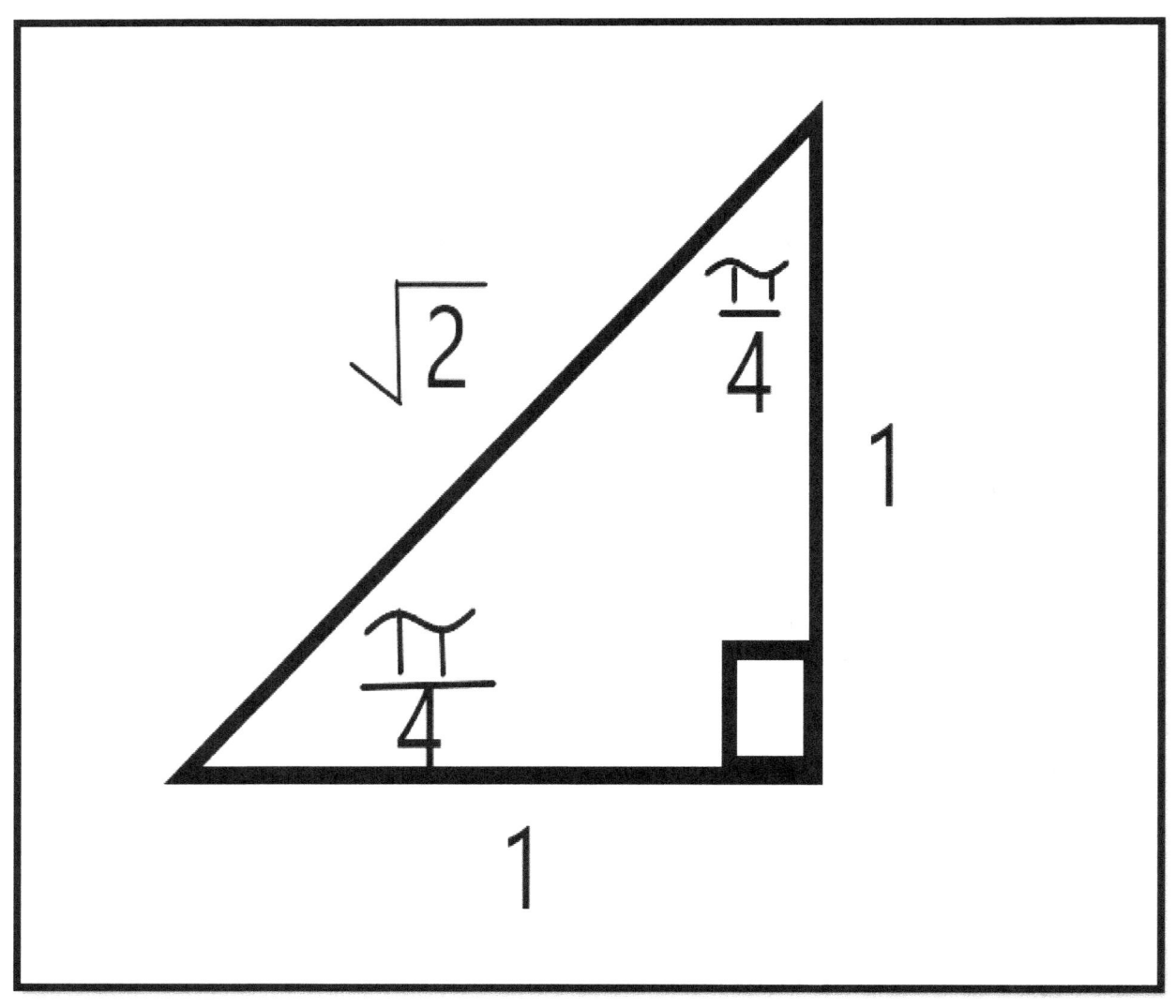

$$\sin\left(\frac{\pi}{4}\right) = \frac{1}{\sqrt{2}} \qquad \cos\left(\frac{\pi}{4}\right) = \frac{1}{\sqrt{2}} \qquad \tan\left(\frac{\pi}{4}\right) = 1$$

On the next 2 pages, there will be 2 large images summing up everything you have learned so far in Chapter 11.

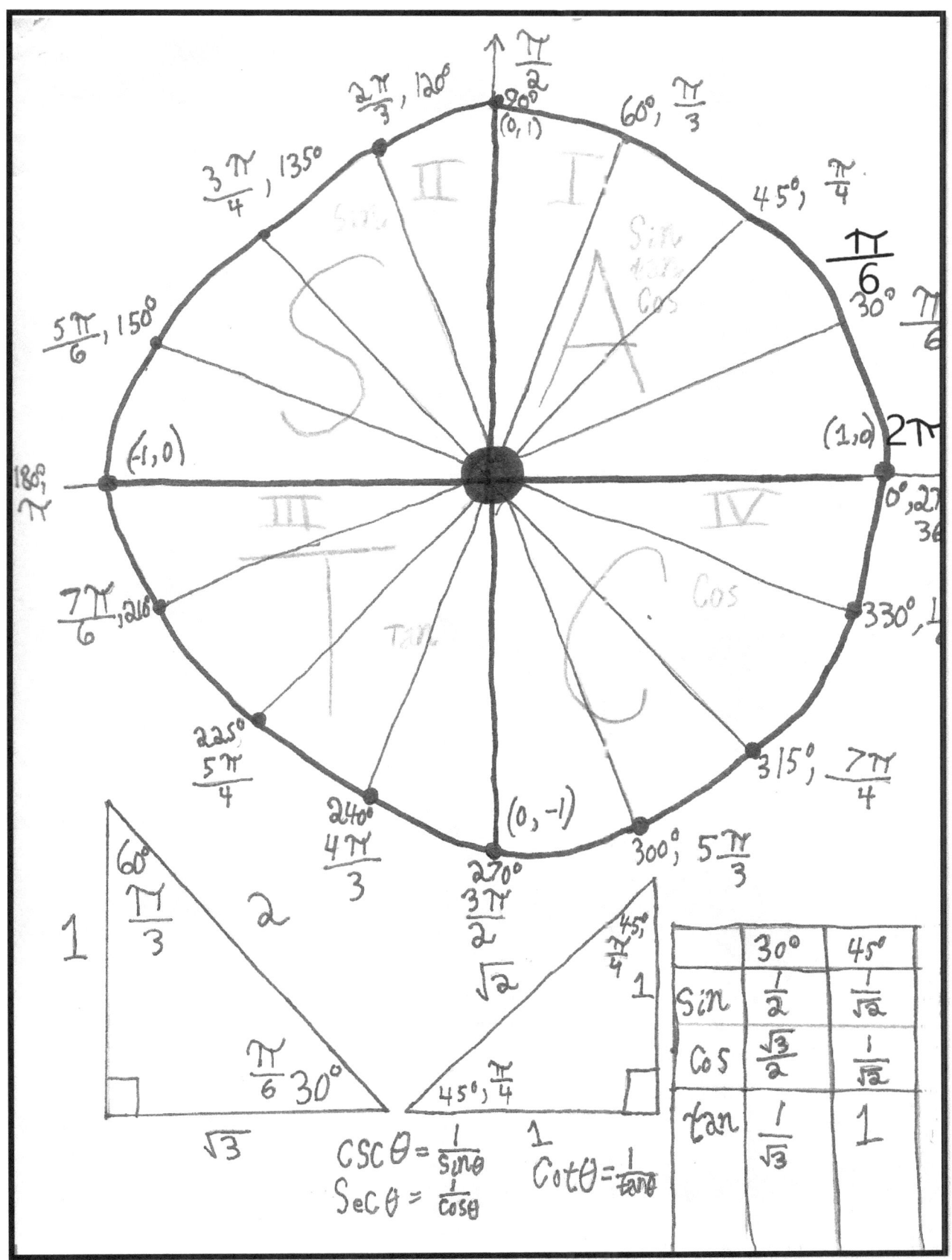

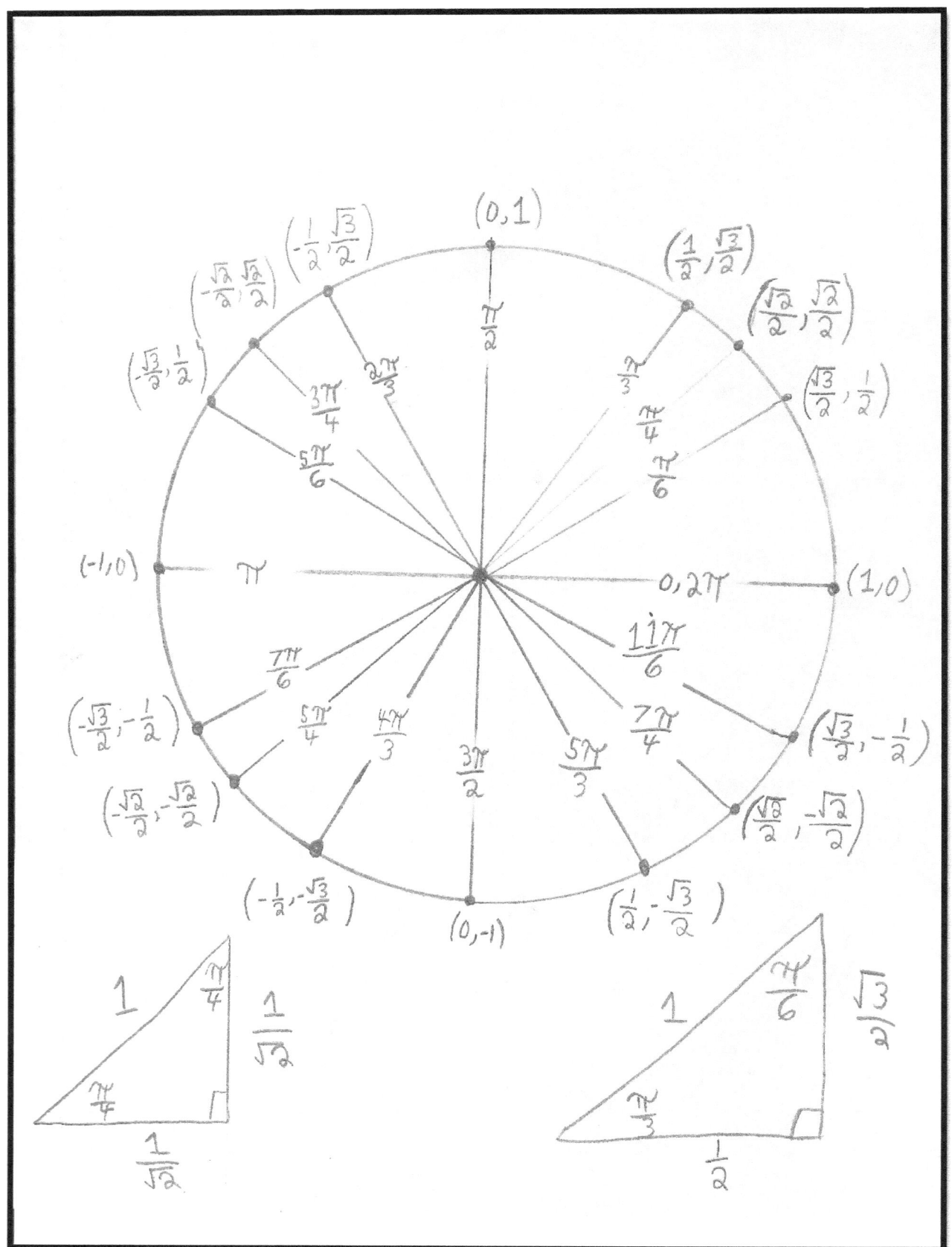

To generate the trig circle:

1. For every 30° you add $\dfrac{\pi}{6}$
2. For every 45° you add $\dfrac{\pi}{4}$

As you probably noticed on the Unit Circle visuals, I was giving the ratios of the reciprocal trig functions, I will repeat them once again:

$$Cosecant = csc\theta = \dfrac{1}{sin\theta}$$

$$Secant = Sec\theta = \dfrac{1}{cos\theta}$$

$$Cotangent = cot\theta = \dfrac{1}{tan\theta}$$

To find the exact values of these functions, first find the ratio of the primary trig function then take its reciprocal.

I think you should go back and review the past few pages before moving on.

Example 1: Find the exact value of $\sin\left(\dfrac{5\pi}{4}\right)$.

Step 1: Draw the angle in standard position,

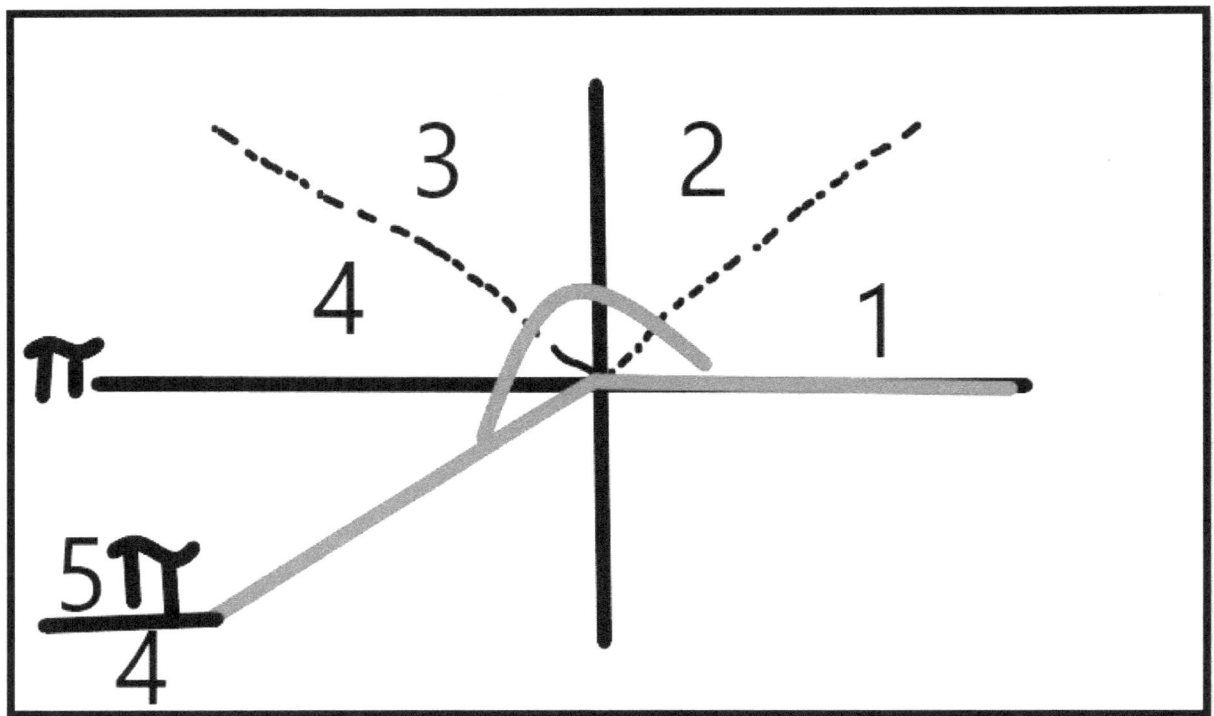

Step 2: Determine the reference angle,

$$\frac{5\pi}{4} - \pi = \frac{5\pi}{4} - \frac{4\pi}{4} = \frac{\pi}{4}$$

Step 3: Find the exact value of this angle using special triangles. Keep in mind that this angle will be negative.

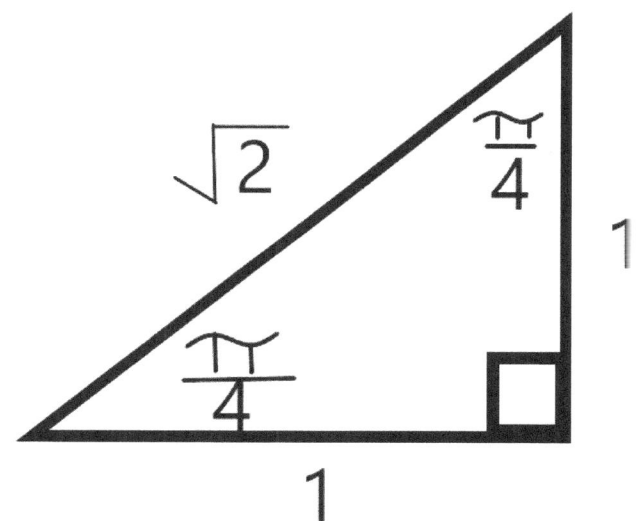

$$\sin\left(\frac{\pi}{4}\right) = \frac{1}{\sqrt{2}}$$

Since this angle lies in Quadrant 3, where sine is negative,

$$\sin\left(\frac{5\pi}{4}\right) = -\frac{1}{\sqrt{2}}$$

This is the exact value of $\sin\left(\dfrac{5\pi}{4}\right)$.

Example 2: $\sec\left(\dfrac{5\pi}{6}\right)$

$$\sec\left(\dfrac{5\pi}{6}\right) = \dfrac{1}{\cos\left(\dfrac{5\pi}{6}\right)}$$

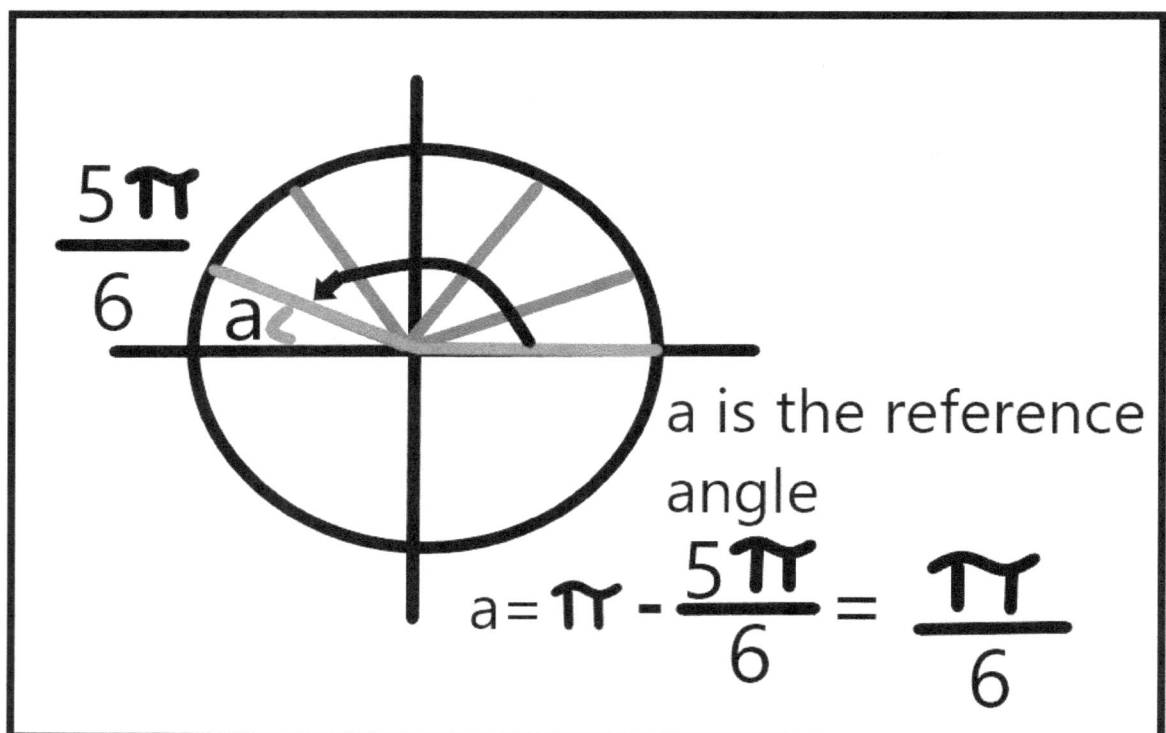

Quadrant 2, so cosine is negative (secant will be negative as well). The angle $\frac{\pi}{6}$ is in the $(\frac{\pi}{6},\frac{\pi}{3},\frac{\pi}{2})$ special triangle.

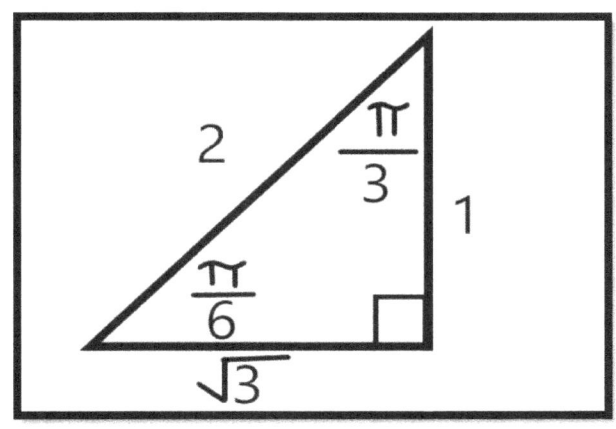

$$\cos\left(\frac{\pi}{6}\right) = \frac{\sqrt{3}}{2}$$

$\cos\left(\frac{5\pi}{6}\right) = -\frac{\sqrt{3}}{2}$ since $\frac{5\pi}{6}$, lies in Quadrant 2 where cosine is negative.

$$\sec\left(\frac{5\pi}{6}\right) = \frac{1}{\cos\left(\frac{5\pi}{6}\right)}$$

$$\sec\left(\frac{5\pi}{6}\right) = \frac{1}{-\frac{\sqrt{3}}{2}}$$ multiply by its reciprocal

$$\sec\left(\frac{5\pi}{6}\right) = 1 * -\frac{2}{\sqrt{3}}$$

$$\sec\left(\frac{5\pi}{6}\right) = -\frac{2}{\sqrt{3}}$$

Example 3: Exact value of $\csc \dfrac{\pi}{6}$

$\csc \theta = \dfrac{1}{\sin \theta}$

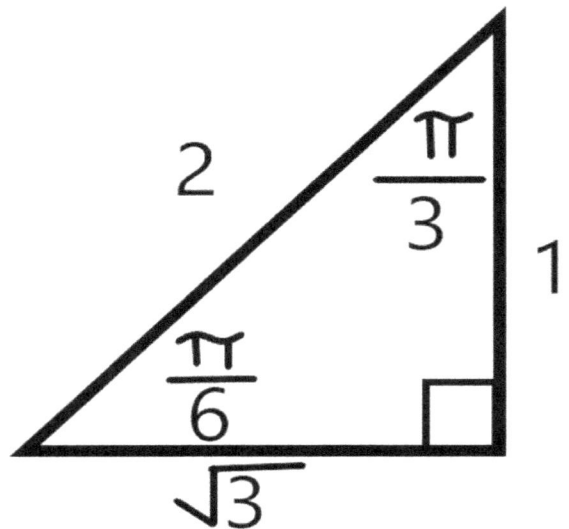

$\sin \left(\dfrac{\pi}{6} \right) = \dfrac{1}{2}$ *this is the exact value of the sine of this angle*

$\csc \left(\dfrac{\pi}{6} \right) = \dfrac{1}{\sin \left(\dfrac{\pi}{6} \right)}$

$\csc \left(\dfrac{\pi}{6} \right) = \dfrac{1}{\dfrac{1}{2}}$

$$\csc\left(\frac{\pi}{6}\right) = 1 * \frac{2}{1}$$ *1 is being multiplied by the reciprocal*

$$\csc\left(\frac{\pi}{6}\right) = 2$$ *exact value of cosecant of $\frac{\pi}{6}$*

Example 4: $\cot\left(\frac{11\pi}{6}\right)$

Where is $\frac{11\pi}{6}$?

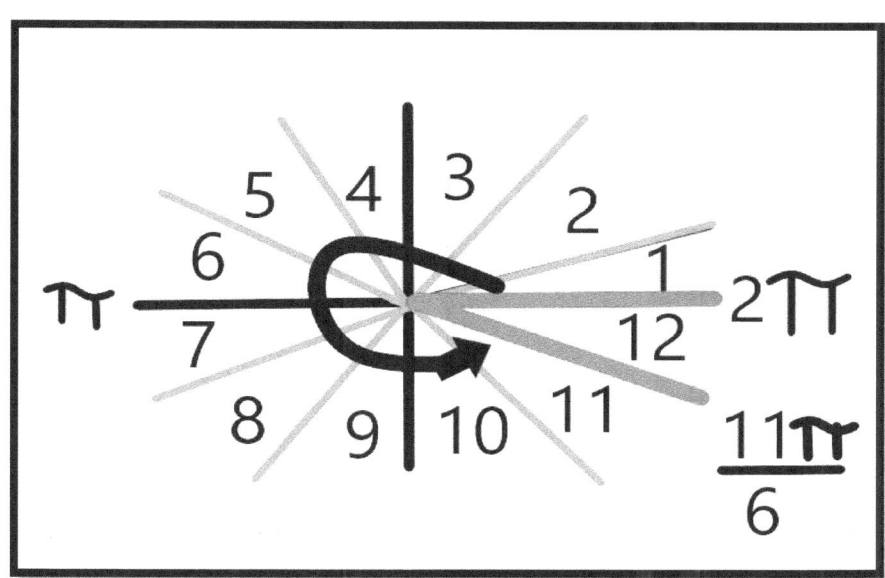

$\cot\theta = \frac{1}{\tan\theta}$ or $\frac{\cos\theta}{\sin\theta}$ you can use whichever one you wish; however, to keep this complicated topic in its most comprehendible form, I will use the $\frac{1}{\tan\theta}$.

What is the reference angle?

$$2\pi - \frac{11\pi}{6} = \frac{12\pi}{6} - \frac{11\pi}{6}$$

$$= \frac{\pi}{6}$$

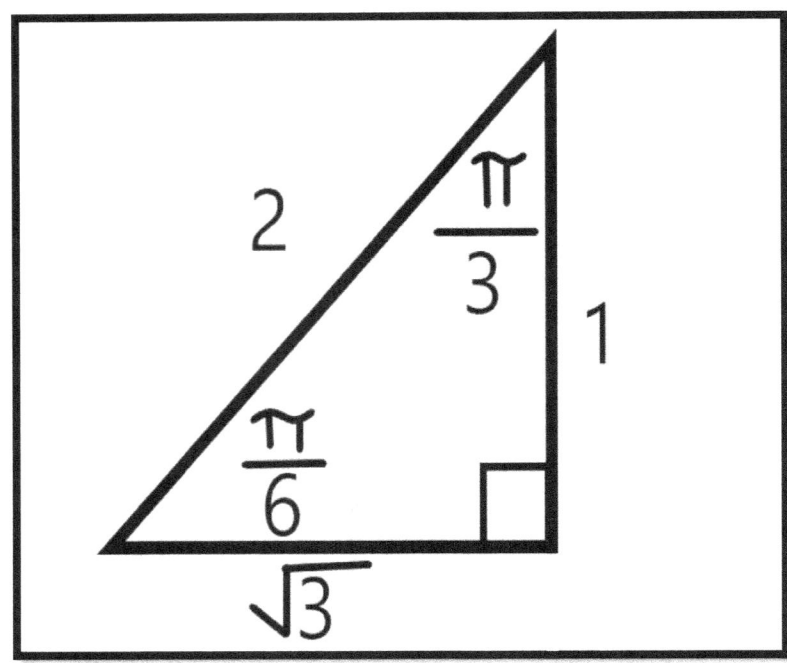

$$\tan\left(\frac{\pi}{6}\right) = \frac{1}{\sqrt{3}}$$

$\frac{11\pi}{6}$ is in quadrant 4, where only cosine is positive; therefore, cotangent and tangent are both negative.

$$\cot\left(\frac{11\pi}{6}\right) = \frac{1}{\tan\left(\frac{11\pi}{6}\right)}$$

$$\cot\left(\frac{11\pi}{6}\right) = \frac{1}{-\frac{1}{\sqrt{3}}}$$

$$\cot\left(\frac{11\pi}{6}\right) = 1 * -\frac{\sqrt{3}}{1}$$ *multiply 1 by the reciprocal*

$$\cot\left(\frac{11\pi}{6}\right) = -\sqrt{3}$$ *this is the exact value*

Example 5: $\cos\left(\frac{3\pi}{2}\right)$

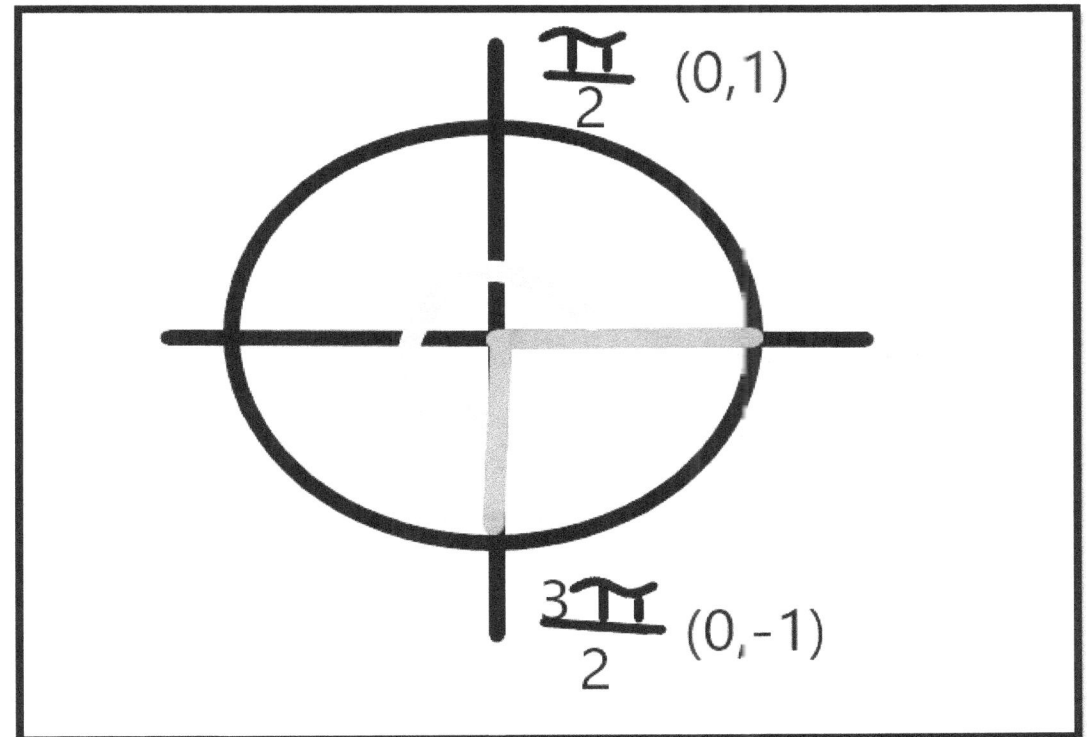

Remember that cosine is the x-coordinate of the point on the unit circle, using this knowledge,

The exact value of $\cos\left(\dfrac{3\pi}{2}\right) = 0$

Example 6: $\log_2\left(\sin\left(\dfrac{\pi}{4}\right)\right)$

Step 1: Deal with $\sin\left(\dfrac{\pi}{4}\right)$ first,

Exact value of $\sin\dfrac{\pi}{4} = \dfrac{1}{\sqrt{2}}$

Step 2: Replace the $\sin\frac{\pi}{4}$ with $\frac{1}{\sqrt{2}}$

$\log_2\left(\frac{1}{\sqrt{2}}\right)$

Step 3: $\frac{1}{\sqrt{2}}$ can become $2^{-\frac{1}{2}}$

$\log_2 2^{-\frac{1}{2}}$

Step 4: Use the power rule,

$-\frac{1}{2}\log_2 2$

$-\frac{1}{2}(1) = -\frac{1}{2}$

The exact value of $\log_2(\sin\frac{\pi}{4})$ is $-\frac{1}{2}$

You can always verify your answer by changing the base,

You can enter $\dfrac{\log\left(\sin\left(\frac{\pi}{4}\right)\right)}{\log 2}$ on your calculator and you will get the same result, as long as your calculator is in radians mode.

"Reciprocal Trig Trick"

In this section we will look at using an easy technique to find the exact value of a reciprocal trig function; instead of having to find the primary trig ratio and the flip it, etc. This is a technique I used,

I basically used their definitions to create their ratios on the special triangles!

$$\cot\theta = \dfrac{adjacent}{opposite} \qquad \csc = \dfrac{hypotenuse}{opposite}$$

$$\sec = \dfrac{hypotenuse}{adjacent}$$

Sin	Csc	Tan Sin Cos	Cot Csc Sec
Tan	Cot	Cos	Sec

These are the quadrants where these trigonometric functions are positive.

Using this idea, we will solve $\sec\dfrac{5\pi}{6}$ for the exact value

Example 1: Find the exact value of $\sec\dfrac{5\pi}{6}$,

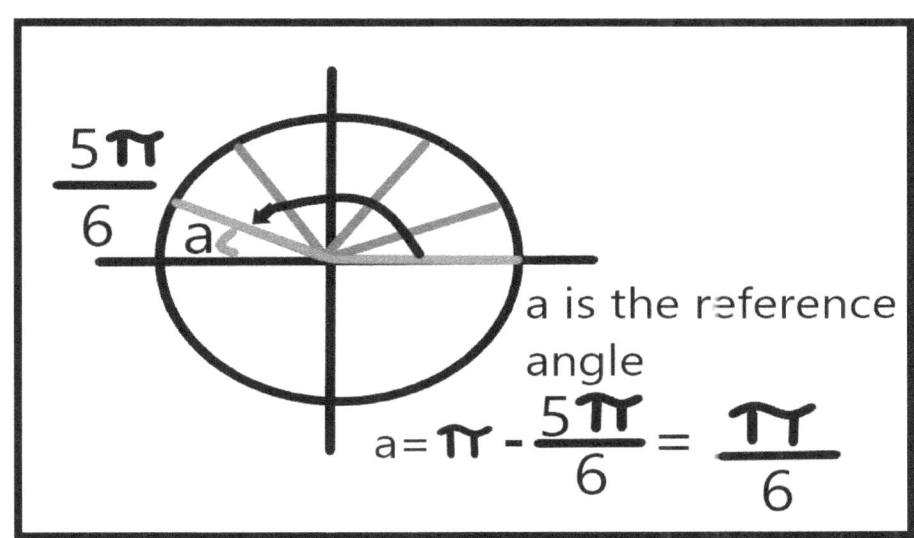

$\dfrac{\pi}{6}$ is the reference angle, draw the special triangle, remember that Secant is negative in Quadrant 2.

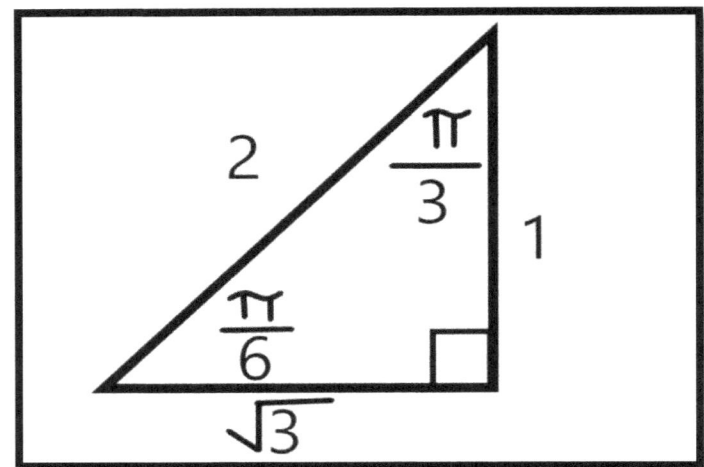

$$\sec\theta = \frac{hypotenuse}{adjacent} \qquad \sec\frac{\pi}{6} = \frac{2}{\sqrt{3}}$$

$$\sec\frac{5\pi}{6} = -\frac{2}{\sqrt{3}}$$

Practice Questions: Chapter 11 Part A

1. Convert the following into radians in terms of π,
   a) $30°$
   b) $240°$
   c) $-10°$

2. Convert the following into degrees exactly,

a) $\dfrac{\pi}{12}$

b) $\dfrac{\pi}{3}$

c) $\dfrac{\pi}{18}$

3. Determine the length of an arc in a circle of radius 5 if it subtends an angle of 3 radians.

4. Draw the following angles in standard position,

a) $\dfrac{3\pi}{4}$

b) $-\dfrac{\pi}{4}$

Circular Functions Section B, Trigonometric Functions:

In this section, you will learn the shape of trigonometric functions as well as the features.

Sine Function

We will start by looking at the graph of a **Sine function,** $y = sinx$

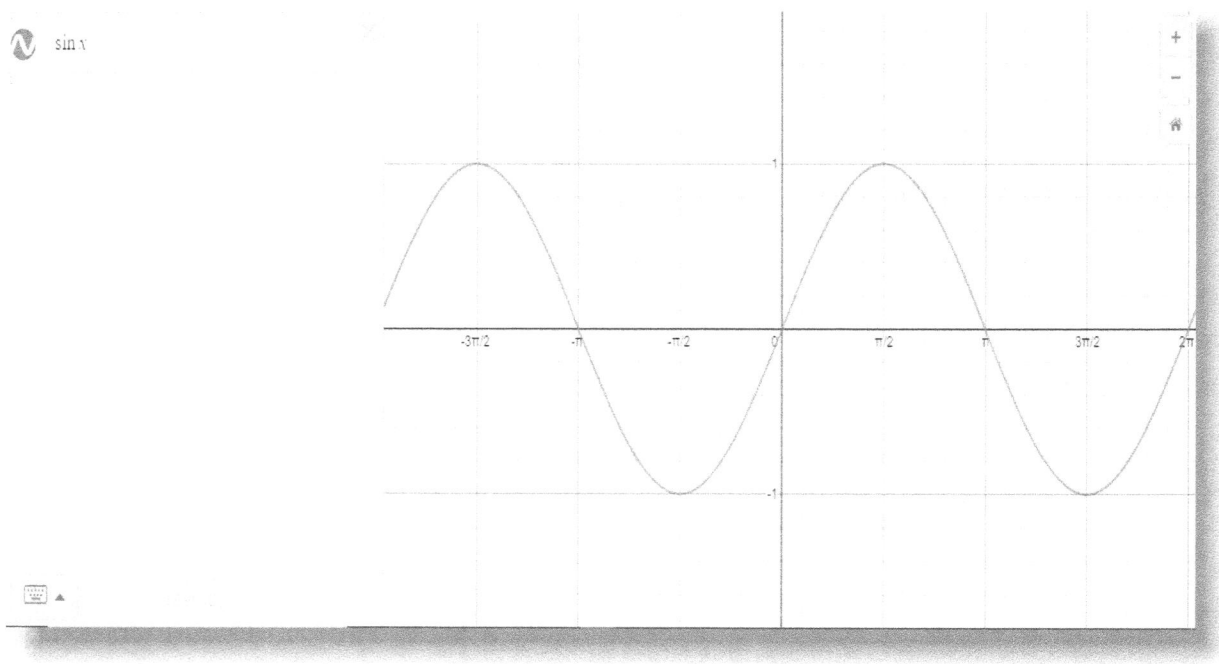

As you can see, this graph is continuous; therefore, its domain is all real numbers. Its maximum value is 1 and its minimum value is -1.

Domain: $\{x | x \in R\}$ Range: $-1 \leq y \leq 1$

When graphing these functions, we only graph them on the interval $-2\pi \leq x \leq 2\pi$

Before we start graphing we divide each side of the x-axis into four parts, $\frac{\pi}{2}, \pi, \frac{3\pi}{2},$ and 2π on the positive side and we divide the negative side of the x-axis into four equal parts as well, $-\frac{\pi}{2}, -\pi, -\frac{3\pi}{2},$ and -2π.

Why is this the scale we use? It is much easier to graph the output values of these inputs. We are saying, "Something interesting happens at each of these four parts"

$\sin 0 = 0$

$\sin\left(\frac{\pi}{2}\right) = 1$

$\sin(\pi) = 0$

$\sin\left(\frac{3\pi}{2}\right) = -1$

$\sin(2\pi) = 0$

(This graph starts at negative infinity on the x-axis) I am just going to start at (0,0) in my example.

The most basic graph of the sine function has a cycle that starts at (0,0) goes up to 1 on the y-axis, goes back down, crosses the x-axis, dips down to -1 on the y-axis, goes back up, crosses the x-axis. And the graph repeats itself infinitely. From 0 to 2π this is one cycle, after that cycle is complete it starts another cycle. 2π **is the period of the function (length of one cycle).** This graph is generated by the unit circle as a point on it rotates.

Our outputs are generated by all of the vertical coordinates of the unit circle as a point rotates from 0 to 2π and so forth. Remember that $\sin(0) = 0$? This is why the graph starts at (0,0). The height of the point at 0 radians is 0 at 0 radians.

$(x,y) = (x,\sin(x))$ y-coordinates are equal to $\sin(x)$

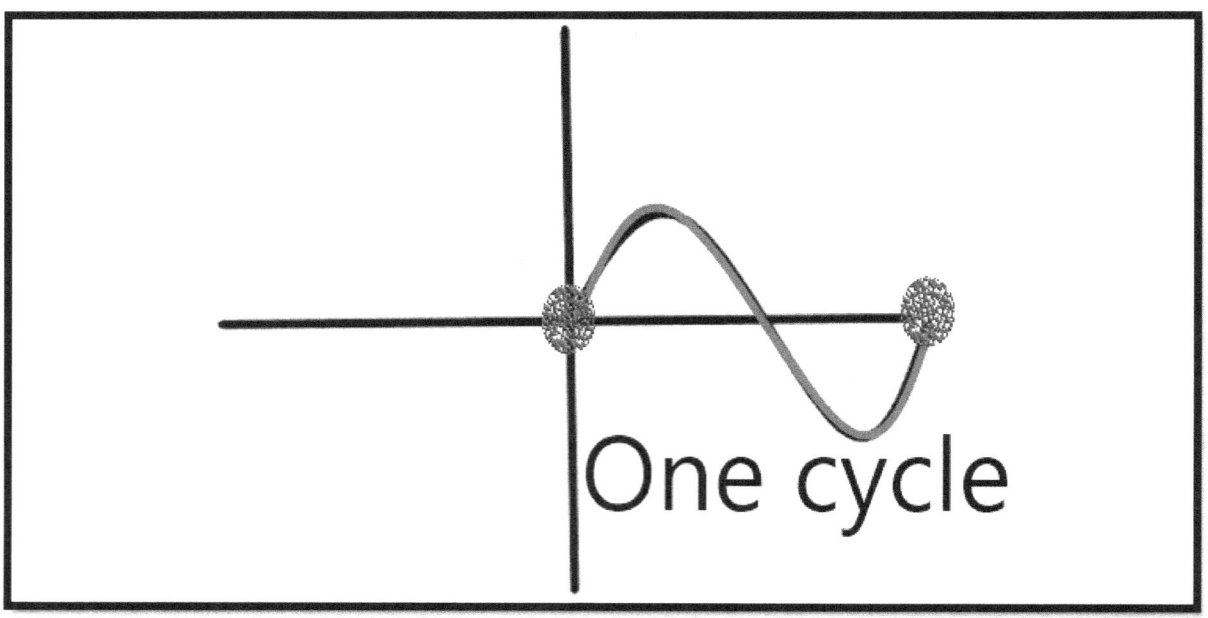

The **amplitude** of this sine function is 1, **Amplitude** is the vertical distance form the **central axis**(in the centre of the graph) to the minimum or maximum value of the periodic function. The example below is an example of a transformed sine function,

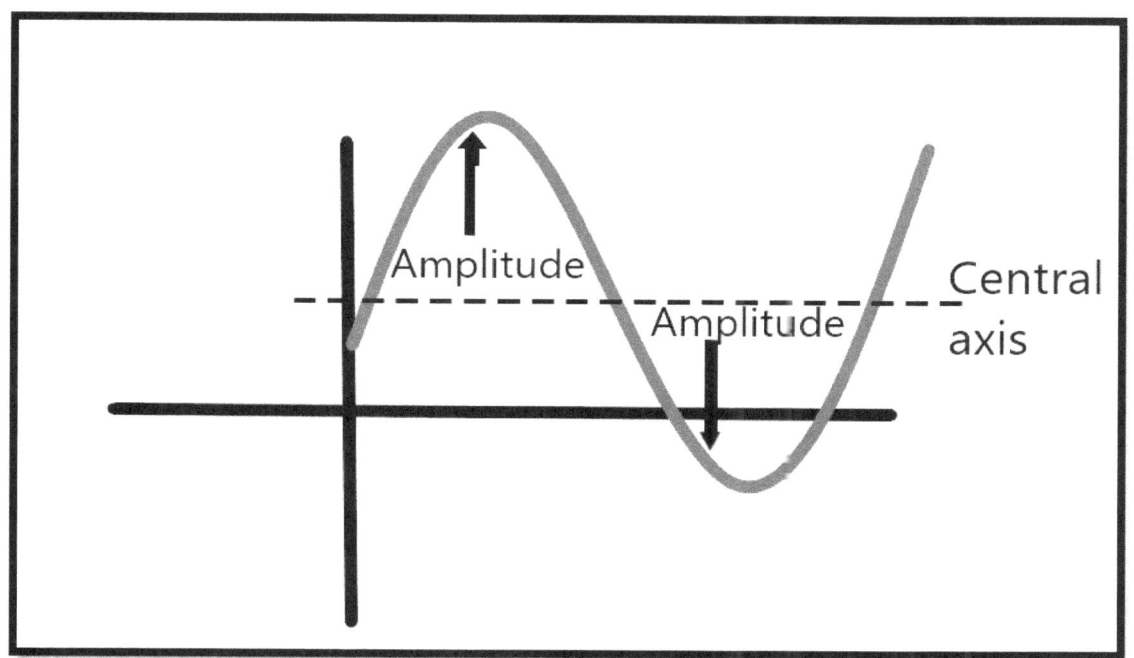

Cosine Function

Now, we will look at the most basic cosine function, $y = \cos x$.

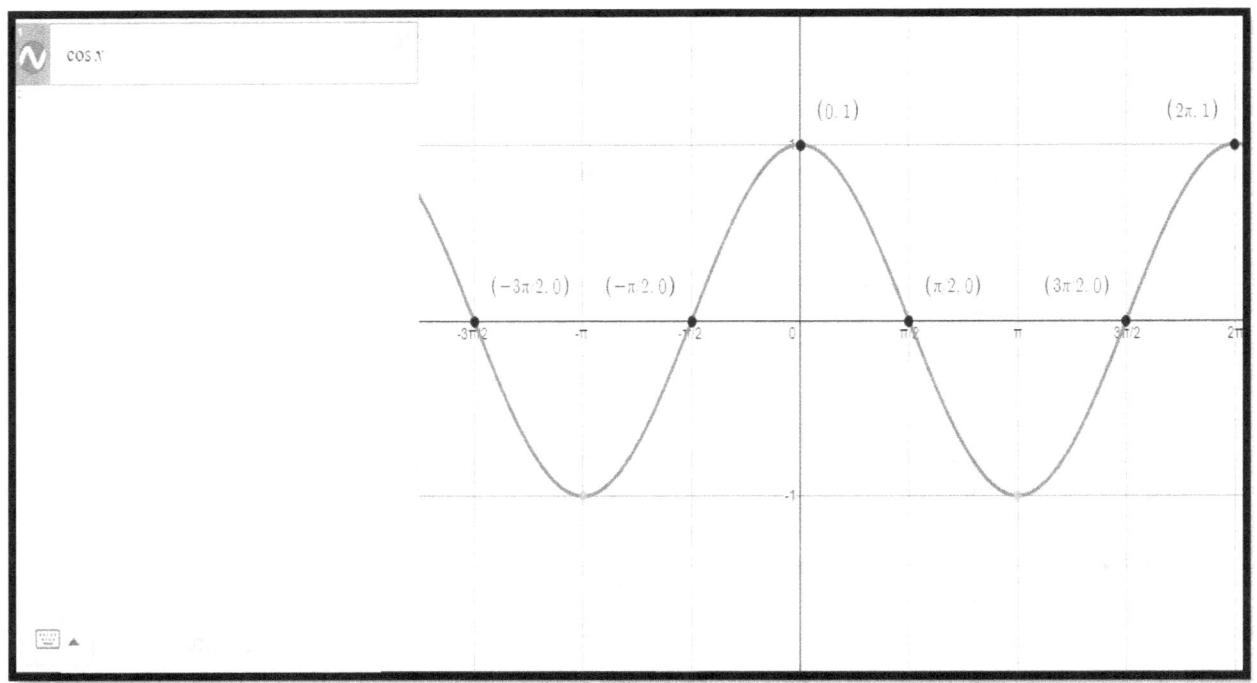

As you can see, this graph is continuous; therefore, its domain is all real numbers. Its maximum value is 1 and its minimum value is -1.

Domain: $\{x | x E R\}$ Range: $-1 \leq y \leq 1$

When graphing these functions, we only graph them on the interval $-2\pi \leq x \leq 2\pi$

Before we start graphing we divide each side of the x-axis into four parts, $\frac{\pi}{2}, \pi, \frac{3\pi}{2},$ and 2π on the positive side and we divide the negative side of the x-axis into four equal parts as well, $-\frac{\pi}{2}, -\pi, -\frac{3\pi}{2}, and -2\pi$.

Why is this the scale we use? It is much easier to graph the output values of these inputs. We are saying, "Something interesting happens at each of these four parts"

$\cos 0 = 1$

$\cos \frac{\pi}{2} = 0$

$\cos \pi = -1$

$\cos \frac{3\pi}{2} = 0$

$\cos 2\pi = 1$

(This graph starts at negative infinity on the x-axis) I am just going to start at (0,0) in my example.

The most basic graph of the cosine function starts at (0,1) goes down, crosses the x-axis $(\frac{\pi}{2})$, dips down to -1 on the y-axis(π), goes back up, crosses the x-axis $(\frac{3\pi}{2})$, goes back up to 1 on the y-axis(2π). And the graph repeats itself infinitely. From 0 to 2π this is one cycle, after that cycle is complete it starts another cycle. 2π **is the period of the function (length**

of one cycle). This graph is generated by the unit circle as a point on it rotates.

Our outputs are generated by all of the horizontal coordinates of the unit circle as a point rotates from 0 to 2π and so forth. Remember that $\cos(0) = 1$? This is why the graph starts at (0,1). The horizontal distance of the point at 0 radians is 1 at 0 radians.

$(x,y) = (x, \cos(x))$ y-coordinates are equal to $\cos(x)$

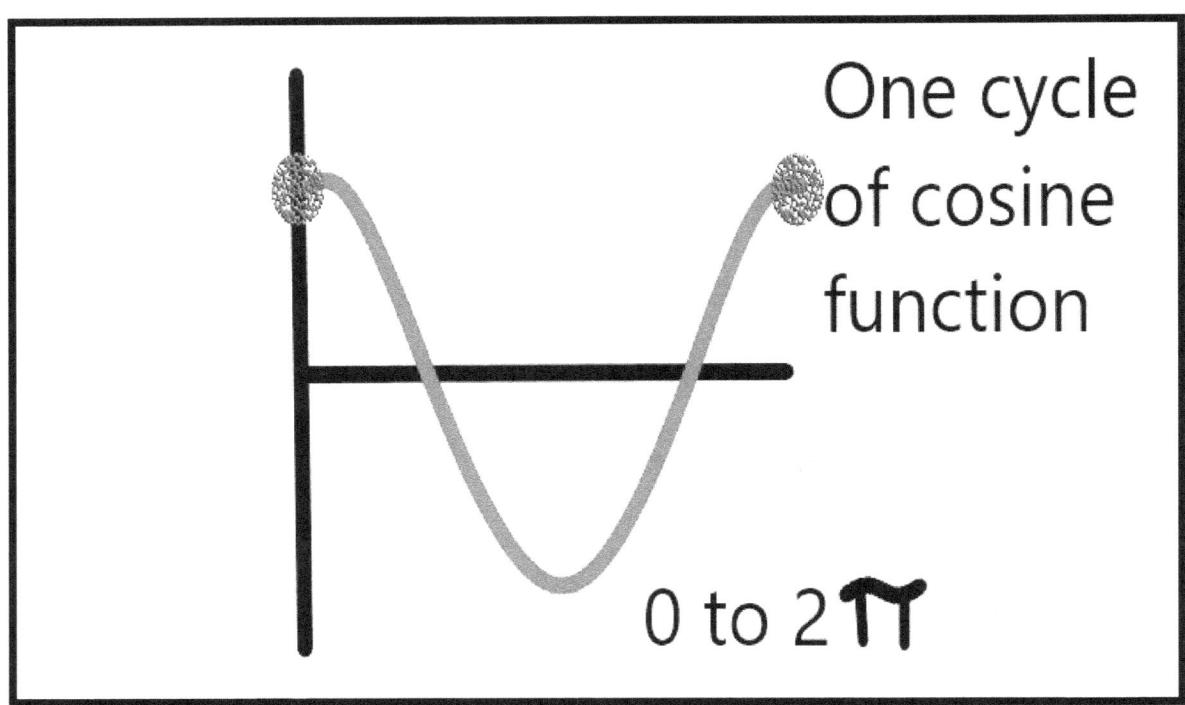

The **amplitude** of this cosine function is 1, **Amplitude** is the vertical distance form the **central axis**(in the centre of the graph) to the minimum or maximum value of the periodic function.

Just like the sine function, there are an infinite number of x-intercepts; however, in the domain we are given, there are 4 x-intercepts $(-\frac{3\pi}{2}, -\frac{\pi}{2}, \frac{\pi}{2}, \frac{3\pi}{2})$.

Transformations of Trigonometric Functions

In this section we will look at transformations of trigonometric functions in the form:

$$y = a\sin b(x - c) + d$$
$$y = a\cos b(x - c) + d$$

Where $|a|$ is the amplitude (vertical height from the central axis), b is the horizontal compression or expansion, c is the phase shift (horizontal translation), and d is the vertical displacement.

First, we will look at vertical factors, a & d

Example 1: Identify the transformations that have been applied to $y = -\frac{1}{3}\cos x$

There has been a reflection in the x-axis, and a vertical compression by a factor of 1/3. Amplitude is 1/3.

Example 2: Identify the transformations that have been applied to $y = 5\sin x + 6$

Vertical expansion by a factor of 5, and a vertical displacement of 6 units up. Amplitude is 5.

Example 3: Graph $y = 2\sin x + 1$

Step 1: Graph $y = \sin x$

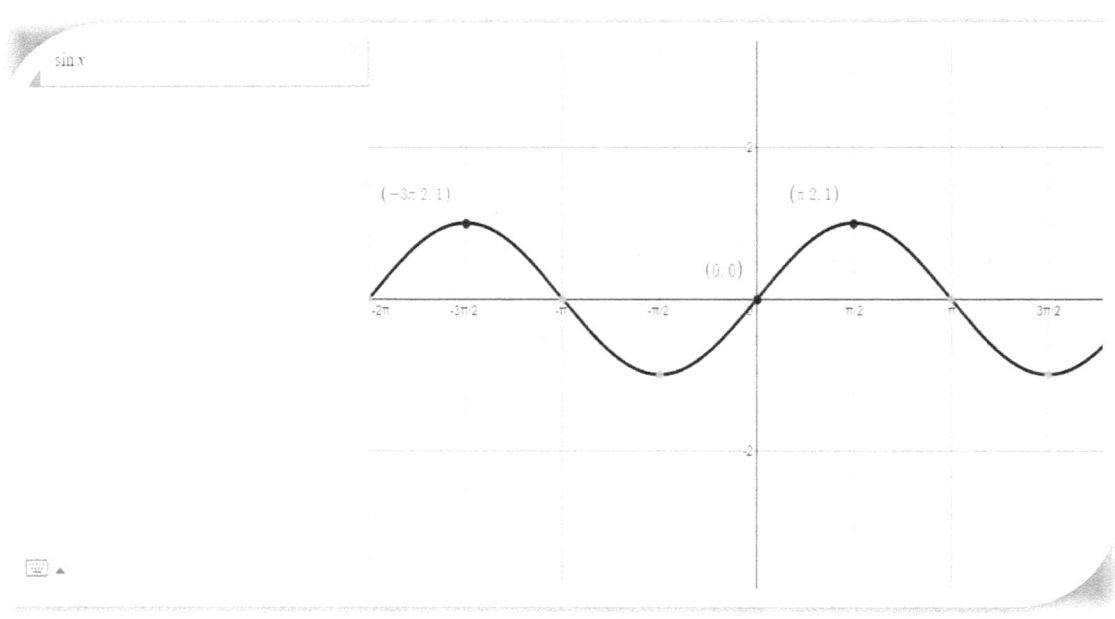

Step 2: Expand it vertically by 2, $y = 2\sin x$

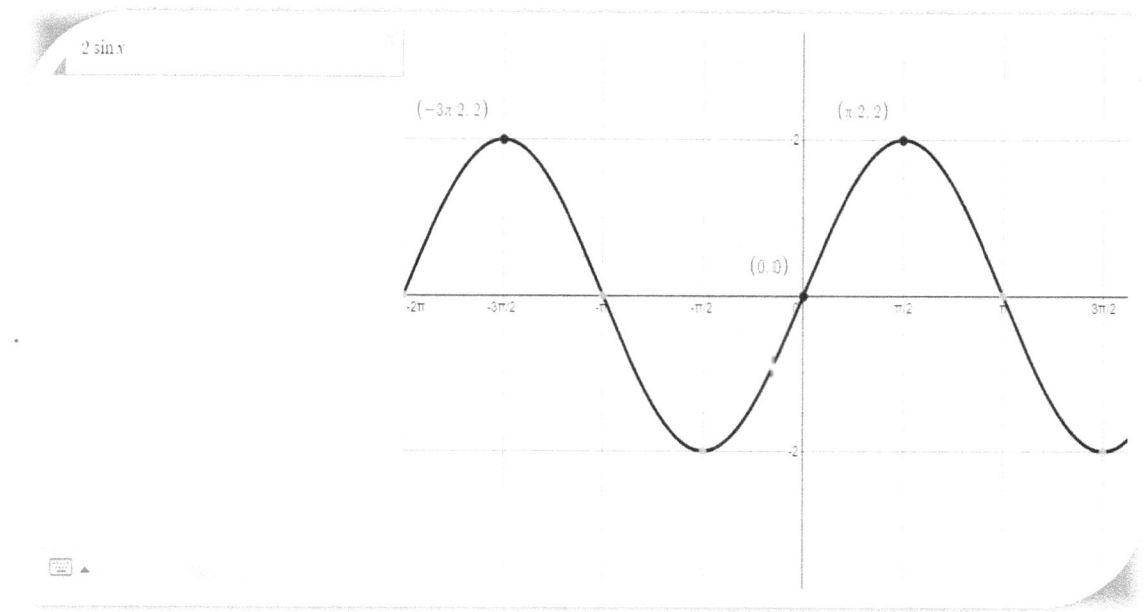

Step 3: Shift the graph 1 unit up, $y = 2\sin x + 1$

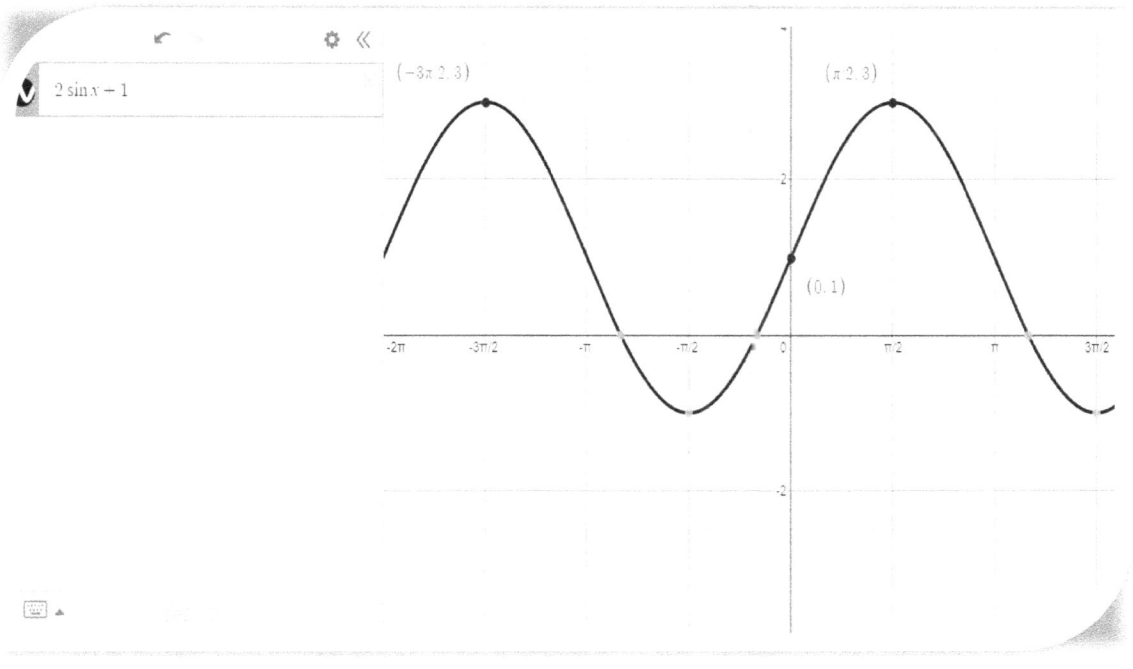

Example 4: Graph $y = -2\cos x + 1$,

Step 1: graph $y = \cos x$

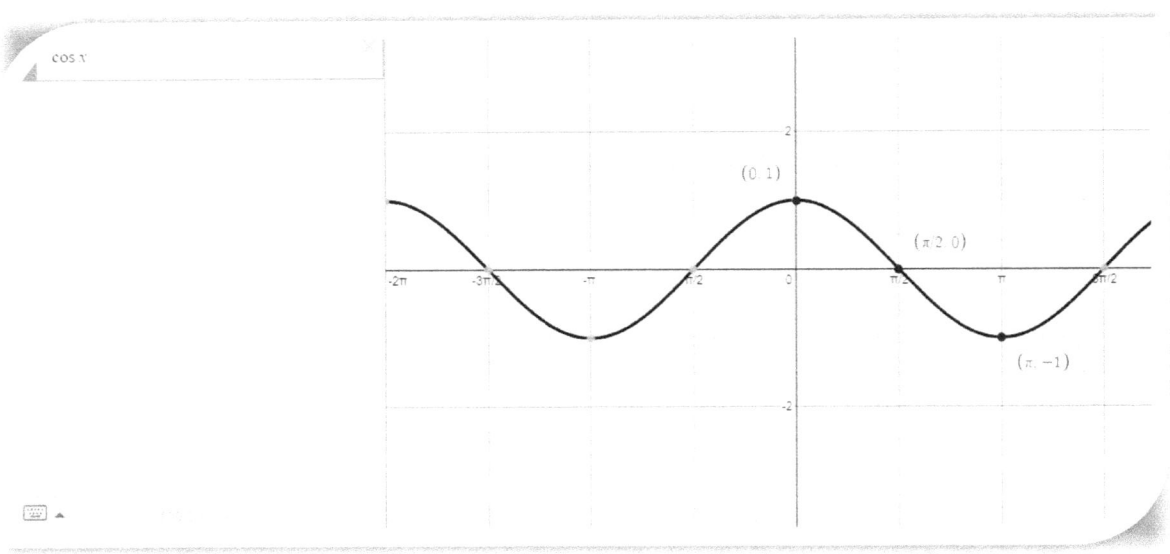

Step 2: Vertically expand it by 2, $y = 2\cos x$,

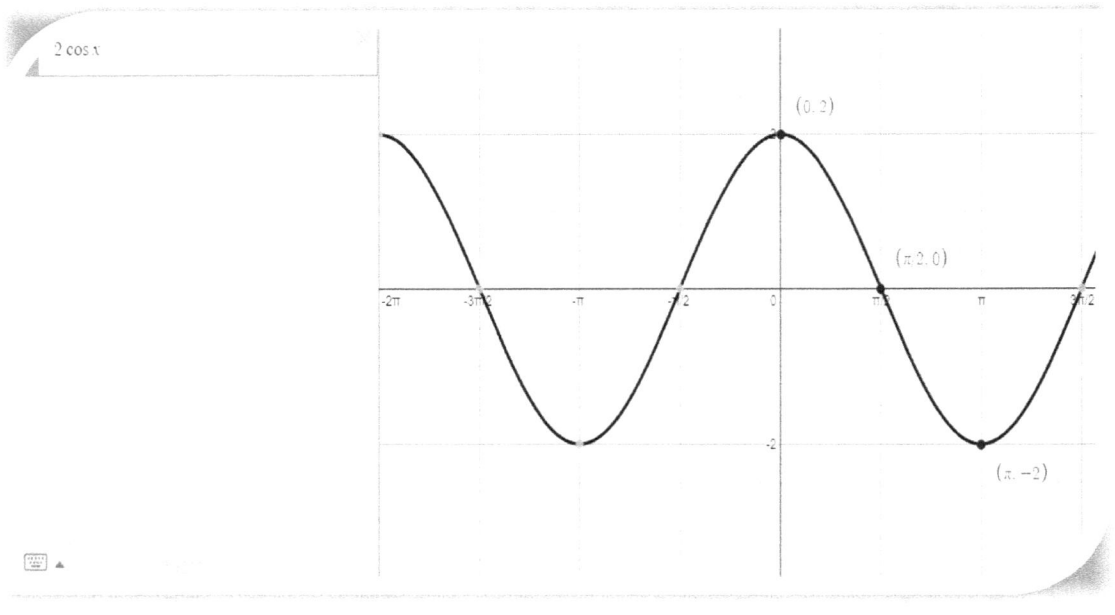

Step 3: Reflect it in the x-axis, $y = -2\cos x$

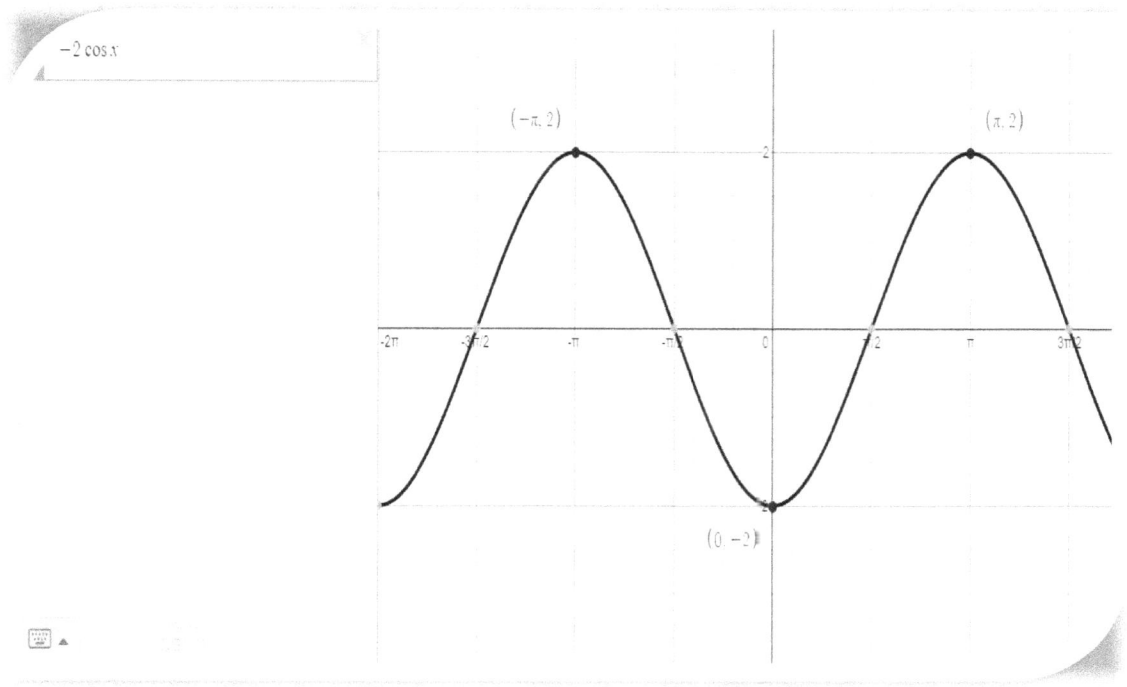

Step 4: Shift the graph one unit up, $y = -2\cos x + 1$

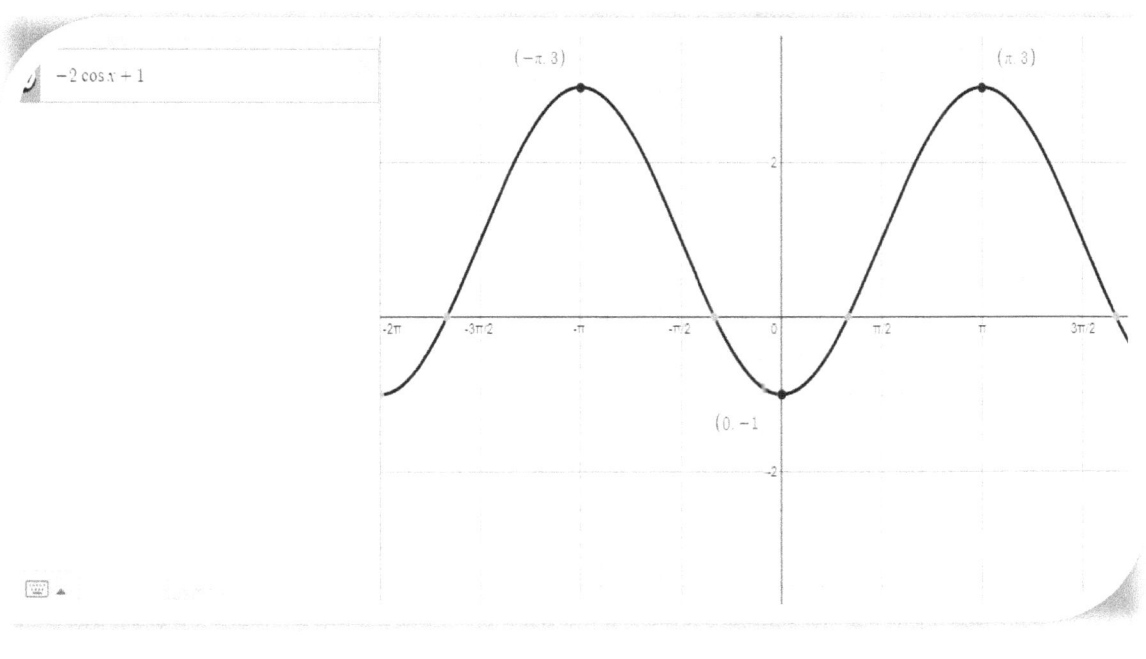

We will now look at an example of writing an equation given the graph, concerning vertical factors only,

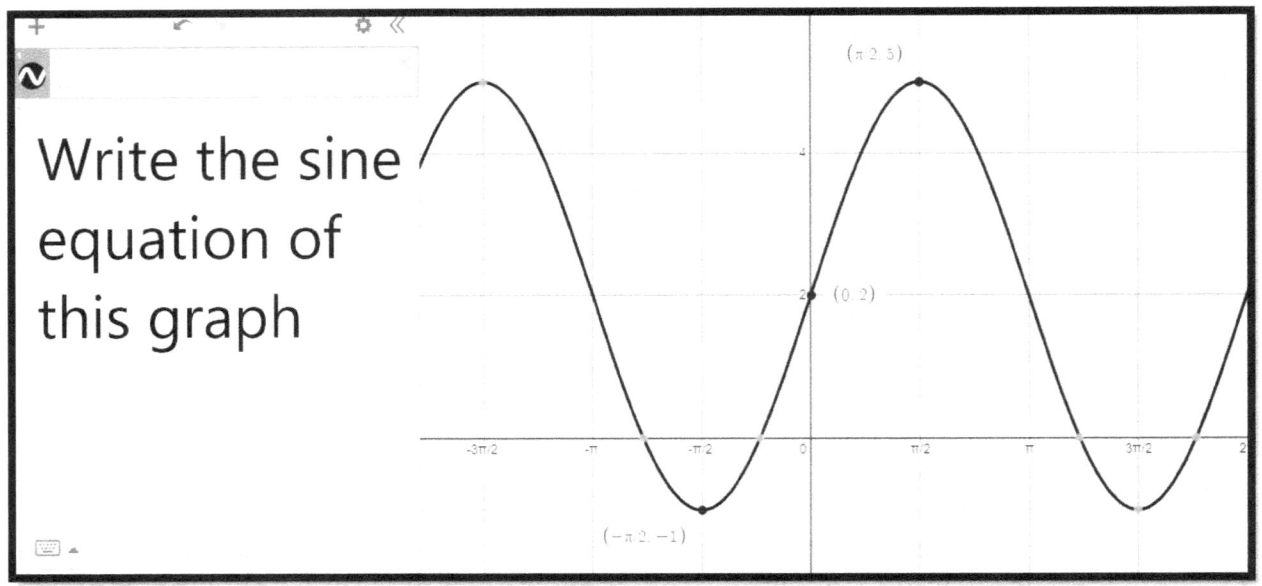

We are writing the equation in the form $y = a\sin x + d$,

To find the amplitude, we must find the maximum point(highest point) and subtract the minimum point(lowest point), and we must divide it by 2.

$$a = \frac{Max - Min}{2} \qquad amplitude = |a|$$

$Max = 5, \ Min = -1$

$$a = \frac{5 - (-1)}{2} = \frac{6}{2} = 3$$

amplitude is 3.

To find the vertical displacement we add the max and the minimum together and we divide it by 2,

$$d = \frac{Max + Min}{2}$$

$$Max = 5, \quad Min = -1$$

$$d = \frac{5 + (-1)}{2} = \frac{5-1}{2} = \frac{4}{2} = 2$$

$d = 2$ and $a = 3$ using this information, the graph of this sine function is:

$$y = 3\sin(x) + 2$$

Now that we have looked at the vertical factors, we will now focus on the horizontal factors and then we will combine all of them together. I am doing it this way, so you get to see what each part of the equation does, instead of having 4 things happen at once, which would leave you confused.

Horizontal Factors:

In the equation, $y = a\sin b(x - c) + d$

We will be looking at the horizontal factors b & c.

WORD OF WARNING: b **is NOT the PERIOD!!!** b plays a role in the period, but it is not the actual period. The formula to determine the period is $\frac{2\pi}{|b|}$ 2 pi divided by the absolute value of b, where $b > 0$.

Example 1: $y = \sin 3(x - \frac{\pi}{3})$

There is a horizontal compression by a factor of 1/3, and a horizontal phase shift of $\frac{\pi}{3}$ radians to the right.

$|b| = 3$

$period = \frac{2\pi}{b} = \frac{2\pi}{3}$

This graph has a period of $\frac{2\pi}{3}$, this is what the graph looks like,

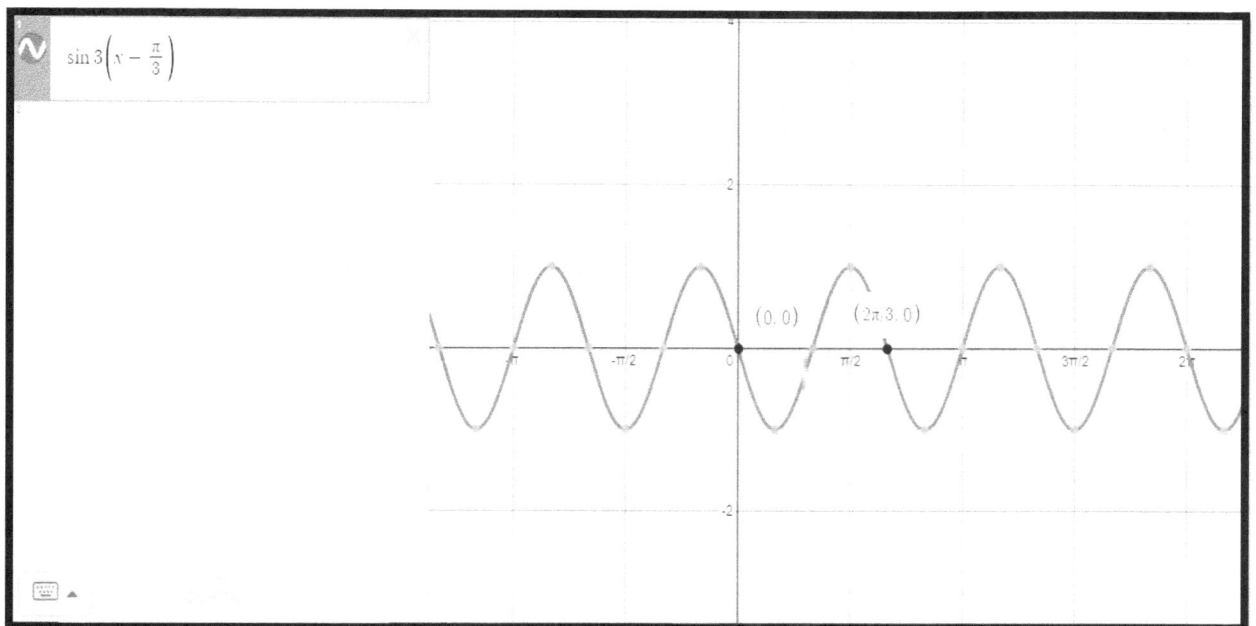

As you can see, every $\dfrac{2\pi}{3}\ radians$ the graph begins another cycle.

Example 2: $y = \cos\left(2x + \dfrac{\pi}{2}\right)$

Before determining the phase shift, we need to factor out the 2,

$y = \cos 2\left(x + \dfrac{\pi}{4}\right)$

There is a horizontal compression by a factor of ½, and there is a horizontal phase shift of $\dfrac{\pi}{4}$ radians to the left.

The period is $\dfrac{2\pi}{|b|} = \dfrac{2\pi}{2} = \pi$

This is the graph of $y = \cos 2\left(x + \dfrac{\pi}{4}\right)$,

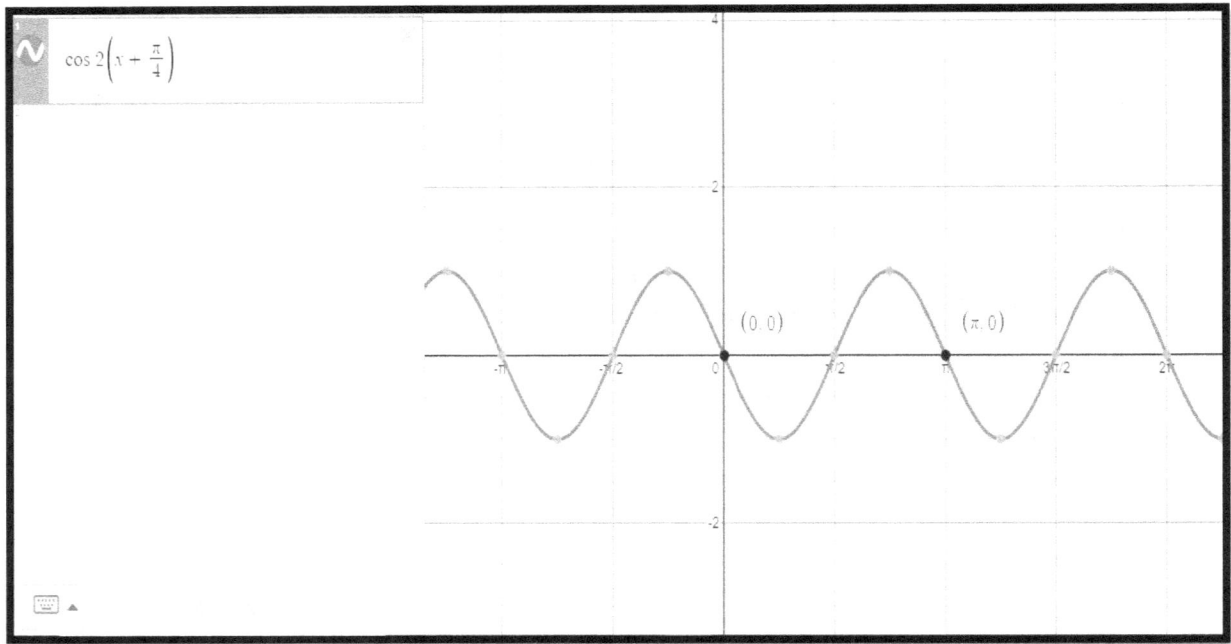

As you can see, the function completes one cycle every π radians and then it repeats itself infinitely.

Why does a horizontal compression change the period? Think about it, since you are pulling this graph towards the y-axis, the cycle gets shorter.

Why would a horizontal expansion change the period?

Think about it, since you are stretching this graph, the cycles become longer, increasing the period.

We will graph a function with a phase shift and a horizontal compression.

Example 3: Graph $y = \cos -2(x - \frac{\pi}{2})$

The period is: $\frac{2\pi}{2} = \pi$

There is a reflection in the y-axis, there is a horizontal compression by a factor of ½ , and there is a horizontal phase shift of $\frac{\pi}{2}$ radians right.

Step 1: Graph $y = \cos x$

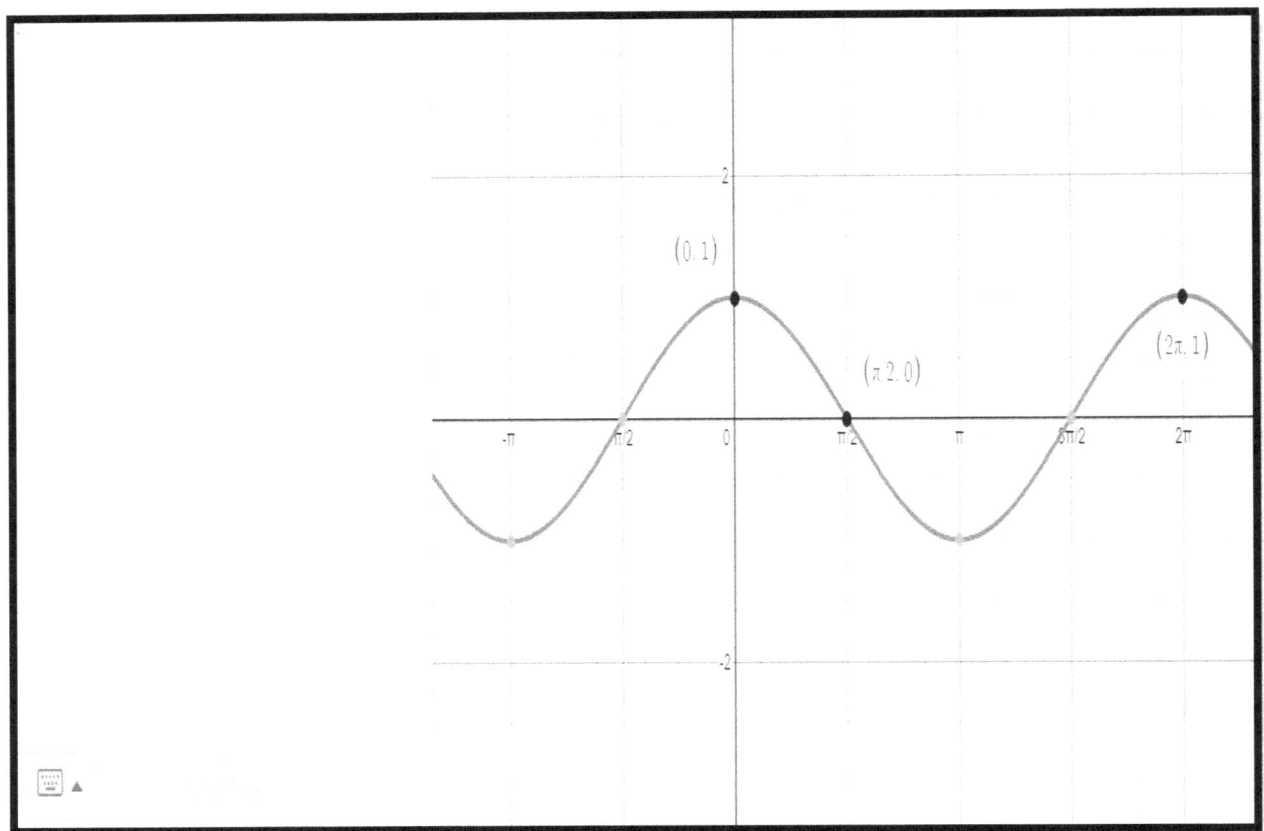

Step 2: Graph $y = \cos 2x$

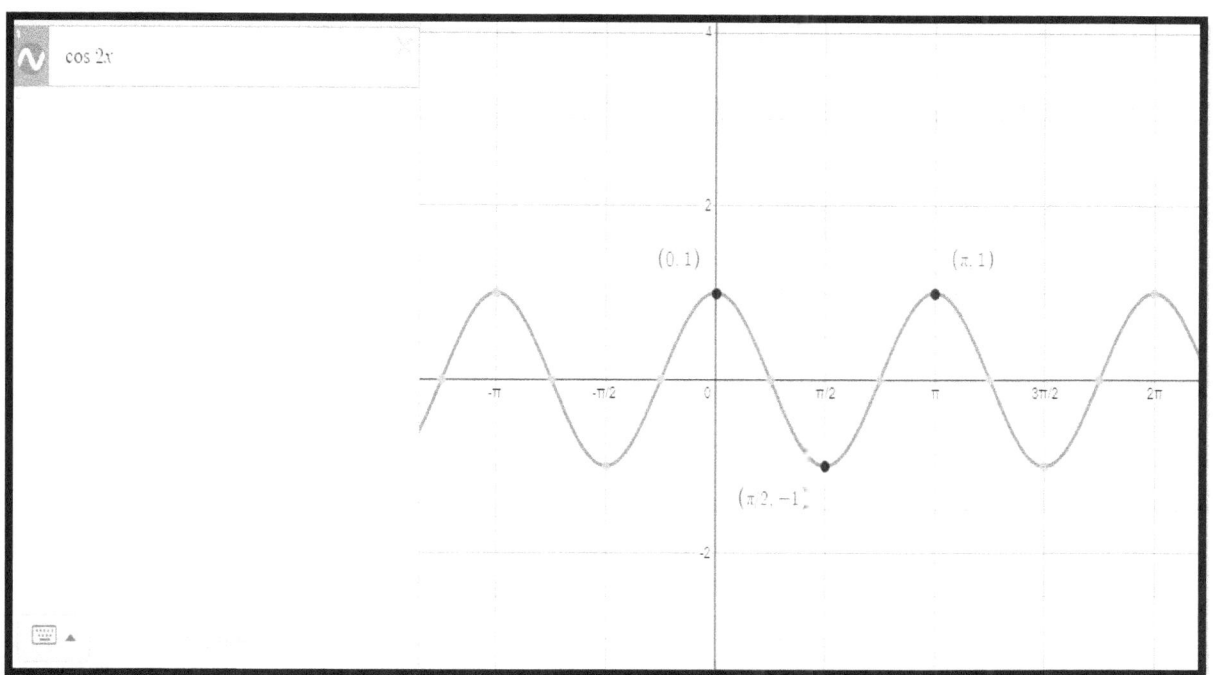

Step 3: Graph $y = \cos -2x$

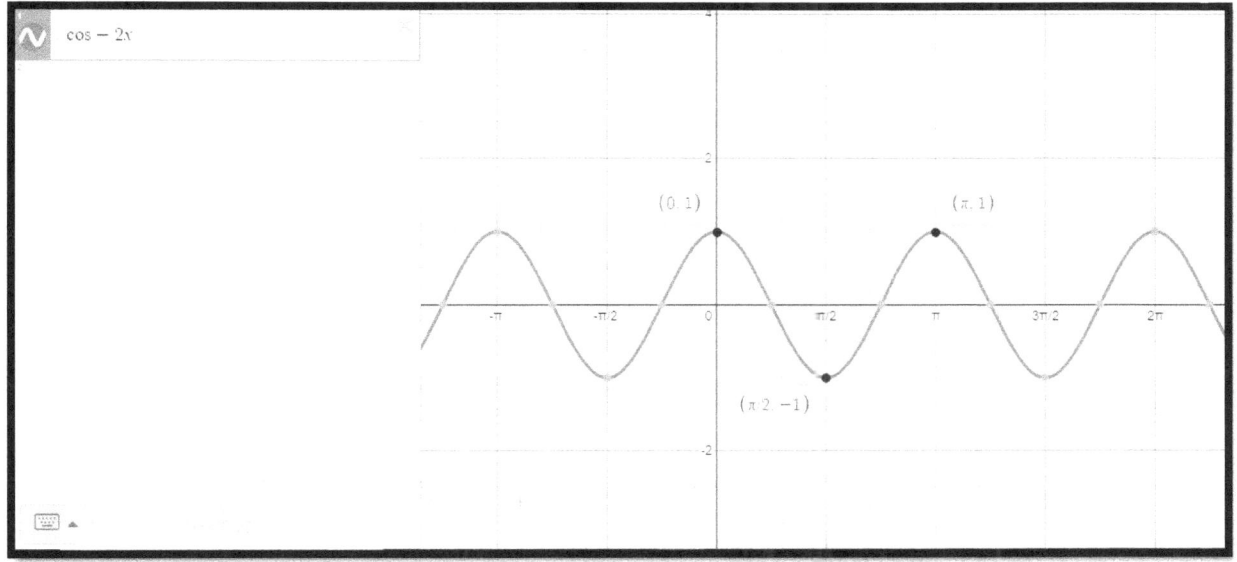

Do you see anything that changed? I do not, why didn't it change? The reason it did not change is because cosine is an even function meaning that $f(x) = f(-x)$ meaning that a reflection in the y-axis will not change the graph since negative x-values wield the same outputs as the positive x-values.

Step 4: Shift the graph $\frac{\pi}{2}$ radians right, $y = \cos -2(x - \frac{\pi}{2})$

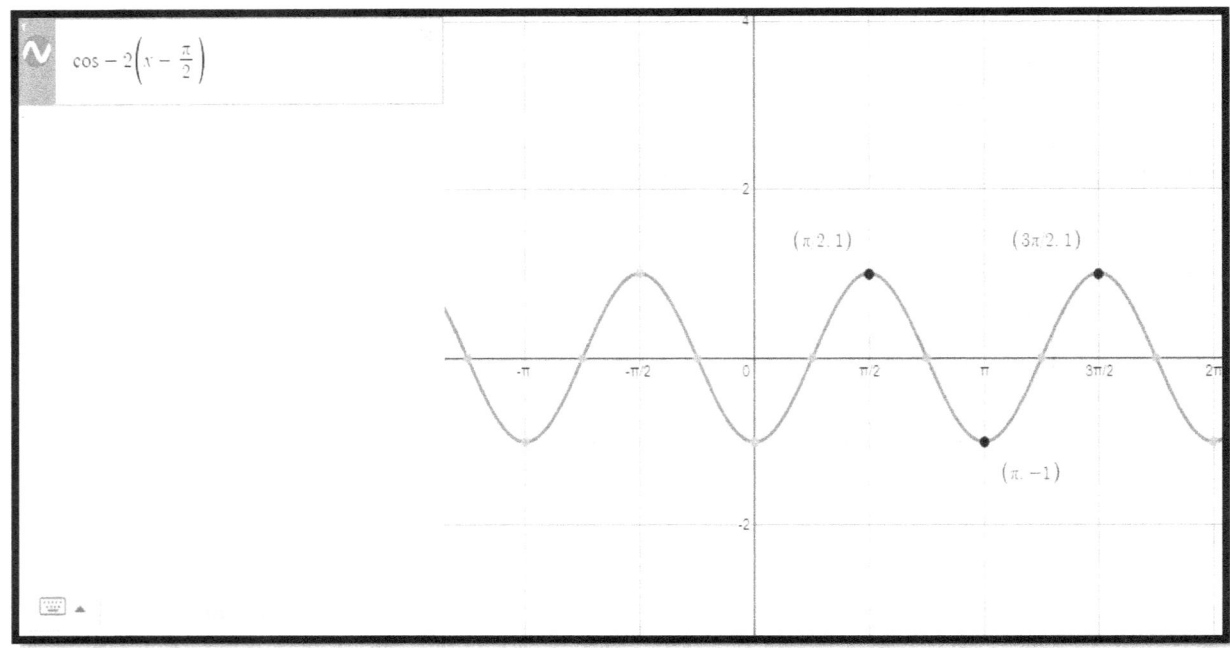

The entire graph has shifted $\frac{\pi}{2}$ radians right.

Trigonometric functions are very tricky to graph, but with practice just like anything else, it will become easier. I suggest you go back and re-read this entire trig graph section and graph the examples yourself. The next section is about my favorite part of Sinusoidal functions, writing the equations of the height of a wave at a harbour at certain times of the day and writing a sinusoidal function to determine the height of a passenger on a Ferris wheel at a certain time. Do not be hard on yourself if you don't understand it right away. Keep on reading it until you understand it or do plenty of practice questions until you understand it. Sometimes students understand something and then they start overthinking about it and then they get all confused about it.

Steve Montgomery, my grade 10 Math and Science Teacher at Fort McMurray Composite High School in Fort McMurray Alberta, said, "Don't overthink about it, just do it." That's a very good point, when you begin to question yourself about something you are doing and why, you might start getting confused because you may tell yourself that it doesn't make any sense what you are doing, even though you are doing it right. Mr. Montgomery said this when we were doing the Factoring Unit in Math 10C. Every time I begin to question the work I am doing; I always think back to what he said in March of 2018 when I was a student at that school. I stop questioning myself and I just do several practice questions on the internet. If I only make a mistake or two, I know that I am doing it right, and if I get several questions wrong then I know that I need to review that concept.

Normally, I forget teachers' words after a few years, but I have never forgotten some of the phrases the teachers at Fort McMurray Composite High School have said. For example, my English teacher at FMCHS, Colleen Schaffner said one time, "Students who take my grade 10-1 (academic) English class, become English Whizzes after." This was actually true, in English 10 I finished with a 75%(B)and then for my Grade 11 and Grade 12 English courses I finished with an A(86+) in both of them. Another thing I would like to mention is that I failed my Grade 10 English Exam, failing that exam was one of the best learning experiences for ELA in High School. It made me realize that there is a lot of room for improvement for some of my English skills. I practiced a lot during my spare time, by reading and writing. There is one more thing I want to mention about this teacher, when we did our Novel Study, she actually read the entire novel with us and explained it as she was reading it. Not many English teachers do that. I aced my Novel Study test because of that. Every Poem we read in class she read to us. I would just like to thank Colleen Schaffner for making English a more amusing class. I did not do my English 11 and 12 with her though, I completed my entire Grade 11 and 12 at GROW centre in Williams Lake, BC. This is a distance education centre, with this being said, I completed all of my courses from home without having to physically attend day classes. I unexpectedly graduated High

School on the 20th of May 2020. I went in town early that day, thinking that I was going to study 2 weeks for the final exam for Pre-Calculus 12, when I got back home, I checked my emails, and I saw that my Math Teacher (Mr. Wiebe) emailed me earlier that day, he said that I was exempt from the Final Exam because of my final mark (96%). What a surprise! I was so shocked and excited that day, it was such a surprise. I must have thanked him 5 times. You should not be hard on yourself when it comes to your grades though. If you try your best, that is all that counts. When I did my Pre-Calculus 11, I only got 64% on one of the Unit tests (Unit 4), because of that, I spent a lot of time reviewing the Quadratic Formula and now I know it by heart and I'm even able to prove the Quadratic Formula now! I am just telling you this so you can see that a 64% doesn't mean anything! Look, I wrote Mathematics books!

Trigonometric Functions with Rational Periods

Before we begin looking at the real-life applications of sinusoidal functions, we will look at writing equations with a period that is not in terms of π. The b value will be, but not the period. I will repeat this once again, the b value is not the period.

This is the form, $$y = a\cos 2\pi \frac{(x-c)}{p} + d$$ where p is the period, the b value of 2π cancels out the irrational period, causing the period to become a rational number.

Example 1: Write the equation of a cosine graph with a period of 1.

$$y = \cos\frac{2\pi(x)}{1} \qquad y = \cos 2\pi(x)$$

$$\text{Period} = \frac{2\pi}{b} = \frac{2\pi}{2\pi} = 1$$

Example 2: Write the equation of a sine graph with a period of 7.

$P = 7 \qquad\qquad b = 2\pi$

$$\sin\frac{b(x)}{p} \qquad \sin\frac{2\pi(x)}{7} \quad \text{or} \quad \sin 2\pi\frac{x}{7}$$

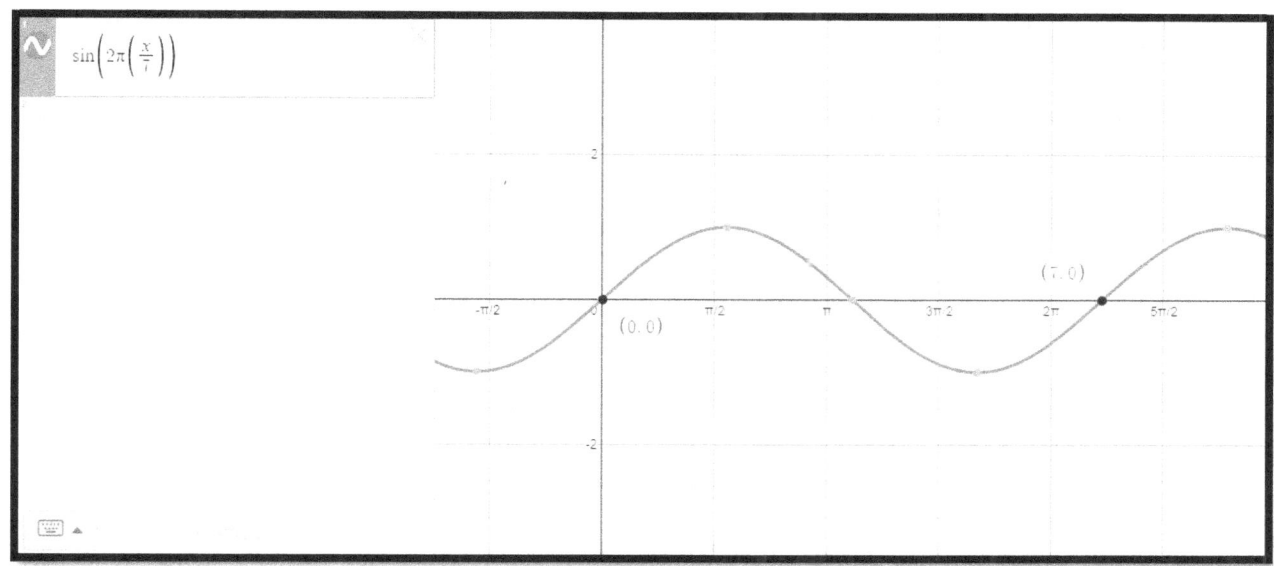

Example 3: Identify the phase shift, amplitude, vertical displacement, and the period of $y = 2\sin 2\pi\dfrac{(x-2)}{4} - 5$

Amplitude: 2

Vertical Displacement: 5 units down

Phase Shift: 2 units right

Period: 4

Example 4: Identify the phase shift, amplitude, vertical displacement, and the period of $y = -3\cos 2\pi\dfrac{(x+3)}{5} - 2$,

$Amplitude = |a| = |-3| = 3$

$Vertical\ Displacement: 2$ units down

$Phase\ Shift: 3$ units left

$Period: 5$

Example 5: Write the equation of a sine function with the following properties:

Amplitude: 8 Period: 17 Phase Shift: -6

Vertical displacement: 4

Using this information, we know that,

$a = 8 \qquad p = 17 \qquad c = -6 \qquad d = 4$

$$y = a\sin 2\pi \frac{(x-c)}{p} + d$$

$$y = 8\sin 2\pi \frac{(x-(-6))}{17} + 4$$

$$y = 8\sin 2\pi \frac{(x+6)}{17} + 4$$ this is the equation of the sine function.

You are probably wondering why we do this in the first place, we do this so we can apply these functions to real life situations. Our x-axis does not have to be in terms of π; if we write the equation like this. All of these things you have been learning are more useful than you think.

Applications of Sinusoidal Functions

In this section, we will look at how sinusoidal functions can be applied in two different situations, Ferris Wheels, and High Tide and Low Tide. Of course, there are numerous other applications; however, all of the real-life applications are beyond the scope of this book.

Example 1: A Ferris Wheel has a radius of 18m. Passengers get on halfway up on the right side. The direction of rotation of the Ferris wheel is counter clockwise. The bottom of the Ferris

Wheel is 2 meters above the ground, and it completes one rotation every 44 seconds.

The first step is to make a sketch of the Ferris Wheel, label it. Radius is 18m, far right corner of circle is where passengers get on, this is where the Ferris Wheel will be every 44 seconds, when it completes one rotation.

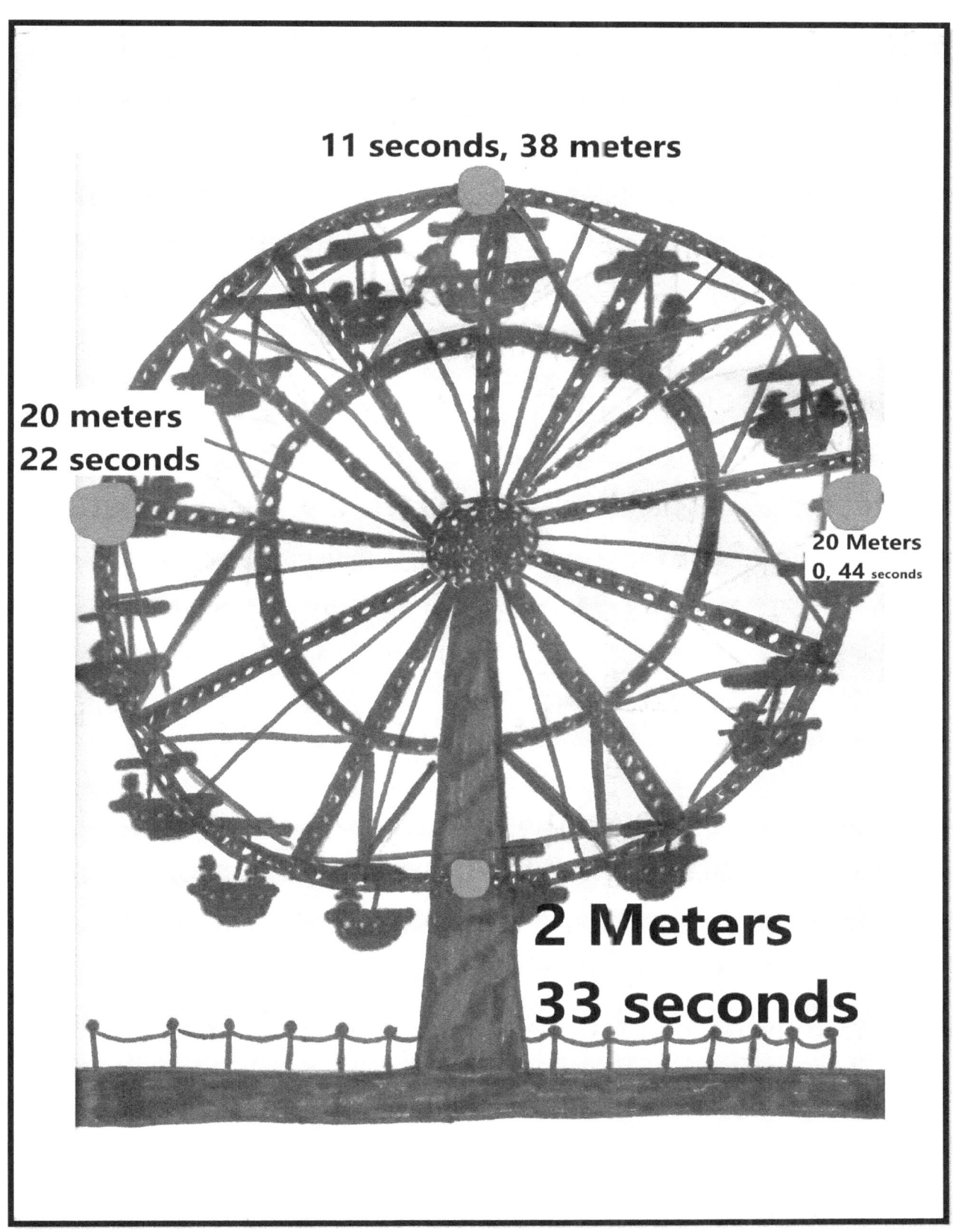

Drawn by: Angelique Martin

So, at 0 and every 44 seconds after that, the passenger is 20 meters in the air. At 11 seconds and every 44 seconds after that the passenger is 38 meters in the air. At 22 seconds and every 44 seconds after that, the passenger is 22 meters in the air. At 33 seconds and every 44 seconds after that, the passenger is only 2 meters above the ground. The Ferris Wheel is spinning constantly at the same speed, there are four important points. The first one is at the far-right corner of the circle. Our dependent variable is the height since the height depends on the time. We will write h (height, meters) as a function of t (time, seconds). The radius as previously mentioned is 18 meters, therefore, the center point is $18 + 2 = 20$ meters above the ground. The highest point is $18 + 18 + 2 = 38$ meters. The lowest point is 2 meters since the Ferris Wheel is suspended an extra 2 meters in the air, or else the poor passengers would bang into the ground and be in for a huge impact and would get severely injured. The next step is to draw a graph of the first 44 seconds of rotation. The x-axis will be labelled as time(t), in seconds. The y-axis will be labelled as height(h), in meters. There are 3 heights to take note of: $2\ meters$, $20\ meters$ and $38\ meters$. There are five important times to take note of: $(0, 11, 22, 33, 44\ seconds)$. Let's graph this using the information we are given.

These are the coordinates we will be graphing,

{(0,20),(11,38),(22,20),(33,2),(44,20)}

Connect these points with a smooth curve passing through them,

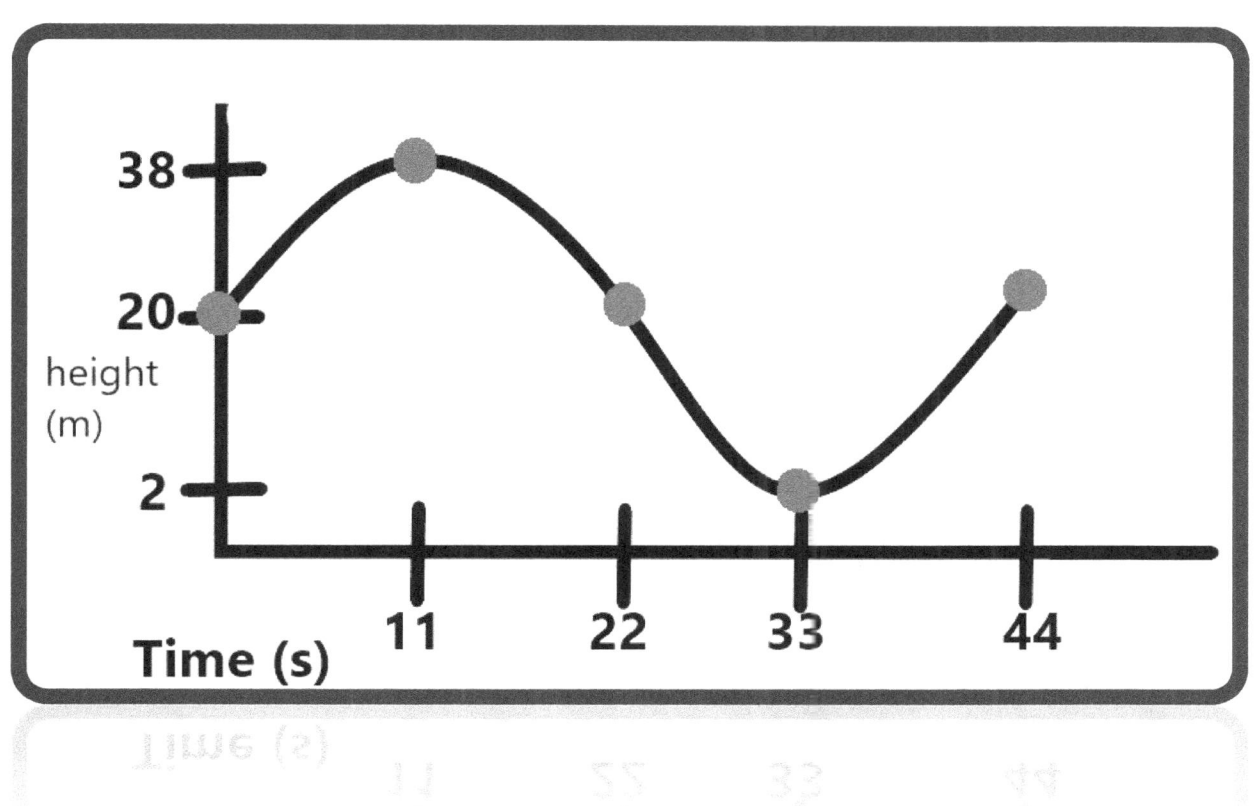

Now we will put all of the information together to write the equation of the Sine function representing this graph.

The amplitude is the radius (18) $a = 18$.

The period is 44 (one rotation every 44 seconds) 44

What is the vertical displacement? The equation of a normal sine function starts at (0,0) this one starts at $(0,20)$. The Vertical displacement is 20. $d = 20$ The radius plus the 2 meters suspension. We will write this function in the form

$$h(t) = a\sin 2\pi\frac{t}{p} + d$$

$$h(t) = 18\sin 2\pi\frac{t}{44} + 20$$ this is the function that models this situation.

How high are you after 25 seconds?

$$h(25) = 18\sin 2\pi\frac{(25)}{44} + 20$$

$$h(25) = 12.52 \; meters$$

The cool thing about this is that since periodic functions are continuous, you can even find out the height of the passenger after 88 seconds, 120 seconds, 400 seconds, etc.

Example 2: The high tide at a harbour is 10.8 meters and 6.4 hours later low tide occurs. The depth at low tide is 1.4 meters. The time that the first high tide is recorded is at 5:30 am.

Drawn by: Angelique Martin

We will write the hours in standard military time. The depth at $5:30\ am\ or\ 5.5$ is 10.8 meters. After adding 6.4 hours we get 11.9 am which is $11:54$ am. We can also plot a point in the middle of that, at 8.7 or $8:42$ am, the depth is assumed to be 6.1 m which is halfway between 1.4 and 10.8. The next high tide must occur 6.4 hours later, $11.9 + 6.4 = 18.3$ or $6:18$ pm. We can plot a point in between 11.9 and 18.3, at 15.1 $(3:06\ pm)$ the depth is assumed to be 6.1 meters.

The depth depends on the time, depth (m) is the y-axis, and time (t) is the x-axis.

These are the coordinates we will plot (d,t):

$\{(5.5,\ 10.8),(8.7,\ 6.1),(11.9,\ 1.4),(15.1,\ 6.1),(18.3,\ 10.8)\}$

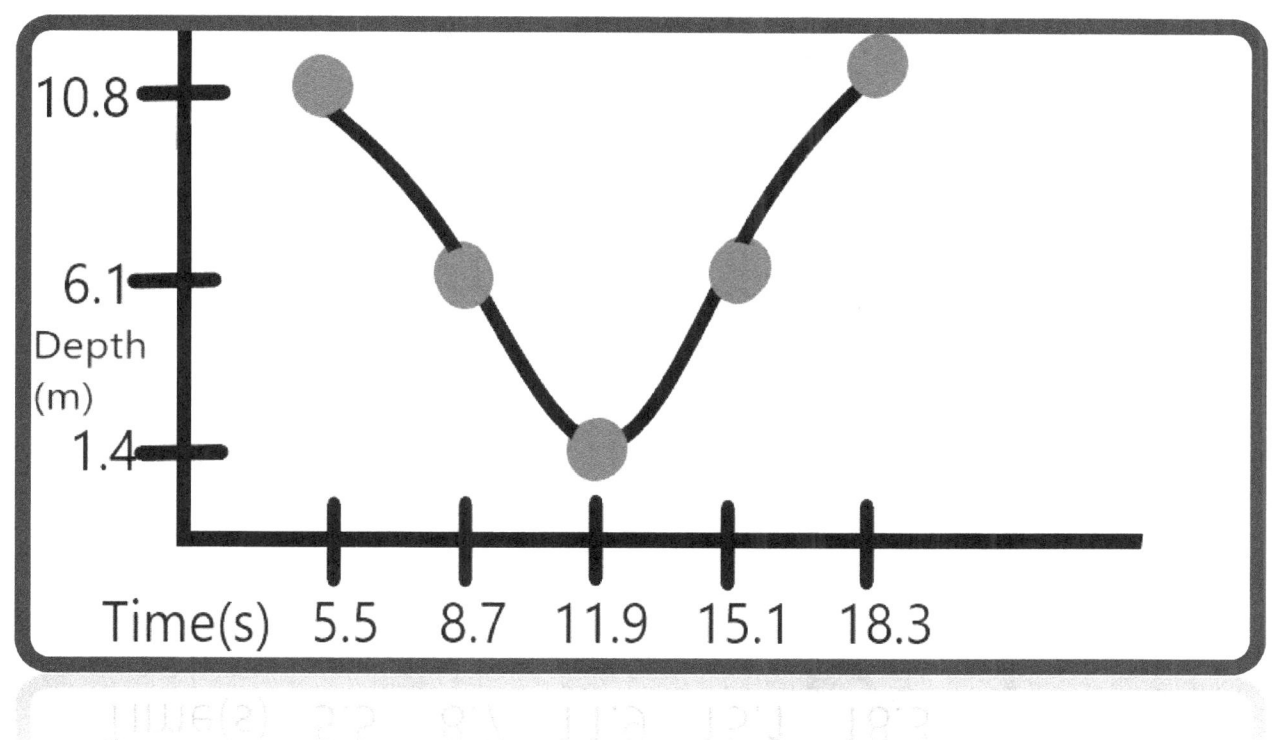

This resembles one wave of a cosine function; therefore, we can write this function in the form:

$$y = a\cos 2\pi \frac{x-c}{p} + d$$

$$Amplitude = \frac{Max - Min}{2} = \frac{10.8 - 1.4}{2} = \frac{9.4}{2} = 4.7$$

$a = 4.7$

$$Vertical\ Displacement = \frac{Max + Min}{2} = \frac{10.8 + 1.4}{2} = \frac{12.2}{2}$$

$$d = 6.1$$

$c = 5.5$ because the graph starts at the time $5{:}30\ am\ or\ 5.5$ in military time.

The period is 12.8 hours since high tide occurs every 12.8 hours, $6.4 + 6.4 = 12.8$ $p = 12.8$

Using this information, we get,

$$y = a\cos 2\pi \frac{x - c}{p} + d$$

$$y = 4.7\cos 2\pi \frac{(x - (5.5))}{12.8} + 6.1$$

$$y = 4.7\cos 2\pi \frac{(x - 5.5)}{12.8} + 6.1$$

Tangent Function

We will look at the most complicated trigonometric graph you will learn in this course, the Tangent function. Of course, the Reciprocal and Hyperbolic Trig functions are more complicated, but these are not usually covered in a High School Math course.

This is the graph of $y = \tan x$,

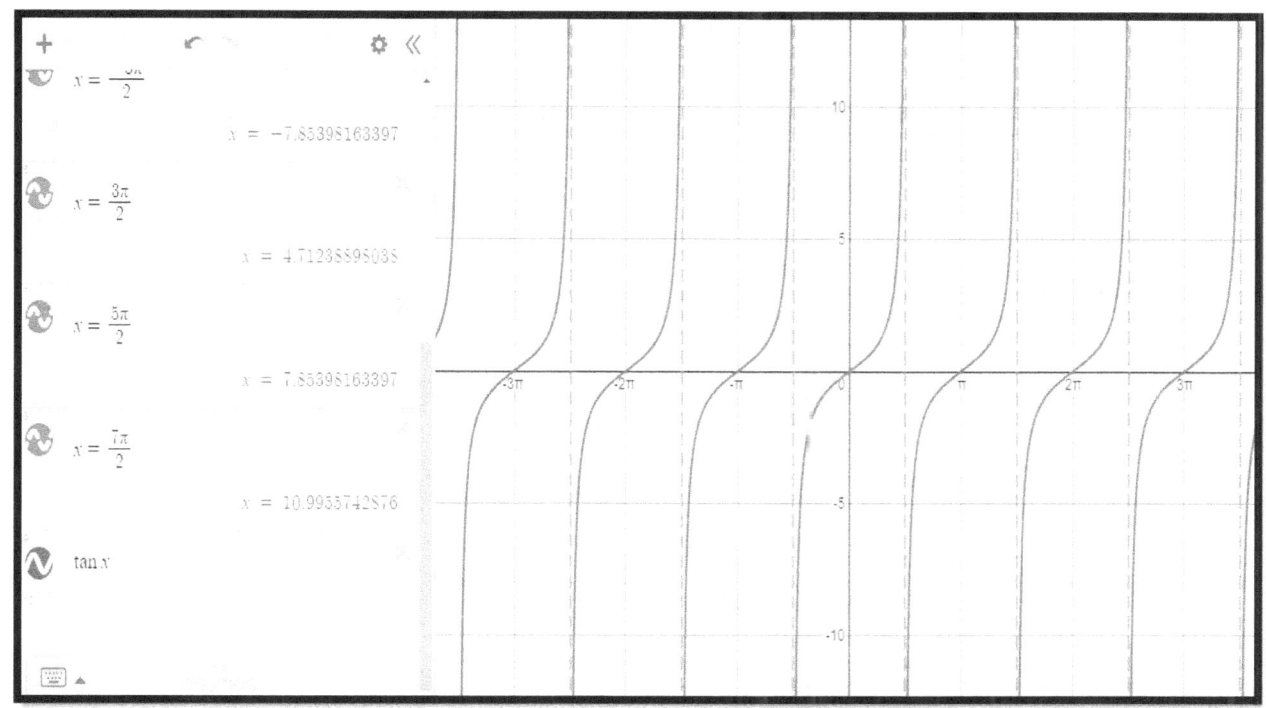

As you can see, every $\frac{\pi}{2} + n\pi$ infinitely (where n is an integer) radians, the function has an asymptote. In this graph, $-\frac{5\pi}{2}, -\frac{3\pi}{2}, -\frac{\pi}{2}, \frac{\pi}{2}, \frac{3\pi}{2}, \frac{5\pi}{2}$ are all non-permissible values. Why are these non-permissible values? They are non-permissible values because $tanx = \frac{sinx}{cosx}$ and the denominator cannot be zero, whenever cosine x equals 0, the function is undefined.

The period of $\tan x$ is π that is why $\frac{\pi}{2}$ plus multiples of π are undefined. The graph completes one cycle every π radians.

We will now change the $x \neq \dfrac{\pi}{2} + n\pi$ to a standard form that we can use to find the asymptotes of all types of tangent functions.

We must factor $\dfrac{\pi}{2}$ out of $\dfrac{\pi}{2}$ and $n\pi$

$x \neq \dfrac{\pi}{2}(1 + 2n)$ 2n so we can cancel the 2 with the 2 in $\dfrac{\pi}{2}$, giving us $n\pi$ once again.

$x \neq \dfrac{\pi}{2}(2n + 1)$

When solving for the domain we use the following form:

$$\left[bx - bc \neq \dfrac{\pi}{2}(2n + 1) \right] b$$

We solve for the domain restrictions using this form because it means, "Whatever is inside the brackets of the tangent function cannot equal $\dfrac{\pi}{2}$ plus multiples of the period."

Example 1: Find the domain and asymptote of
$y = \tan\left(\dfrac{1}{8}x\right) - 1$

We are only concerned with what is inside of the brackets of the tangent function. $(\frac{1}{8}x)$,

$b = \dfrac{1}{8}, \quad c = 0$

$\left[bx - bc \neq \dfrac{\pi}{2}(2n + 1)\right]b$

$\left[\left(\dfrac{1}{8}\right)x - \left(\dfrac{1}{8}\right)0 \neq \dfrac{\pi}{2}(2n + 1)\right]\left(\dfrac{1}{8}\right)$

$\left[\dfrac{1}{8}x \neq \dfrac{\pi}{2}(2n + 1)\right]\left(\dfrac{1}{8}\right)$ Multiply both sides by $\dfrac{1}{8}$

$\left[\dfrac{1}{64}x \neq \dfrac{\pi}{16}(2n + 1)\right]$ Multiply both sides by 64 to solve for x,

$64\left(\dfrac{1}{64}x\right) \neq 64\left(\dfrac{\pi}{16}(2n + 1)\right)$

$x \neq 4\pi(2n + 1)$ expand it to finish simplifying,

$x \neq 8n\pi + 4\pi$

Equation of the asymptote: $x = 8n\pi + 4\pi$

Example 2: Find the domain and asymptote of $tan3\left(x - \dfrac{\pi}{2}\right)$

We can expand $3\left(x - \dfrac{\pi}{2}\right)$ to get $\left(3x - \dfrac{3\pi}{2}\right)$

$b = 3 \qquad c = \dfrac{\pi}{2}$

$\left[(3)x - (3)\left(\dfrac{\pi}{2}\right) \neq \dfrac{\pi}{2}(2n + 1)\right]3$

$\left[3x - \dfrac{3\pi}{2} \neq \dfrac{\pi}{2}(2n + 1)\right]3$ Multiply both sides by 3,

$\left[9x - \dfrac{9\pi}{2} \neq \dfrac{3\pi}{2}(2n + 1)\right]$ Add $\dfrac{9\pi}{2}$ to both sides,

$9x - \dfrac{9\pi}{2} + \dfrac{9\pi}{2} \neq \dfrac{3\pi}{2}(2n + 1) + \dfrac{9\pi}{2}$ expand the right side

$9x \neq \dfrac{6n\pi}{2} + \dfrac{3\pi}{2} + \dfrac{9\pi}{2}$ Collect like terms,

$9x \neq 3n\pi + \dfrac{12\pi}{2}$

$9x \neq 3n\pi + 6\pi$ Solve for x, divide both sides by 9,

$\dfrac{9x}{9} \neq \dfrac{3n\pi + 6\pi}{9}$

$$x \neq \frac{3(n\pi + 2\pi)}{3(3)}$$

$$x \neq \frac{n\pi + 2\pi}{3}$$

Equation of asymptote: $x = \frac{n\pi + 2\pi}{3}$

Practice Questions: Chapter 11 Part B

(Graph these functions over the domain $-2\pi \leq x \leq 2\pi$)

1. Graph $y = 3\sin 2x$

2. Graph $y = 2\sin\left(x + \frac{\pi}{2}\right)$

3. Graph $y = 2\cos 2\left(x - \frac{\pi}{2}\right)$

4. Graph $y = 3\cos x + 1$

5. Identify the transformations that have been applied to, find the period as well,

   a) $y = 5\sin 3\left(x - \frac{\pi}{6}\right) + 4$

b) $y = -2\cos 5\left(x - \dfrac{\pi}{7}\right) - 1$

6. A Ferris Wheel has a radius of 20m. Passengers get on halfway up on the right side. The direction of rotation of the Ferris wheel is counter clockwise. The bottom of the Ferris Wheel is 3 meters above the ground, and it completes one rotation every 48 seconds. Write the equation.

7. Determine the domain of $\tan 2\left(x - \dfrac{\pi}{6}\right)$

Trigonometric Equations & Identities

Chapter 12

In this section we will focus on solving Trigonometric equations to find the x-intercepts over a certain domain, and a general solution. After that, we will look at some

Trigonometric identities and we will take them as they are and use them to prove other identities.

Solving Graphically

In this section, we will learn how to solve these graphically, by sketching 2 functions and seeing how many solutions there are.

Example 1: $\sin x = 0.6$ *how many solutions over the domain* $0 \leq x \leq 2\pi$?

Sketch $y = \sin x$ and $y = 0.6$ count how many times they intersect. There are 2 solutions,

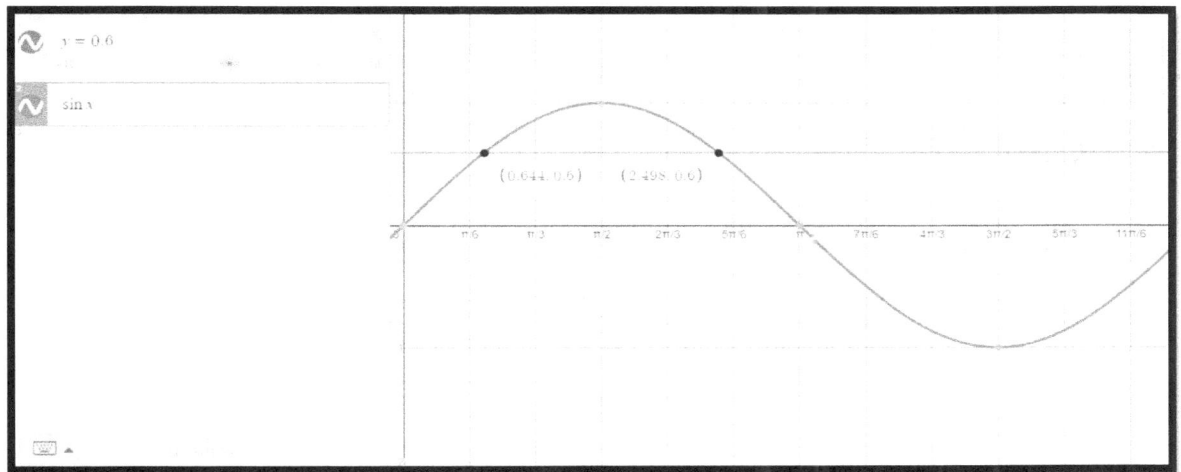

Example 2: $\cos 2x = -0.3$ *how many solutions over the domain* $0 \leq x \leq 2\pi$? There are 4 solutions,

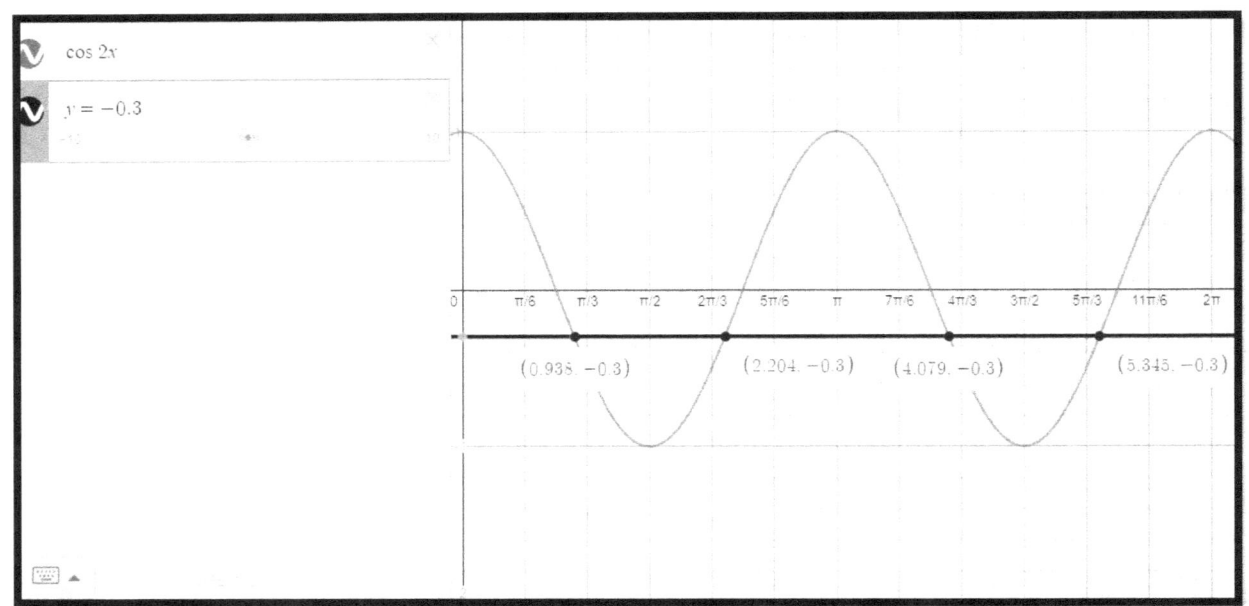

Example 3: $\sin 4x = 0.25$

how many solutions over the domain $0 \leq x \leq 2\pi$?

There are 8 solutions,

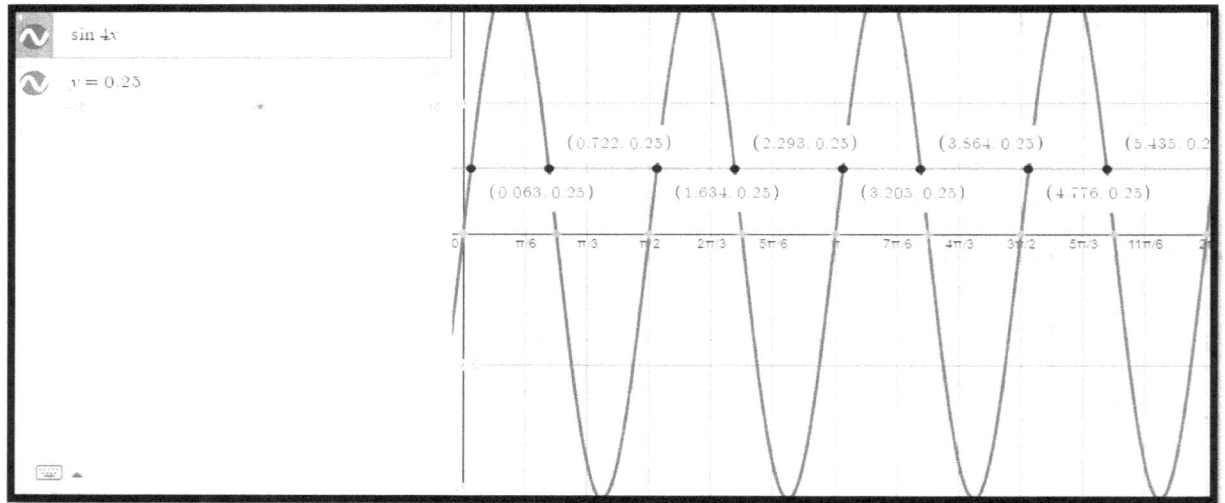

There will always be $2b$ solutions, which means *2 multiplied by The b value* as long as long as the value on the other side is in the range of the function.

$\cos x$ – 2 solutions

$\cos 2x$ – 4 solutions

$\cos 3x$ – 6 solutions

Sometimes the graph will have no solution.

For example, $sinx = 7$ has no solution since there is no x-value that will result in 7.

We will now go back to our first example, $\sin x = 0.6$, write a general solution that represents all of the roots. To do this we need to find the period of $\sin x$ which is 2π and we must add multiples of this period to the two roots we found, x=0.64 and x=2.5,

$x = 0.64 + 2n\pi$

$x = 2.5 + 2n\pi$ $\quad nEI \quad$ where n is an integer

This represents all of the roots of $\sin x = 0.6$.

We will now go back to our second example, $\cos 2x = -0.3$, how many solutions did this graph have? This graph has 4 solutions. The period is $\frac{2\pi}{2} = \pi$.

The solutions were,

$x = 0.94, 2.2, 4.08, 5.35$

This does not mean that we will have 4 general solutions, we only need two since the period is π and there are two solutions for every π radians.

$x = 0.94 + n\pi$

$x = 2.2 + n\pi$

These are the two general solutions for $\cos 2x = -0.3$, you add multiples of the period to the roots. This will give you an expression for all of the roots, there are an infinite number of roots, too many to count.

Solving Trigonometric Equations Algebraically

Remember how we solved Trigonometric Expressions for exact values? In this section, we will solve more complex trigonometric equations using a combination of skills you have already learned.

Example 1: Solve $\sin x = -\dfrac{1}{2}$ over the domain $0 \leq x \leq 2\pi$,

- Step 1: Determine the reference angle using special triangles. Let a represent the reference angle. Which triangle has a $\dfrac{opposite}{hypotenuse}$ ratio that gives $\dfrac{1}{2}$?

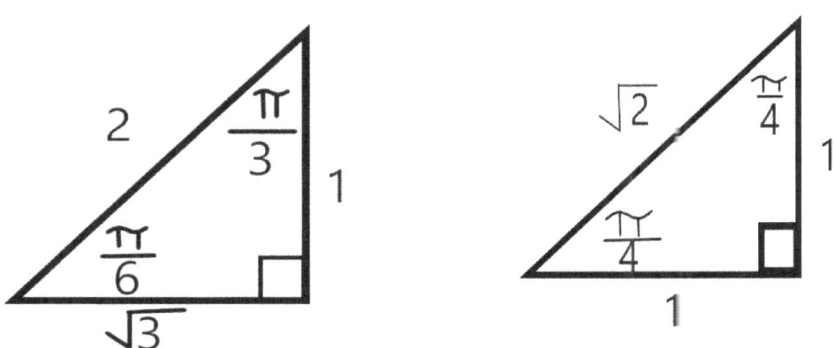

The $\frac{\pi}{6}, \frac{\pi}{3}, \frac{\pi}{2}$ special triangle does, what is the reference angle? How do we get $\sin \theta = \frac{1}{2}$?

We get this by using the reference angle of $\frac{\pi}{6}$, since $\sin \frac{\pi}{6} = \frac{1}{2}$ $a = \frac{\pi}{6}$

- Step 2: Look at the sign (positive or negative) of the original equation. Determine in which quadrant the angles lie. Draw the reference angle in the quadrant. Since the period is 2π and $b = 1$; there are only two distinct solutions

$\sin x = -\frac{1}{2}$ In which Quadrants is Sine negative? Sine is negative in Quadrants 3 and 4.

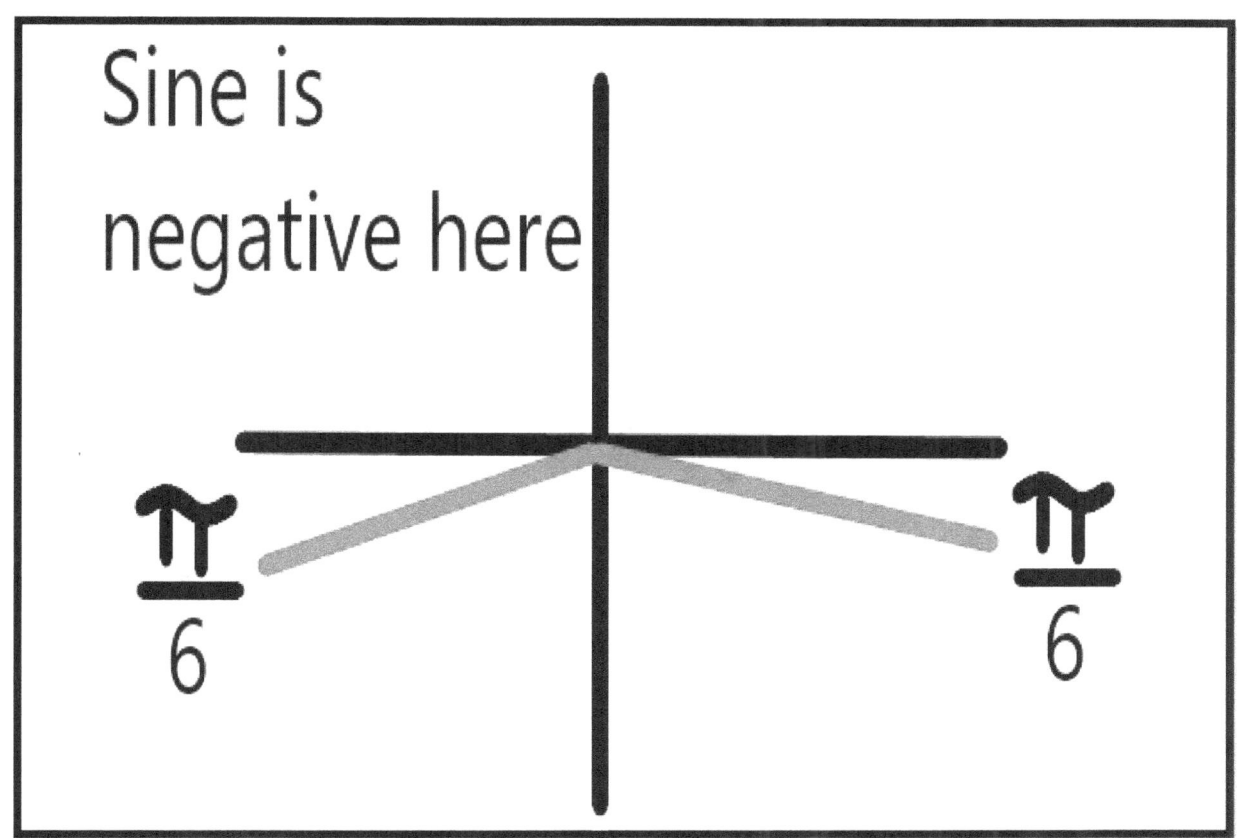

- Step 3: Determine the angles in standard position. These are the solutions.

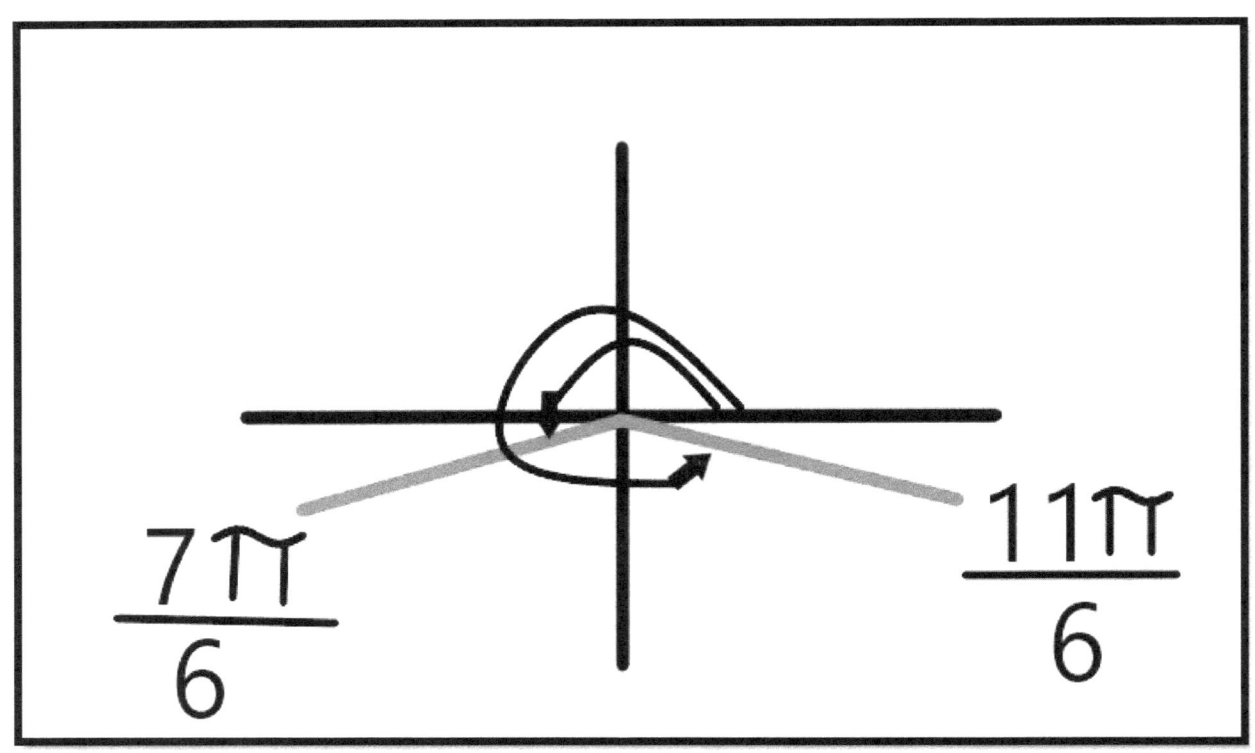

Here is how I got $\dfrac{7\pi}{6}$ and $\dfrac{11\pi}{6}$:

I got $\dfrac{7\pi}{6}$ because the reference angle was $\dfrac{\pi}{6}$ and it was underneath π; therefore, it is greater than π,

$$Solution\ 1 = \pi + \frac{\pi}{6} = \frac{6\pi}{6} + \frac{\pi}{6} = \frac{7\pi}{6}$$

I got $\dfrac{11\pi}{6}$ because the reference angle was $\dfrac{\pi}{6}$ and it was underneath 2π; therefore, it is $\dfrac{\pi}{6}$ less than 2π,

$$Solution\ 2 = 2\pi - \frac{\pi}{6} = \frac{12\pi}{6} - \frac{\pi}{6} = \frac{11\pi}{6}$$

The solutions over the domain $0 \leq x \leq 2\pi$ are $\frac{7\pi}{6}$ and $\frac{11\pi}{6}$. What would the general solution be for all roots?

The period of $\sin x$ is 2π meaning that the general solution is the roots that we found plus multiples of the period.

General solution:

$$x = \frac{7\pi}{6} + 2n\pi$$

$$x = \frac{11\pi}{6} + 2n\pi$$

Example 2: Solve $\csc x - \sqrt{2} = 0$ over the domain $0 \leq x \leq 2\pi$.

- Step 1: Rearrange the equation,
  $\csc x - \sqrt{2} + \sqrt{2} = \sqrt{2}$
  $\csc x = \sqrt{2}$ this is a reciprocal trig function; therefore, we can put it in terms of the primary trig function and solve for the exact value,

$$\csc x = \frac{1}{\sin x}$$

$$\frac{1}{\sin x} = \sqrt{2}$$

This means that the reciprocal of $\sin x$ is $\sqrt{2}$, what would be the value of $\sin x$? We can flip both sides.

$$\sin x = \frac{1}{\sqrt{2}}$$

- Step 2: Step 1: Determine the reference angle using special triangles. Let a represent the reference angle.

Which triangle has a $\frac{opposite}{hypotenuse}$ ratio that gives $\frac{1}{\sqrt{2}}$?

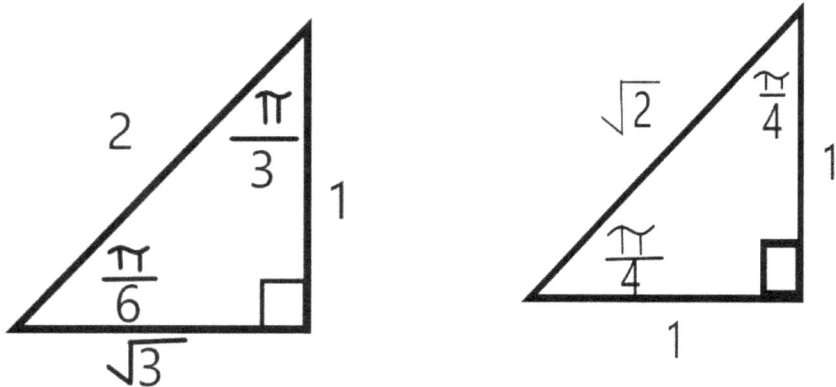

As you can see, the $\frac{\pi}{4}, \frac{\pi}{4}, \frac{\pi}{2}$ special triangle does.

The reference angle $\dfrac{\pi}{4}$ gives an exact value of $\dfrac{1}{\sqrt{2}}$?

- Step 3: Look at the sign (positive or negative) of the original equation. Determine in which quadrant the angles lie. Draw the reference angle in the quadrant. Since the period is 2π and $b = 1$; there are only two distinct solutions. $a = \dfrac{\pi}{4}$

The solutions should result in a positive output of $\sqrt{2}$ since $\dfrac{1}{\sin\left(\dfrac{1}{\sqrt{2}}\right)} = \sqrt{2}$.

In which quadrants is Sine positive? Sine is positive in Quadrants 1 and 2.

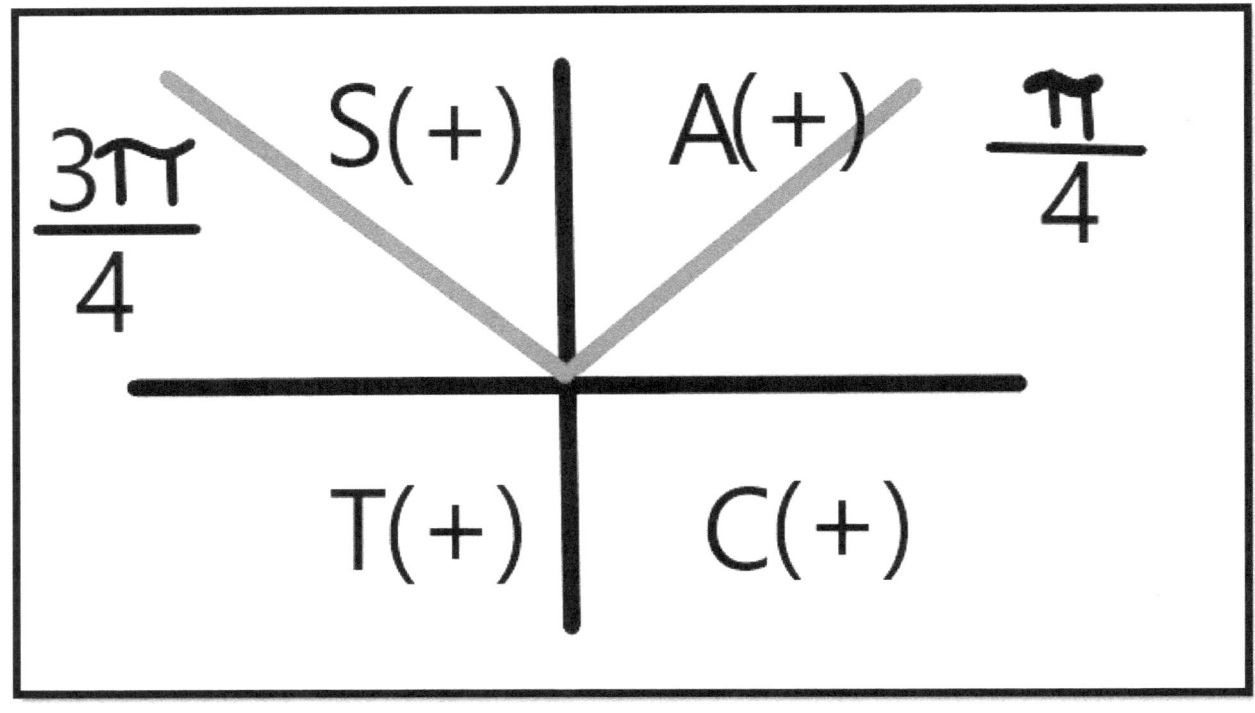

I got $\dfrac{\pi}{4}$ as one of the solutions because $\dfrac{\pi}{4}$ is an angle in quadrant 1. I got $\dfrac{3\pi}{4}$ because sine needs to be positive and a positive angle in Quadrant 2 with a reference angle of $\dfrac{\pi}{4}$ is $\dfrac{3\pi}{4}$.

Solving Second Degree Equations

In this section you will learn how to solve second degree equations such as $\sin^2 x - 3\sin x = 0$ using factoring to determine solutions on a domain of 2π radians. You will also find the general solution over the domain of real numbers.

To factor trig expressions, you factor them the same way you factor other equations:

- Common Factor,

    Example: $\sin^2 x - \sin x$ $\sin^2 x = \sin(x)\sin(x)$

    this means $\sin(x)$ multiplied by itself

    $\sin^2 x - \sin x$
    $= \sin x(\sin x - 1)$ since $\sin x$ is a common factor of both terms.

- Difference of two squares

    $9\sin^2 x - 1$
    $= (\sqrt{9\sin^2 x} - \sqrt{1})(\sqrt{9\sin^2 x} + \sqrt{1})$
    $= (3\sin x - 1)(3\sin x + 1)$

- Factoring Trinomials of the form $ax^2 + bx + c$. For example,

$\sin^2 x + 2\sin x - 3$ factor this just like you would with

$x^2 + 2x - 3$ and then just put a $\sin$ in front of both of the x-terms inside of the brackets.

Which two numbers multiply to -3 and add up to 2?

$-3 \,\&\, 1$ satisfy both of these conditions,

$x^2 + 2x - 3 = (x - 1)(x + 3)$

$\sin^2 x + 2\sin x - 3 = (\sin x - 1)(\sin x + 3)$

You can always verify your answer on your graphing the original function and the factored function on your graphing calculator, the graphs should be exactly the same. Or you can plug in several values and you should get the same output.

We will look at several other example before moving on,

Example 1: $\csc^2 x + \csc x = 0$

$= \csc x(\csc x + 1)$

Example 2: $2\tan^2 x + 4\tan x = 0$

$= 2\tan x(\tan x + 2)$

Example 3: $4\tan^2 x - 16 = 0$, *difference of squares,*

$(2\tan x - 4)(2\tan x + 4)$

Example 4: $5\cot^2 x - 10\cot x - 15 = 0$ *Factor out 5,*

$5(\cot^2 x - 2\cot x - 3) = 0$

Which numbers multiply to -3 and add up to -2? -3 and 1 satisfy these conditions.

$x^2 - 2x - 3 = (x+1)(x-3)$

Add cot to the x-terms,

$5(\cot x + 1)(\cot x - 3) = 0$ *this is the fully factored form.*

Example 5: $\sin^2 x + \sin x \cos x = 0$, factor out the sinx,

$\sin x(\sin x + \cos x) = 0$

What about factoring a messy trinomial such as $2\cos^2 x + 7\cos x - 4$? We can factor this just like we would with a normal messy trinomial and add $\cos$ to the x-terms.

Which two numbers multiply to -8 (2*-4) and add up to 7? $8\ \&\ -1$ satisfy these conditions.

$2x^2 + 8x - x - 4$

$= 2x(x+4) - 1(x+4)$

$= (2x-1)(x+4)$

Add cos(x) to the x-terms,

$= (2\cos x - 1)(\cos x + 4)$

To solve these expressions for the roots, we deal with each factor separately just like with any other equation.

Example 1: Solve $4\cos^2 x - 1 = 0$ *algebraically for the roots.*

- Step 1: Factor $4\cos^2 x - 1 = 0$, this is a difference of squares,

  $(2\cos x - 1)(2\cos x + 1) = 0$

Deal with $(2\cos x - 1)$ & $(2\cos x + 1)$ separately, solve for $\cos x$

$2\cos x - 1 = 0$	$2\cos x + 1 = 0$
$2\cos x = 1$	$2\cos x = -1$
$\dfrac{2\cos x}{2} = \dfrac{1}{2}$	$\dfrac{2\cos x}{2} = -\dfrac{1}{2}$
$\cos x = \dfrac{1}{2}$	$\cos x = -\dfrac{1}{2}$
$a = \dfrac{\pi}{3}$ since $\cos\dfrac{\pi}{3} = \dfrac{1}{2}$	$a = \dfrac{\pi}{3}$ since $\cos\dfrac{\pi}{3} = \dfrac{1}{2}$
Positive, Quadrant 1 and 4,	Negative, Quad 2 and 3

$$x = \frac{\pi}{3}, \frac{5\pi}{3} \qquad\qquad\qquad x = \frac{2\pi}{3}, \frac{4\pi}{3}$$

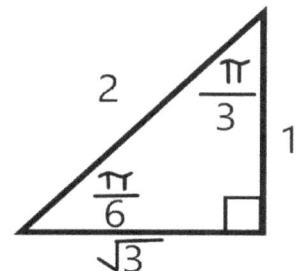

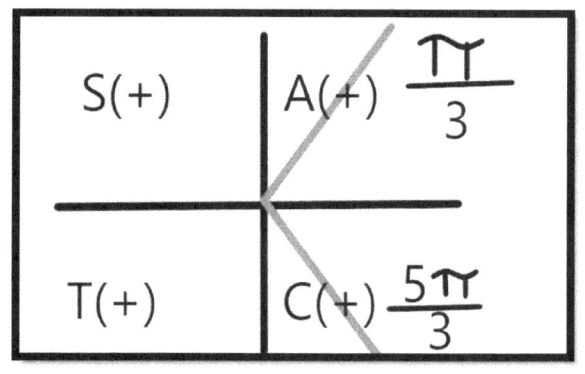

$\frac{5\pi}{3}$ because $2\pi - \frac{\pi}{3} = \frac{6\pi}{3} - \frac{\pi}{3} = \frac{5\pi}{3}$

$Period = \frac{2\pi}{b} = \frac{2\pi}{1} = 2\pi$ there are 4 distinct solutions,

$$x = \frac{\pi}{3}, \frac{2\pi}{3}, \frac{4\pi}{3}, \frac{5\pi}{3},$$

The general solutions are:

$$x = \frac{\pi}{3} + 2n\pi \qquad\qquad x = \frac{4\pi}{3} + 2n\pi$$

$$x = \frac{2\pi}{3} + 2n\pi \qquad x = \frac{5\pi}{3} + 2n\pi$$

Example 2: Solve $\cos^2 x = -\cos x$ *and give the general solution.*

First, make the right side equal 0.

$\cos^2 x + \cos x = 0$ factor it,

$\cos x(\cos x + 1) = 0$ deal with both factors separately,

$\cos x = 0$ we do not need to use a special triangle

Which x-values result in 0,

$x = \frac{\pi}{2}, \quad x = \frac{3\pi}{2}$ Every rotation on the unit circle it happens twice.

$\cos x + 1 = 0$

$\cos x = -1$

$\cos \pi = -1$

Cosine is only equal to negative 1 once per rotation (cycle).

$$x = \frac{\pi}{2}, \frac{3\pi}{2}, \pi$$

General solution is:

$$x = \frac{\pi}{2} + n\pi \qquad\qquad x = \pi + 2n\pi$$

The reason $x = \frac{\pi}{2} + n\pi$ is because the first root occurs at $\frac{\pi}{2}$ and every π after that. The reason $x = \pi + 2n\pi$ is because this root only occurs once a cycle $(2\pi\ radians)$.

Example 3: Solve $\sqrt{3}\tan x = -1$ over the domain $0 \leq x \leq 2\pi$

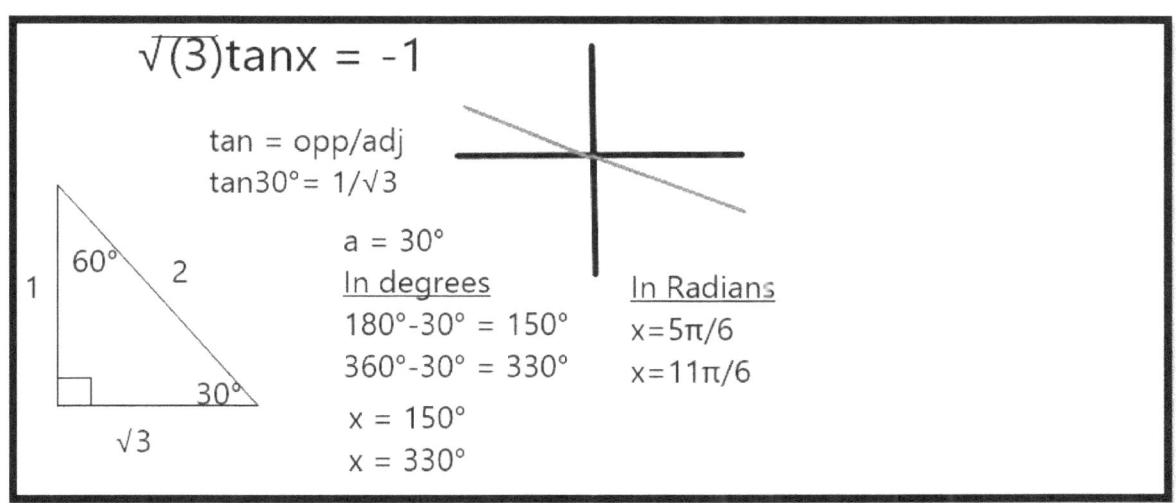

Example 4: Solve $\sin 2x = \frac{\sqrt{3}}{2}$ over the domain $0 \leq x \leq 2\pi$.

b is greater than 1; therefore, start by solving for $\sin x$

- Step 1: $\sin x = \dfrac{\sqrt{3}}{2}$ use a special triangle,

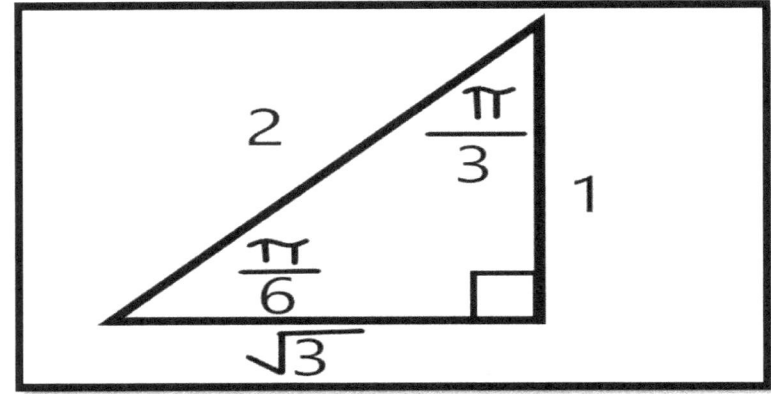

$$\sin \dfrac{\pi}{3} = \dfrac{\sqrt{3}}{2}$$

- Draw the two solutions in standard position, $\dfrac{\pi}{3}$ & $\dfrac{2\pi}{3}$,

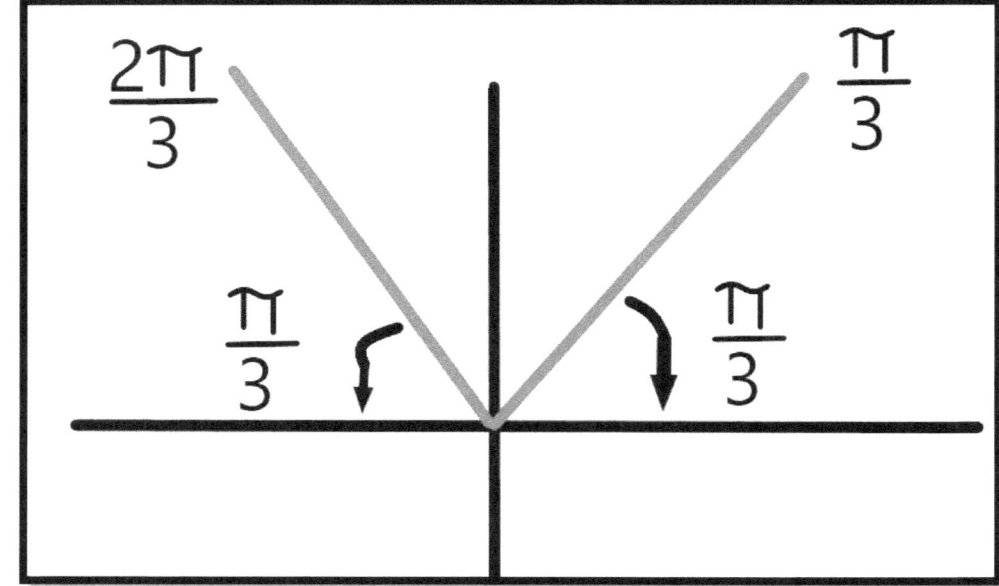

$$x = \dfrac{\pi}{3}, \dfrac{2\pi}{3}$$

- We want the solution for $\sin 2x = \frac{\sqrt{3}}{2}$

$\sin \frac{\pi}{3} = \frac{\sqrt{3}}{2}$ should be the same as $\sin 2x = \frac{\sqrt{3}}{2}$

Therefore,

$$\sin 2x = \frac{\sqrt{3}}{2}$$

$$2x = \frac{\pi}{3}$$

$$\frac{2x}{2} = \frac{\left(\frac{\pi}{3}\right)}{2}$$

$$x = \frac{\pi}{3} * \frac{1}{2} \qquad x = \frac{\pi}{6}$$

What is the period?

$b = 2$

$$period = \frac{2\pi}{b} = \frac{2\pi}{2} = \pi$$

$p = \pi$

Add multiples of the period to $\frac{\pi}{6}$ the solutions will be everything above 0 and less than 2π.

$$x = \frac{\pi}{6} + \pi = \frac{\pi}{6} + \frac{6\pi}{6} = \frac{7\pi}{6}$$

$$x = \frac{\pi}{6}, \frac{7\pi}{6}$$ these are two of the solutions, we are not done yet.

$$\sin\frac{2\pi}{3} = \frac{\sqrt{3}}{2}$$

$$2x = \frac{2\pi}{3} \qquad\qquad \frac{2x}{2} = \frac{\left(\frac{2\pi}{3}\right)}{2}$$

$$x = \frac{2\pi}{3} * \frac{1}{2} \qquad\qquad x = \frac{2\pi}{6} = \frac{\pi}{3}$$

Add multiples of the period to $\frac{\pi}{3}$ the solutions will be everything above 0 and less than 2π.

$$x = \frac{\pi}{3} + \pi = \frac{\pi}{3} + \frac{3\pi}{3} = \frac{4\pi}{3}$$

Solution for $\sin 2x = \frac{\sqrt{3}}{2}$

$$x = \frac{\pi}{6}, \frac{7\pi}{6}, \frac{\pi}{3}, \frac{4\pi}{3}$$

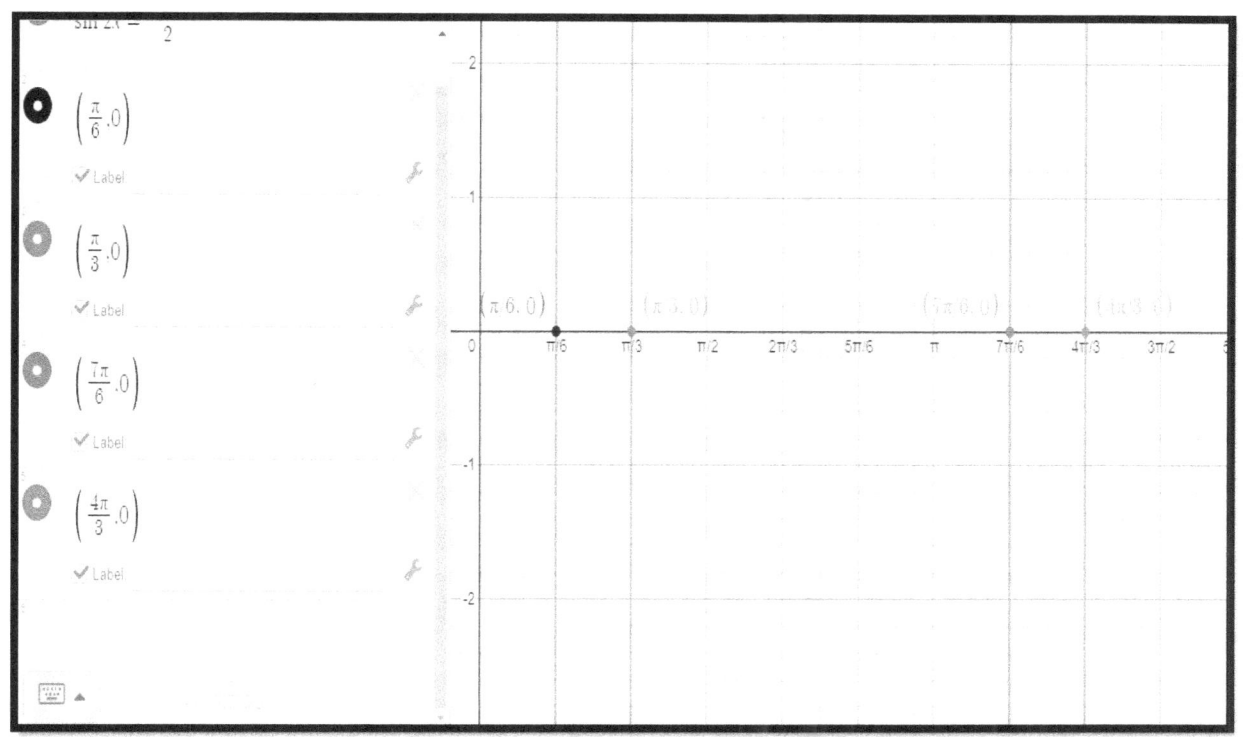

We will look at another example that is similar to the one we just looked at, it may look very complicated at first, but with another example, it should start to be clearer to you.

Example 5: $\cos 2x = -\dfrac{1}{\sqrt{2}}$

What is the period? $b = 2$,

$Period = \dfrac{2\pi}{b} = \dfrac{2\pi}{2} = \pi$

Deal with cos(x) first,

$\cos x = -\dfrac{1}{\sqrt{2}}$ This is negative, solutions are in Quadrant 2 and 3. Use a special triangle to find the reference angle and then sketch both of the angles. $\cos\dfrac{\pi}{4} = \dfrac{1}{\sqrt{2}}$

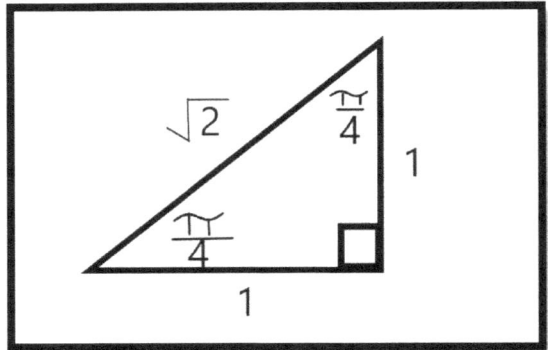

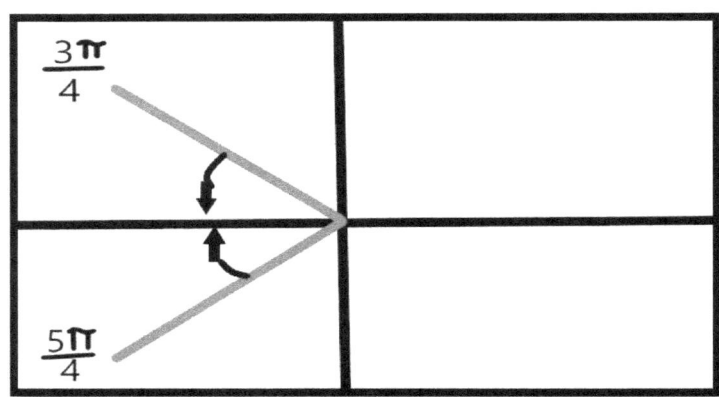

$\cos\dfrac{3\pi}{4} = -\dfrac{1}{\sqrt{2}}$

$\cos 2x = -\dfrac{1}{\sqrt{2}}$ $2x = \dfrac{3\pi}{4}$

$\dfrac{2x}{2} = \dfrac{\frac{3\pi}{4}}{2}$ solve for x

$$x = \frac{3\pi}{4} * \frac{1}{2} \qquad x = \frac{3\pi}{8}$$ add multiples of the period as long as $0 \leq x \leq 2\pi$

$$x = \frac{3\pi}{8} + \pi = \frac{3\pi}{8} + \frac{8\pi}{8} = \frac{11\pi}{8}$$

$$x = \frac{3\pi}{8}, \frac{11\pi}{8}$$ we are not done yet, we must look at the other solution as well,

$$\cos\frac{5\pi}{4} = -\frac{1}{\sqrt{2}}$$

$$2x = \frac{5\pi}{4} \quad \text{solve for x,} \quad \frac{2x}{2} = \frac{\frac{5\pi}{4}}{2}$$

$$x = \frac{5\pi}{4} * \frac{1}{2}$$

$$x = \frac{5\pi}{8}$$ add multiples of the period as long as $0 \leq x \leq 2\pi$

$$x = \frac{5\pi}{8} + \pi = \frac{5\pi}{8} + \frac{8\pi}{8} = \frac{13\pi}{8}$$ Solution is,

$$x = \frac{3\pi}{8}, \frac{5\pi}{8}, \frac{11\pi}{8}, \frac{13\pi}{8}$$

Reciprocal Identities

What is an identity, an identity is a special type of equation that is true for all values, both sides are always equal.

An identity is something such as,

$x^2 - 4x + 4$		$(x - 2)^2$
$x^2 - 4x + 4$		$(x - 2)(x - 2)$
$x^2 - 4x + 4$		$x^2 - 4x + 4$
Left side	=	Right side

When verifying an identity treat the Left side(LS) and Right side(RS) separately. Never bring something from one side to the other.

An equation is when only a few values make it true.

For example, $x - 2 = 7$

$x = 9$ and only 9, because watch what happens if I plug in another value,

$(12) - 2 = 7$ $\qquad 10 \neq 7$ this does not make sense.

Here are the **Reciprocal Identities:**

$$\sec \theta = \frac{1}{\cos \theta} \qquad \csc \theta = \frac{1}{\sin \theta} \qquad \cot \theta = \frac{1}{\tan \theta}$$

Here are the **Quotient Identities:**

$$\tan \theta = \frac{\sin \theta}{\cos \theta} \qquad \cot \theta = \frac{\cos \theta}{\sin \theta}$$

We will review some methods for simplifying rational expressions before moving on,

Example 1: $\frac{1}{x} * \frac{z}{y} * \frac{x}{z} * y$ you can cancel either vertically or diagonally,

$$= \frac{1}{x} * \frac{z}{y} * \frac{x}{z} * y$$

$$= 1$$

Example 2: $\frac{1}{\frac{1}{b}}$ To simplify this, you have two options,

Option 1: Multiply 1 by the reciprocal of the denominator,

Option 2: Multiply the two denominators by a common denominator.

Option 1: $\dfrac{\frac{1}{1}}{b}$

$= 1 * \dfrac{b}{1} = b$

Option 2: $\dfrac{\frac{1}{1}}{b}$

$= \dfrac{\frac{1}{1}}{\frac{1}{b}}$

CD=$(1 * b) = b$

$= \dfrac{\frac{1}{1}(b)}{\frac{1}{b}(b)}$

$= \dfrac{b}{1} = b$

Example 3: $\dfrac{\dfrac{a}{b}}{\dfrac{c}{d}}$

$CD = (b * d) = (bd)$

$$\dfrac{\left(\dfrac{a}{b}\right)(bd)}{\left(\dfrac{c}{d}\right)(bd)} = \dfrac{\left(\dfrac{a}{b}\right)bd}{\left(\dfrac{c}{d}\right)bd} = \dfrac{ad}{cb}$$

Prove $\quad cos\theta csc\theta tan\theta = 1$

$\csc \theta = \dfrac{1}{sin\theta} \qquad \cos \theta \left(\dfrac{1}{\sin \theta}\right) tan\theta$

$\tan \theta = \dfrac{sin\theta}{cos\theta} \qquad \cos \theta \left(\dfrac{1}{\sin \theta}\right)\left(\dfrac{sin\theta}{\cos \theta}\right)$

$$1 = 1$$

$$Left\ Side = Right\ Side$$

Prove $\quad \cot\theta \sec\theta = \csc\theta$

$\cot\theta = \dfrac{\cos\theta}{\sin\theta}$ $\quad \left(\dfrac{\cos\theta}{\sin\theta}\right)\sec\theta \quad \Big| \quad \dfrac{1}{\sin\theta} \quad\quad \csc\theta = \dfrac{1}{\sin\theta}$

$\sec\theta = \dfrac{1}{\cos\theta}$ $\quad \left(\dfrac{\cos\theta}{\sin\theta}\right)\left(\dfrac{1}{\cos\theta}\right)$

$$\dfrac{1}{\sin\theta} = \dfrac{1}{\sin\theta}$$

$$Left\ Side = Right\ Side$$

Prove $\quad \tan\theta \csc\theta = \sec\theta$

$\tan\theta = \dfrac{\sin\theta}{\cos\theta}$ $\quad \left(\dfrac{\sin\theta}{\cos\theta}\right)\csc\theta \quad \Big| \quad \dfrac{1}{\cos\theta}$

$\csc\theta = \dfrac{1}{\sin\theta}$ $\quad \left(\dfrac{\sin\theta}{\cos\theta}\right)\left(\dfrac{1}{\sin\theta}\right)$

$$\dfrac{1}{\cos\theta} = \dfrac{1}{\cos\theta}$$

Pythagorean Identities

There are three main Pythagorean identities,

- $\sin^2\theta + \cos^2\theta = 1$
- $1 + \cot^2\theta = \csc^2\theta$

- $1 + \tan^2\theta = \sec^2\theta$

Remember the unit circle? Remember how the radius was 1 (hypotenuse was one)? Remember how sine was equal to the y-coordinate and cosine was equal to the x-coordinate. The Pythagorean Identity is based off of the Pythagorean Theorem ($c^2 = a^2 + b^2$)

Where $c = 1$, $a = \sin\theta$, & $b = \cos\theta$

We obtain $1 + \cot^2\theta = \csc^2\theta$ by dividing $\sin^2\theta + \cos^2\theta = 1$ by $\sin^2\theta$

We obtain $1 + \tan^2\theta = \sec^2\theta$ by dividing $\sin^2\theta + \cos^2\theta = 1$ by $\cos^2\theta$

These three Pythagorean Identities can be written several ways,

- $\sin^2\theta = 1 - \cos^2\theta$
- $\cot^2\theta = \csc^2\theta - 1$
- $\tan^2\theta = \sec^2\theta - 1$

$\cos^2\theta = 1 - \sin^2\theta$
$1 = \csc^2\theta - \cot^2\theta$
$1 = \sec^2\theta - \tan^2\theta$

These identities are very useful when proving trigonometric Identities, they will save you a lot of time, as you can immediately substitute its equivalent without having to expand and simplify just to get to the same result. Write these identities in your notebook as they will be very useful. We will prove some Trig Identities using the Pythagorean Identities and then after that, we will look at proving Trigonometric Identities using factoring or even its conjugate as a last resort.

Example 1: Prove, $\sin x + \cot x \cos x = \csc x$

$\cot x = \dfrac{\cos x}{\sin x}$ $\quad\quad \sin x + \left(\dfrac{\cos x}{\sin x}\right)\cos x$

$\quad\quad\quad\quad\quad\quad\quad \sin x + \dfrac{\cos^2 x}{\sin x}$

Find the common denominator to

Continue simplifying, $\quad \dfrac{\sin x}{1}\left(\dfrac{\sin x}{\sin x}\right) + \dfrac{\cos^2 x}{\sin x}\left(\dfrac{1}{1}\right)$

$\quad\quad\quad\quad\quad\quad\quad \dfrac{\sin^2 x}{\sin x} + \dfrac{\cos^2 x}{\sin x}$

$\sin^2 x + \cos^2 x = 1 \quad\quad \dfrac{\sin^2 x + \cos^2 x}{\sin x}$

$$\frac{1}{\sin x} = \frac{1}{\sin x}$$

Example 2:

$$\frac{(\csc x - 1)(\csc x + 1)}{} \stackrel{?}{=} \frac{1}{\tan^2 x}$$

Expand,

$$\csc x(\csc x + 1) - 1(\csc x + 1)$$

$$\csc^2 x + \csc x - \csc x - 1$$

$$\csc^2 x - 1$$

$\csc^2 x - 1 = \cot^2 x \qquad \cot^2 x$

$\cot^2 x = \dfrac{1}{\tan^2 x} \qquad \dfrac{1}{\tan^2 x} = \dfrac{1}{\tan^2 x}$

$$Left\ Side = Right\ Side$$

Proving Trig Identities Using Factoring

Here are some more Trigonometric Identities you should take note of:

$$\cot x = \frac{\cos x}{\sin x} \qquad \frac{\sin^2 x}{\cos^2 x} = \tan^2 x$$

$$\cot x \sin x = \cos x$$

$$\sin^2 x - 1 = -\cos^2 x \qquad \tan^2 x \cot x = \tan x$$

$$\csc x \tan x = \sec x \qquad \cos^2 x + \sin^2 x + \tan^2 x = \sec^2 x$$

Before looking at how to use factoring to prove identities we will look at simplifying complex fractions,

Example 1: Solve this Complex Fraction,

$$\frac{\frac{7}{12}+1}{\frac{1}{3}+\frac{3}{4}}$$

Find the common denominator, (3*4)=12 multiply the numerator and denominator by 12,

$$\frac{\left(\frac{7}{12}+1\right)12}{\left(\frac{1}{3}+\frac{3}{4}\right)12} = \frac{7+12}{4+9} = \frac{19}{13}$$

Prove $\dfrac{1+\sin x}{1+\csc x} = \sin x$

$\csc x = \dfrac{1}{\sin x}$ $\qquad \dfrac{1+\sin x}{1+\dfrac{1}{\sin x}}$

Find the common denominator, $\sin x$

$$\dfrac{\sin x(1+\sin x)}{\sin x\left(1+\dfrac{1}{\sin x}\right)}$$

$$\dfrac{\sin x(1+\sin x)}{\sin x + 1}$$

$\qquad \sin x = \sin x$

$\qquad \text{Left Side} = \text{Right Side}$

Prove $\dfrac{1+\sec x}{1+\cos x} = \sec x$

$\qquad \dfrac{1+\left(\dfrac{1}{\cos x}\right)}{1+\cos x} \qquad\qquad \dfrac{1}{\cos x}$

Find the common denominator

$$\frac{\left(cosx\left(1+\dfrac{1}{\cos x}\right)\right)}{\cos x(1+\cos x)}, \cos x$$

$$\frac{\cos x + 1}{\cos x(1+\cos x)}$$

$$\frac{1}{\cos x} = \frac{1}{\cos x}$$

Proving Trig Identities Using Conjugates

When you are trying to prove a trigonometric identity and all of the other methods have failed, use the denominator's conjugate. We multiply the denominator and the numerator by the denominator's **conjugate**. For example, the conjugate of $(a+b)$ is $(a-b)$ and vice versa.

Example 1: **Prove** $\dfrac{\cos x}{1+\sin x} = \dfrac{1-\sin x}{\cos x}$

Multiply the left side by

Its conjugate $(1-\sin x)$

$$\frac{(1-\sin x)(\cos x)}{(1+\sin x)(1-\sin x)}$$

$$\frac{(1-\sin x)(\cos x)}{1-\sin^2 x}$$

$\cos^2 x = 1 - \sin^2 x \qquad \dfrac{(1-\sin x)(\cos x)}{\cos^2 x}$

$$\frac{(1-\sin x)(\cos x)}{\cos x(\cos x)}$$

$$\frac{1-\sin x}{\cos x} = \frac{1-\sin x}{\cos x}$$

Sine and Cosine Addition and Difference Identities

Trigonometric functions do not distribute over a sum or difference inside parenthesis. For example, $\sin(A+B)$

Cannot be distributed,

We will verify the Sine Sum Identity,

$\sin(a+b) = \sin a\cos b + \cos a\sin b$

We will verify it for, $a = \dfrac{\pi}{4}$ and $b = \dfrac{\pi}{2}$

$$\sin(a+b) = \sin a \cos b + \cos a \sin b$$

$$\sin\left(\dfrac{\pi}{4} + \dfrac{\pi}{2}\right) \quad \sin\left(\dfrac{\pi}{4}\right)\cos\left(\dfrac{\pi}{2}\right) + \cos\left(\dfrac{\pi}{4}\right)\sin\left(\dfrac{\pi}{2}\right)$$

$$\sin\left(\dfrac{\pi}{4} + \dfrac{2\pi}{4}\right) \quad \left(\dfrac{1}{\sqrt{2}}\right)(0) + \left(\dfrac{1}{\sqrt{2}}\right)(1)$$

$$\sin\left(\dfrac{3\pi}{4}\right) \quad \dfrac{1}{\sqrt{2}}$$

The exact value of

$$\sin\left(\dfrac{3\pi}{4}\right) \text{ is } \dfrac{1}{\sqrt{2}} \qquad \dfrac{1}{\sqrt{2}} \qquad \dfrac{1}{\sqrt{2}}$$

$$Left\ Side = Right\ Side$$

Sine Addition Identity:

$\sin(a+b) = \sin a \cos b + \cos a \sin b$

Trigonometric Sum and Difference Identities:

$\sin(a+b) = \sin a \cos b + \cos a \sin b$

$\sin(a-b) = \sin a \cos b - \cos a \sin b$

$\cos(a+b) = \cos a \cos b - \sin a \sin b$

$\cos(a-b) = \cos a \cos b + \sin a \sin b$

$\tan(a+b) = \dfrac{\tan a + \tan b}{1 - \tan a \tan b}$

$\tan(a-b) = \dfrac{\tan a - \tan b}{1 + \tan a \tan b}$

You basically, expand it and find the exact values of the trig ratios inside of the equation and then simplify. When you work backwards, you find the a & b value inside of the expanded identity and put them both inside of the parenthesis and finish simplifying.

Example 1: $\cos 4\theta \cos 3\theta + \sin 4\theta \sin 3\theta$

which identity is this? It is the Cosine difference Identity,

$\cos(a-b) = \cos a \cos b + \sin a \sin b$

$\qquad a = 4\theta \qquad b = 3\theta$

$\cos((4\theta) - (3\theta)) = \cos 4\theta \cos 3\theta + \sin 4\theta \sin 3\theta$

$\cos(4\theta - 3\theta) = \cos \theta$

Example 2: $\cos 3x \cos 7x - \sin 3x \sin 7x$

Which identity is this? This is the cosine sum identity,

$$\cos(a+b) = \cos a \cos b - \sin a \sin b$$

$$a = 3x \qquad b = 7x$$

$$\cos((3x)+(7x)) = \cos 3x \cos 7x - \sin 3x \sin 7x$$

$$= \cos(3x + 7x)$$

$$= \cos 10x$$

Example 3: $\dfrac{\tan \dfrac{\pi}{3} - \tan \dfrac{\pi}{12}}{1 + \tan \dfrac{\pi}{3} \tan \dfrac{\pi}{12}}$

Which identity is this? It is the tangent difference identity,

$$\tan(a-b) = \frac{\tan a - \tan b}{1 + \tan a \tan b}$$

$$a = \frac{\pi}{3} \qquad b = \frac{\pi}{12}$$

$$\tan\left(\left(\frac{\pi}{3}\right)-\left(\frac{\pi}{12}\right)\right) = \frac{\tan\frac{\pi}{3} - \tan\frac{\pi}{12}}{1 + \tan\frac{\pi}{3}\tan\frac{\pi}{12}}$$

$$= \tan\left(\frac{4\pi}{12} - \frac{\pi}{12}\right)$$ Simplify everything inside of the brackets,

$$= \tan\left(\frac{3\pi}{12}\right)$$ this can be reduced further...

$$= \tan\left(\frac{\pi}{4}\right)$$ Find the exact value of $\frac{\pi}{4}$

The exact value of $\tan\frac{\pi}{4}$ is 1,

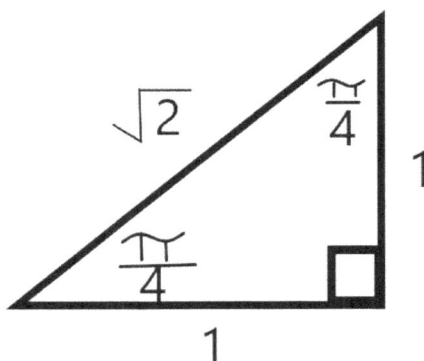

$\tan\frac{\pi}{4} = 1$

as you can see, these identities are very useful.

Example 4: $\sin 3y \cos 5y + \cos 3y \sin 5y$

Which Identity is this? This is the Addition Identity of Sine,

$a = 3y \qquad b = 5y$

$\sin(a + b) = \sin a \cos b + \cos a \sin b$

$\sin((3y) + (5y))$ simplify what is inside of the brackets

$= \sin(3y + 5y)$ collect like terms,

$= \sin 8y$

Example 5: $\cos(\pi - \theta)$

We can use the cosine difference identity to simplify this expression,

$\cos(a - b) = \cos a \cos b + \sin a \sin b$

$a = \pi \qquad b = \theta$

$\cos(\pi - \theta) = \cos \pi \cos \theta + \sin \pi \sin \theta$

$= (-1)\cos \theta + (0)\sin \theta$

$= -1\cos \theta$

$= -\cos \theta$

Example 6: $\sin\left(\dfrac{3\pi}{2} - \theta\right)$

We can use the Sine difference identity to simplify this expression,

$\sin(a - b) = \sin a \cos b - \cos a \sin b$

$a = \dfrac{3\pi}{2} \qquad b = \theta$

$= \sin\left(\dfrac{3\pi}{2}\right)\cos\theta - \cos\left(\dfrac{3\pi}{2}\right)\sin\theta$

$= (-1)\cos\theta - (0)\sin\theta$

$= -\cos\theta$

Example 7: $\tan(2\pi + \theta)$

We can use the tangent sum identity to simplify this expression,

$\tan(a + b) = \dfrac{\tan a + \tan b}{1 - \tan a \tan b}$

$a = 2\pi \qquad b = \theta$

$$= \frac{\tan(2\pi) + \tan\theta}{1 - \tan(2\pi)\tan\theta}$$ find the exact value of $\tan 2\pi$, (0)

$$= \frac{0 + \tan\theta}{1 - (0)\tan\theta}$$

$$= \frac{\tan\theta}{1}$$

$$= \tan\theta$$

This is actually some pretty neat stuff, whether you like math or not.

Example 8: $\cos\theta = \dfrac{3}{5}$ **where θ is in quadrant 4, evaluate** $\cos\left(\theta + \dfrac{\pi}{3}\right)$

Questions like these might sound very difficult, but if you pay close attention to this example, it should not actually be that difficult after all,

In the solution, we will make use of the cosine addition identity,

$\cos(a + b) = \cos a \cos b - \sin a \sin b$

$a = \theta \qquad b = \dfrac{\pi}{3}$

$\cos\left(\theta + \dfrac{\pi}{3}\right) = \cos\theta\cos\left(\dfrac{\pi}{3}\right) - \sin\theta\sin\left(\dfrac{\pi}{3}\right)$ Something is missing here, what is the exact value of $\sin\theta$? We can use the fact that "$\cos\theta = \dfrac{3}{5}$ where θ is in quadrant 4" This means that θ is in quadrant 4, and has a ratio of $\dfrac{x}{r} = \dfrac{3}{5}$

$x = 3 \qquad r = 5$

We need to find the 'y' since $\sin\theta = \dfrac{y}{r} = \dfrac{y}{5}$

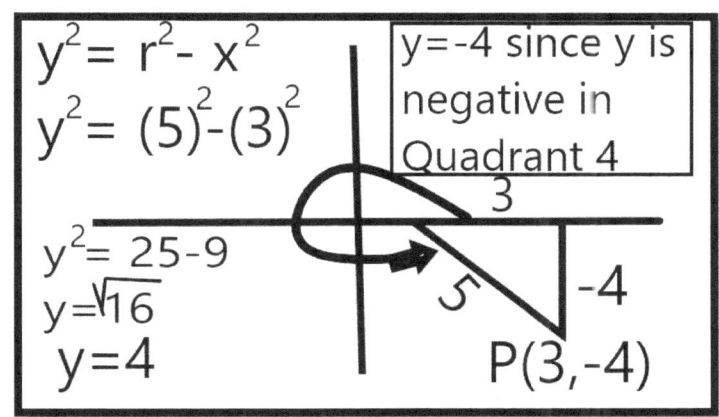

$\sin\theta = \dfrac{y}{r} = -\dfrac{4}{5}$

What is the sine and cosine of $\frac{\pi}{3}$? Let's use a special triangle for that,

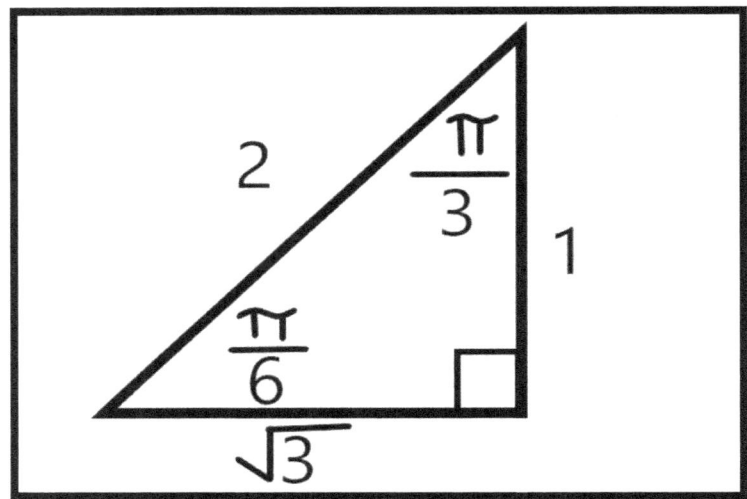

$$\cos\frac{\pi}{3} = \frac{1}{2} \qquad \sin\frac{\pi}{3} = \frac{\sqrt{3}}{2} \qquad \cos\theta = \frac{3}{5} \qquad \sin\theta = -\frac{4}{5}$$

We can put all of this information to get,

$$\cos\left(\theta + \frac{\pi}{3}\right) = \cos\theta\cos\left(\frac{\pi}{3}\right) - \sin\theta\sin\left(\frac{\pi}{3}\right)$$

$$= \left(\frac{3}{5}\right)\left(\frac{1}{2}\right) - \left(-\frac{4}{5}\right)\left(\frac{\sqrt{3}}{2}\right) \quad \text{simplify,}$$

$$= \frac{3}{10} - \left(-\frac{4\sqrt{3}}{10}\right) \quad \text{Denominators are conveniently the same,}$$

$$= \frac{3}{10} + \frac{4\sqrt{3}}{10} = \frac{3+4\sqrt{3}}{10}$$ This is the exact value!

Double Angle Identities

We will look at double angle identities, double angle identities are when you have two identical angles, such as

$\sin(\theta + \theta) = \sin\theta\cos\theta + \cos\theta\sin\theta$

$= \sin\theta\cos\theta + \sin\theta\cos\theta$

$\sin 2\theta = 2\sin\theta\cos\theta$ this is how the first double angle identity is obtained

Let's verify this double angle identity by making $\theta = \frac{\pi}{2}$

$$\sin 2\theta \quad \neq \quad 2\sin\theta\cos\theta$$

$$\sin 2\left(\frac{\pi}{2}\right) \quad \Big| \quad 2\sin\frac{\pi}{2}\cos\frac{\pi}{2}$$

$$\sin \pi \qquad 2(1)(0)$$
$$0 \qquad 0$$
$$Left\ Side\ =\ Right\ Side$$

We will now use the cosine addition identity to obtain a double angle identity for cosine,

$\cos(\theta + \theta) = \cos\theta\cos\theta - \sin\theta\sin\theta$

$\cos 2\theta = \cos^2\theta - \sin^2\theta$ the special thing about the cosine double angle identity is that it can also equal,

$\cos 2\theta = \cos^2\theta - \sin^2\theta$

$\cos 2\theta = 1 - 2\sin^2\theta$

$\cos 2\theta = 2\cos^2\theta - 1$

Example 1: Simplify $1 - 2\sin^2\dfrac{\pi}{4}$

$\theta = \dfrac{\pi}{4}$

$1 - 2\sin^2\theta = \cos 2\theta$ plug in $\theta = \dfrac{\pi}{4}$

$$1 - 2\sin^2\frac{\pi}{4} = \cos 2\left(\frac{\pi}{4}\right)$$

$$= \cos\frac{2\pi}{4}$$ Simplify this fraction, remove common factor, 2.

$$= \cos\frac{\pi}{2}$$

$$= 0$$

Example 2: Simplify $10\cos^2\theta - 5$

It is this identity $\cos 2\theta = 2\cos^2\theta - 1$; however, it has been multiplied by 5. To simplify this, we simply multiply both sides by 5,

$$5(\cos 2\theta) = 5(2\cos^2\theta - 1)$$

$$5\cos 2\theta = 10\cos^2\theta - 5$$

$$= 5\cos 2\theta$$ by using the double angle identity, we got something even easier to work with!

Example 3: Simplify, $2\sin 5y\cos 5y$

This is a double angle identity where $\theta = 5y$,

$$2\sin\theta\cos\theta = \sin 2\theta$$

$2\sin 5y \cos 5y = \sin 2(5y)$

$= \sin 10y$

We can even use the double angle identities to express a trigonometric expression as a **single trigonometric function**!

Example 4: Write $\cos^2 \frac{1}{2}x - \sin^2 \frac{1}{2}x$ as a single trigonometric expression.

This is the $\cos 2\theta = \cos^2 \theta - \sin^2 \theta$ double angle identity where in this case, $\theta = \frac{1}{2}x$ let's plug this into the identity,

$\cos 2\left(\frac{1}{2}x\right) = \cos^2 \left(\frac{1}{2}x\right) - \sin^2 \left(\frac{1}{2}x\right)$

$= \cos x$ because the 2 cancels the ½ inside of the brackets.

Example 5: Write $\sin 3 \cos 3$ as a single trigonometric function.

This seems to be the double angle identity

$\sin 2\theta = 2\sin\theta\cos\theta$ where the right side has been halved,

We simply multiply both sides by ½ and plug in $\theta = 3$,

$\left(\dfrac{1}{2}\right)\sin 2(3) = \left(\dfrac{1}{2}\right)2\sin 3\cos 3$

$\dfrac{1}{2}\sin 6 = \sin 3\cos 3$

$= \dfrac{1}{2}\sin 6$

As you can see, it is not that bad, just take your time and follow the equivalent identity and you should be fine.

On the next page, we will prove a Trigonometric Identity using some double angle identities.

Prove $\quad \dfrac{1 - \cos 2\theta}{\sin 2\theta} = \tan\theta$

Before we begin, do you

See anything familiar?

$\cos 2\theta = \cos^2\theta - \sin^2\theta$

Or $= 2\cos^2\theta - 1$

Or $= 1 - 2\sin^2\theta$ I will use this identity since, there is a $\sin$ in the denominator.

$$\frac{1-(1-2sin^2\theta)}{sin2\theta}$$

$$\frac{1-1+2\sin^2\theta}{sin2\theta}$$

$\sin 2\theta = 2\sin\theta\cos\theta$ Substitute,

and factor

$$\frac{2\sin\theta(\sin\theta)}{2\sin\theta(\cos\theta)}$$

$$\frac{sin\theta}{cos\theta}$$

$$\tan\theta \quad = \quad \tan\theta$$

As you can see, even the double angle identities can be used to prove trigonometric identities, now how cool is that!

What about using the double angle identities to solve a problem like this one:

If $\cos\theta = -\frac{2}{7}$ and θ is in Quadrant 3, evaluate $\cos 2\theta$

$$\cos 2\theta = 2\cos^2 \theta - 1 \qquad \cos \theta = -\frac{2}{7}$$

$$\cos 2\theta = 2\left(-\frac{2}{7}\right)\left(-\frac{2}{7}\right) - 1$$

$$\cos 2\theta = \frac{8}{49} - 1$$

$$\cos 2\theta = \frac{8}{49} - \frac{49}{49}$$

$$\cos 2\theta = -\frac{41}{49}$$

Restrictions for Trigonometric Expressions

We will now look at restrictions for Trigonometric Expressions. Restrictions occur because the denominator cannot equal 0. $denominator \neq 0$

Example 1: find the restrictions of $\csc 2\theta$,

$$\csc 2\theta = \frac{1}{\sin 2\theta}$$ we can expand the denominator using the double angle identity in order to find all of the restrictions,

$$\csc 2\theta = \frac{1}{2\sin\theta \cos\theta}$$

$\sin\theta \neq 0 \qquad \cos\theta \neq 0$

Example 2: Find the restrictions of $\dfrac{\sec x}{\sin x}$,

$\sec x = \dfrac{1}{\cos x}$, $\qquad \dfrac{\left(\dfrac{1}{\cos x}\right)}{\sin x}$

$\cos x \neq 0 \qquad \sin x \neq 0$

Example 3: Find the restrictions of $\dfrac{\tan x}{2\cos x - 1}$

$\tan x = \dfrac{\sin x}{\cos x} \qquad \dfrac{\left(\dfrac{\sin x}{\cos x}\right)}{2\cos x - 1}$

$\cos x \neq 0 \qquad\qquad 2\cos x - 1 \neq 0$

$$2\cos x \neq 1$$
$$\frac{2\cos x}{2} \neq \frac{1}{2}$$
$$\cos x \neq \frac{1}{2}$$

Practice Questions: Chapter 12

1. How many solutions does $\cos 3x$ have over the domain $0 \leq x \leq 2\pi$?

2. Factor the following,
   a) $3\sec^2 x + 6\sec x$
   b) $4\cot^2 x - 9$
   c) $\cos^2 x + 3\cos x + 2$
   d) $2\sin^2 x + 4\sin x - 6$
   e) $\csc^2 x - 16$
   f) $\cos^2(j)\sin(j) - \cos(j)\sin^2(j) = 0$
   g) $4\cos^2(f) - 25 = 0$
   h) $2\sin^2(x) + 7\sin(x) + 3 = 0$
   i) $v^2\csc^2(b) + v\csc(b)\csc(g) = 0$
   j) $9\sec^2(x) - 4\cot^2(x) = 0$
   k) $\cot^2(x) + 6\cot(x) + 8$

l) $\tan^2(x) - \tan(x) - 12$

m) $\sin^2(m) - 7$

n) $4\csc^4(v) - 12\csc^3(v) = 0$

o) $3\sin^4(x) - 3\sin^3(x) - 36\sin^2(x) = 0$

3. Solve the following and give the general solution,

   a) $\sin x \cos x = \sin^2 x$

4. Solve for the exact values of θ, for $0 \leq \theta \leq 2\pi$

   a) $2\sin\theta = 1$
   b) $2\cos\theta - \sqrt{3} = 0$
   c) $3\tan\theta + 3 = 0$
   d) $4\cos^2\theta - 1 = 0$

5. Prove the following identity

   a) $\cot\theta + \tan\theta = \csc\theta \sec\theta$

6. Determine the exact value or simplify using the sum or difference identity,

   a) $\cos\left(\dfrac{\pi}{4} + \dfrac{\pi}{6}\right)$
   b) $\sin(x - \pi)$

c) $\cos\left(\dfrac{\pi}{2} - x\right)$

d) $\sin\left(x - \dfrac{\pi}{4}\right)$

e) $\dfrac{\tan 7z - \tan 2z}{1 + \tan 7z \tan 2z}$

7. Write the following as a single trigonometric function,
   a) $4\sin 5x \cos 5x$
   b) $2 - 4\sin^2 \dfrac{1}{2}x$
   c) $6\sin 6x \cos 6x$
   d) $8\sin\theta \cos\theta$
   e) $14\cos^2 7x - 7$

8. Prove the following Trigonometric Identity,
   a) $\dfrac{\sin 2\theta}{2} = \cot\theta \sin^2\theta$

Intervals

Chapter 13

In this section, I will go over some notation you will see, I will not go in depth, but I will show you what you can expect and how to prepare for your first few weeks of University Calculus.

When a is an element of S, it is written as: $a \in S$

When a is not an element of S: $a \notin S$

If you see something like, $S \cup T$ this means a set consisting of all elements in S or T.

If there is an intersection of S and T is the set, it is written as $S \cap T$.

If there is an empty set which contains no elements, it is denoted by ∅.

Set example,

$A = \{1,2,3,4,5,6\}$

$A = \{x | x \text{ is an integer and } 0 < x < 7\}$

Intervals,

$(a,b) = \{x | \ a < x < b\}$

Endpoints, a & b are excluded, denoted by "(" brackets,

$[a,b] = \{x | \ a \leq x \leq b\}$

Endpoints, a & b is included, denoted by the square brackets "["

<u>There are 9 possible types of intervals:</u>

Interval	Inequality	Number Line
(a, b)	$\{x \mid a < x < b\}$	○——○→ a b
$[a, b]$	$\{x \mid a \leq x \leq b\}$	●——●→ a b
$[a, b)$	$\{x \mid a \leq x < b\}$	●——○→ a b
$(a, b]$	$\{x \mid a < x \leq b\}$	○——●→ a b
(a, ∞)	$\{x \mid x > a\}$	○——→ a
$[a, \infty)$	$\{x \mid x \geq a\}$	●——→ a
$(-\infty, b)$	$\{x \mid x < b\}$	←——○ b
$(-\infty, b]$	$\{x \mid x \leq b\}$	←——● b
$(-\infty, \infty)$	$\mathbb{R}$ (set of real numbers)	←——→

∞ means positive infinity, $-\infty$ means negative infinity

Here are some more rules to take note of:

a) If $a < b$, then $a + c < b + c$
b) If $a < b$ and $c < d$, then $a + c < b + d$
c) If $a < b$ and $c > 0$, then $ac < bc$
d) If $a < b$ and $c < 0$, then $ac > bc$
e) If $0 < a < b$, then $\dfrac{1}{a} > \dfrac{1}{b}$

Rule d) is true because the closer the negative number is to 0, the larger it is.

Rule e) is true because the smaller the denominator, the larger the resulting value is, test these out.

Intervals Example 1: Give the interval and inequality that corresponds to "All real numbers x so that
$0 < x \leq 9$ *and* $x \leq 3$

Step 1: Sketch this on the number line

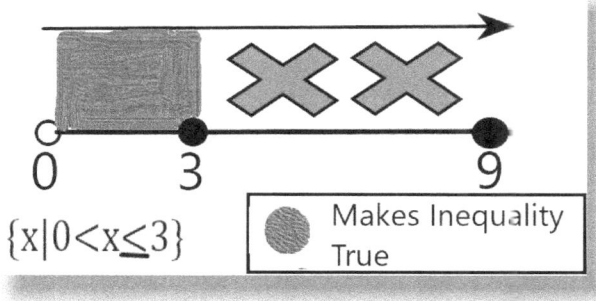

Interval: $(0, 3]$ greater than 0 and less than or equal to 3,
Domain: $\{x | 0 < x \leq 3\}$

Before starting Calculus, I strongly recommend that you review everything in this book (especially Trigonometry). In the University course Study Guides, Textbooks, etc. they assume you already know everything and skip several steps in a solution, causing confusion if you do not remember some things. This concludes the brief chapter on Introductory University Calculus.

Practice Questions: Intervals

1. Write the corresponding Interval,
    a) $-2 < x \leq 6$
    b) $3 \leq x < 9$
    c) $x > \pi$
    d) $x < 7$
    e) $x > 6$ & $x \leq 18$

2. Write the corresponding Inequality,
    a) $(3, \infty)$
    b) $[\pi, 3\pi)$
    c) $(-\infty, \sqrt{3})$
    d) $[2\sqrt{2}, 7]$

Conclusion

Congratulations on finishing reading this book, you did it! I hope this book helped you throughout your entire High School Math Years. Remember, always practice, and learn from your mistakes, if you have any questions to ask your teacher, don't hesitate. I am wishing you the best of luck in your future, always try your best, and follow your dreams. Don't sell yourself short when it comes to choosing a career. I am planning on getting a career in the Computer Science field (most likely will be a Computer Programmer/Software Developer). Please make sure to take the time to appreciate the help that the teachers give. They chose that career to help students (even though they could pursue higher paying jobs with the same degree).

One of my greatest memories from High School was being a Judge at the 2018 Fort McMurray Composite High School Science Fair! It was such a fun experience, being the one grading the kids, and not being the one graded, it was a different perspective.

Wishing you the best of luck in the future,

SINCERELY, JEREMY MARTIN

About the Author

Jeremy Martin (18 years old) is an aspiring Computer Programmer, who is currently studying the **Bachelor of Science in Computing and Information Systems** with Athabasca University. This is the third book he published. He graduated High School with Honours in May of 2020.

He was even a volunteer judge at the 2018 Fort McMurray Composite High School Science Fair

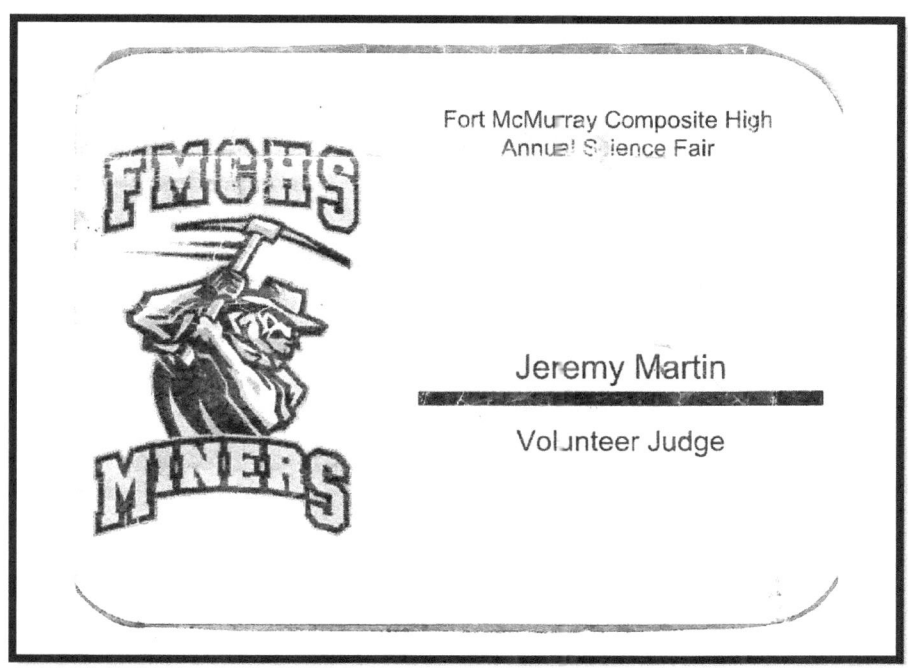

Here are Jeremy's Life Mottos:

Motto #1: "Do what you love doing, don't do or choose a career just because that's what another human being wanted you to do with your life. It's your life, besides, what's better? Living a happy life with a job you love doing every day? Or, living a miserable life hating and dreading your job everyday, just because it pays more, or you think you have to choose that career to please other people? "

Motto #2: "There is a blurred boundary between reality and dreams, believe in yourself and your dreams, and one day that boundary will become defined, and the dream may become reality. "

Answers & Solutions

Part 1: Math 11

Chapter 1: Trigonometry

1.

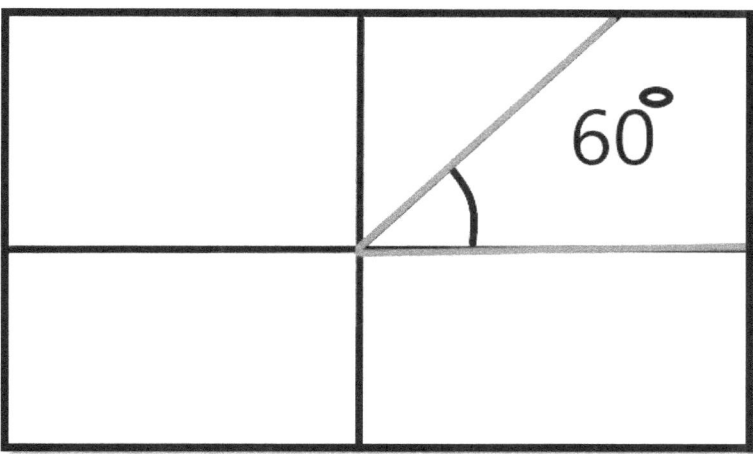

2.

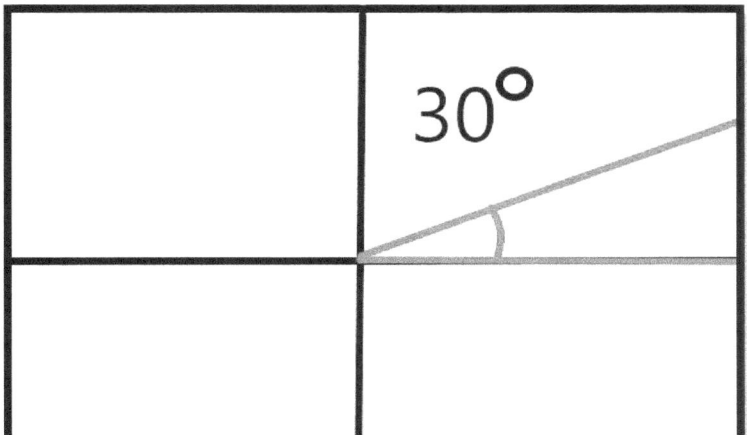

3.

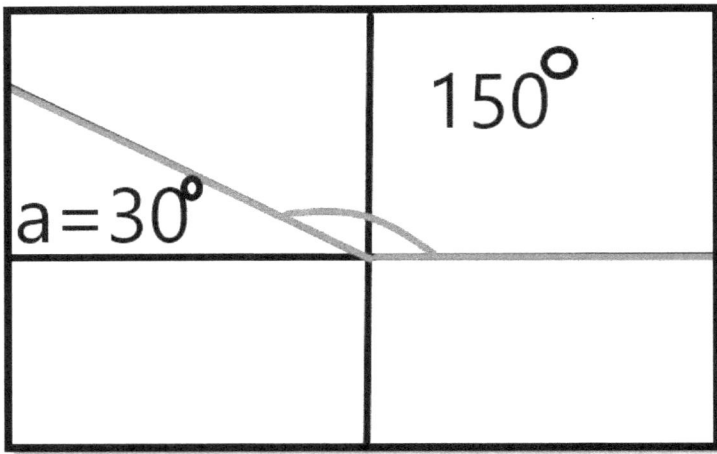
a = 30°, 150°

4.

Sinθ = y/r = √3/2
Cosθ = x/r = 1/2
Tanθ = y/x = 3/1 = √3

5. $\dfrac{23}{sin50} = \dfrac{b}{sin74}$

$$\frac{23 \sin 74}{\sin 50} = b$$

$$b \approx 28.9 \ cm$$

6. $$\frac{18}{\sin 36} = \frac{a}{\sin 120}$$

$$\frac{18 \sin 120}{\sin 36} = a$$

$$a \approx 26.5$$

7. $$Reference \ angle = 180° - 120° = 60°$$

Chapter 2: Quadratic Functions

1.

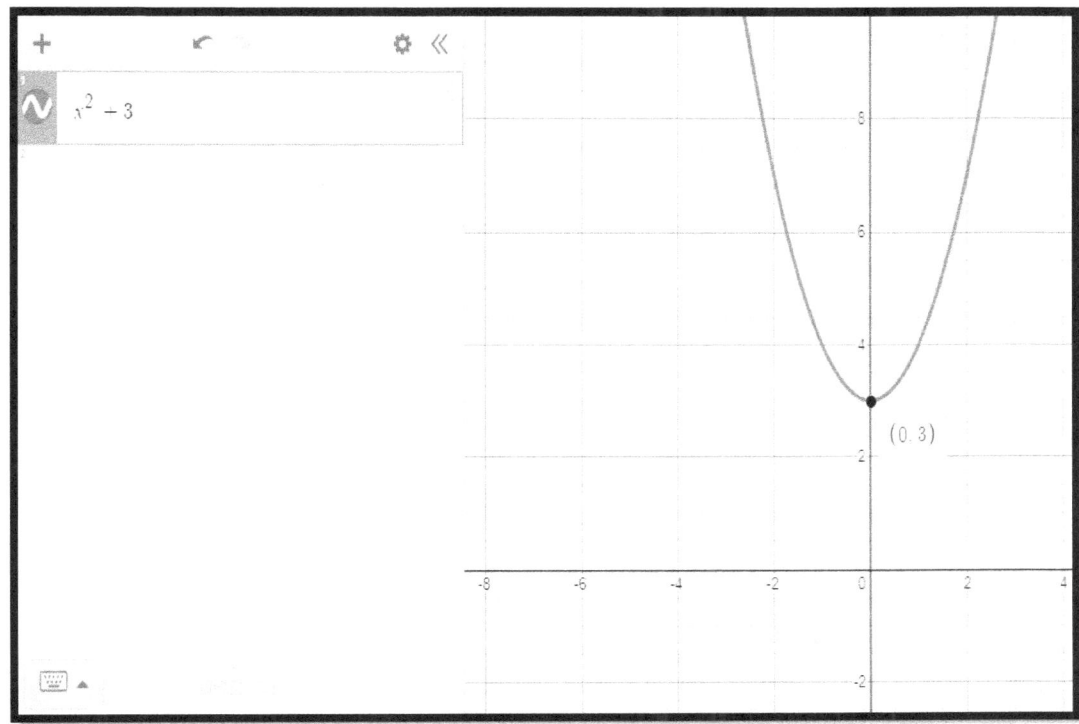

2. A) (-2,-3)

   b) (0,4)

   c) (-w,-h)

   d) (1,80)

   e) (200,-300)

3. a) $x^2 + 4x + 3$

   $(x+3)(x+1)$

   $x = -3 \quad x = -1$

   b) $x^2 + 4x - 21$

   $(x+7)(x-3)$

   $x = -7 \quad x = 3$

4. Range: $y \geq 8$

5. A) $x = \dfrac{-b \pm \sqrt{b^2 - 4ac}}{2a} = \dfrac{-(3) \pm \sqrt{(3)^2 - 4(2)(-7)}}{2(2)}$

   $x = \dfrac{-3 \pm \sqrt{9 + 56}}{4} = \dfrac{-3 \pm \sqrt{65}}{4}$

   b) $x = \dfrac{-b \pm \sqrt{b^2 - 4ac}}{2a} = \dfrac{-(17) \pm \sqrt{(17)^2 - 4(-6)(5)}}{2(-6)}$

   $x = \dfrac{-17 \pm \sqrt{289 + 120}}{-12} = \dfrac{-17 \pm \sqrt{409}}{-12}$

6. $x^2 + 4x + 5 = 0$
   $x^2 + 4x = -5$
   $x^2 + 4x + 4 = -5 + 4$
   $x^2 + 4x + 4 = -1$
   $(x + 2)^2 = -1$

7.

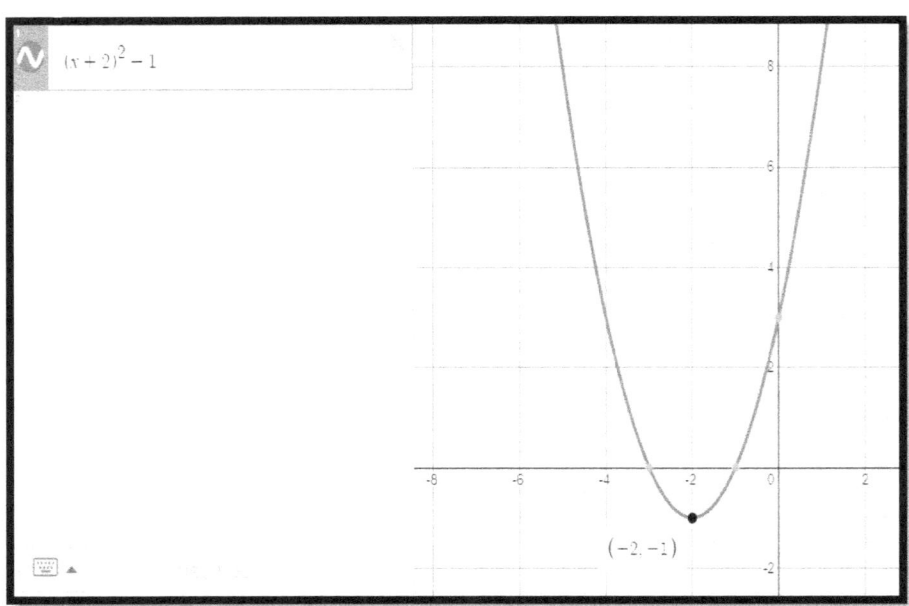

Chapter 3: Radical Expressions and Equations

1.a) $3\sqrt{7} + 92\sqrt{7} = 95\sqrt{7}$
b) $2\sqrt{5} + 3\sqrt{5} = 5\sqrt{5}$
c) $\sqrt{28} + \sqrt{63} = \sqrt{(2)(2)(7)} + \sqrt{(3)(3)(7)}$
$2\sqrt{7} + 3\sqrt{7} = 5\sqrt{7}$
d) $\sqrt{xz^2} + \sqrt{xy^2} = z\sqrt{x} + y\sqrt{x} = (z+y)\sqrt{x}$

2. A) $9\sqrt{3} - 2\sqrt{12} = 9\sqrt{3} - 4\sqrt{3} = 5\sqrt{3}$
b) $25\sqrt{7} - 16\sqrt{7} = 9\sqrt{7}$
3. a) $8\sqrt{6}$
b) $15\sqrt{15}$

4. A) $\dfrac{a}{\sqrt{b}} = \dfrac{a}{\sqrt{b}}\left(\dfrac{\sqrt{b}}{\sqrt{b}}\right) = \dfrac{a\sqrt{b}}{b}$

b) $\dfrac{a}{y+\sqrt{x}} = \dfrac{a(y-\sqrt{x})}{(y+\sqrt{x})(y-\sqrt{x})} = \dfrac{ay - a\sqrt{x}}{y^2 + x}$

Chapter 4: Rational Expressions and Equations

1. A) $\dfrac{1}{j+7}$ $j \neq -7$

b) $\dfrac{1}{x}$ $x \neq 0$

c) $\dfrac{x}{y^2 - 4} = \dfrac{x}{(y+2)(y-2)}$ $y \neq \pm 2$

2. A) $\dfrac{3x+6}{3y+9} = \dfrac{3(x+2)}{3(y+3)} = \dfrac{x+2}{y+3}$ $y \neq 3$

b) $\dfrac{x^2+x-2}{x^2+2x-3} = \dfrac{(x+2)(x-1)}{(x+3)(x-1)} = \dfrac{x+2}{x+3}$

$x \neq 1$ $x \neq -3$

c) $\dfrac{a^2b+ab+b}{b^2+b} = \dfrac{b(a^2+a+1)}{b(b+1)} = \dfrac{a^2+a+a}{b+1}$

$b \neq 0$ $b \neq -1$

d) $\dfrac{4xy+8x}{12x^2+4x} = \dfrac{4x(y+2)}{4x(3x+1)} = \dfrac{y+2}{3x+1}$

$x \neq 0$ $x \neq -\dfrac{1}{3}$

Chapter 5: Absolute Value and Reciprocal Functions

1. A) $|x+2| = 3$

$\quad\quad\quad\quad\quad x+2 = 3 \quad\quad\quad -x-2 = 3$

$x = 3-2 \quad\quad -x = 3+2$

$x = 1 \quad\quad\quad\quad -\dfrac{x}{-1} = \dfrac{5}{-1}$

$\quad\quad\quad\quad\quad\quad\quad\quad\quad x = -5$

b) $|x-4| = 2$

$\quad x-4 = 2 \quad\quad -x+4 = 2$

$\quad x = 2+4 \quad\quad -x = 2-4$

$\quad x = 6 \quad\quad\quad\quad -\dfrac{x}{-1} = -\dfrac{2}{-1}$

$$x = 2$$

c) $|x - 1| = 5$

$x - 1 = 5$ $-x + 1 = 5$

$x = 5 + 1$ $-x = 5 - 1$

$x = 6$ $-\dfrac{x}{-1} = \dfrac{4}{-1}$

                    $x = -4$

2.

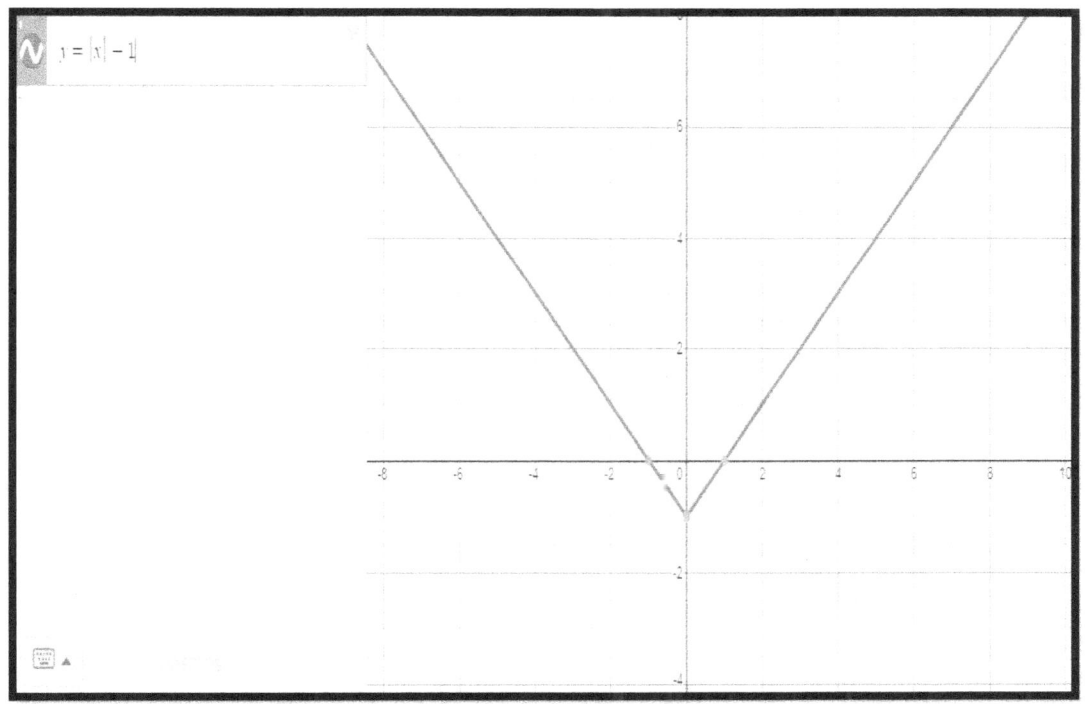

3.

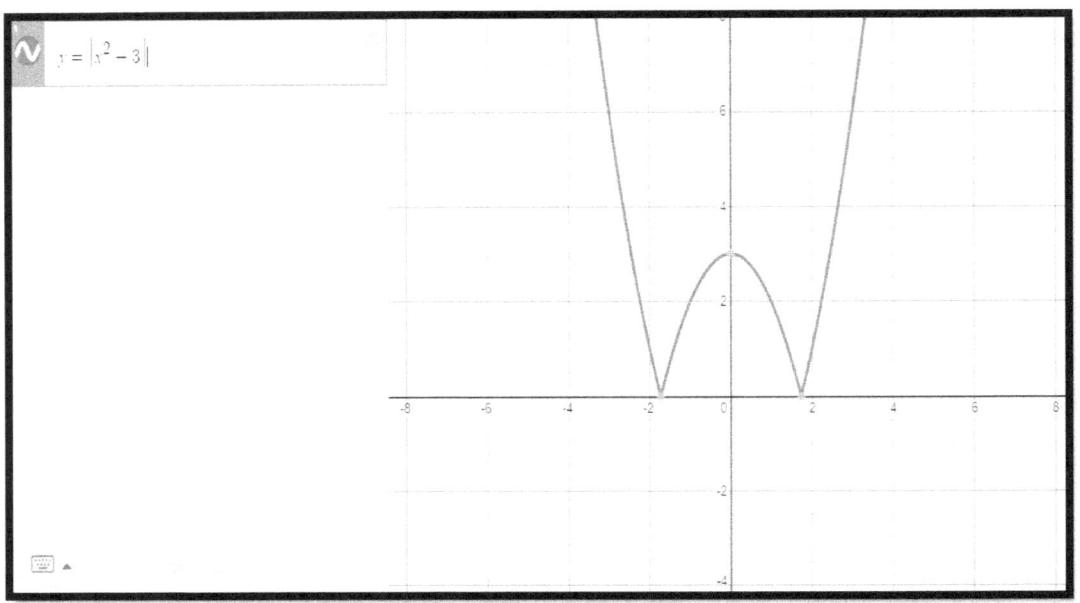

4. I have given you a graph as well, so you can visualize it.

$$y = |x-2|$$

$$y = \begin{cases} x-2 & \text{if } x \geq 2 \\ -x+2 & \text{if } x < 2 \end{cases}$$

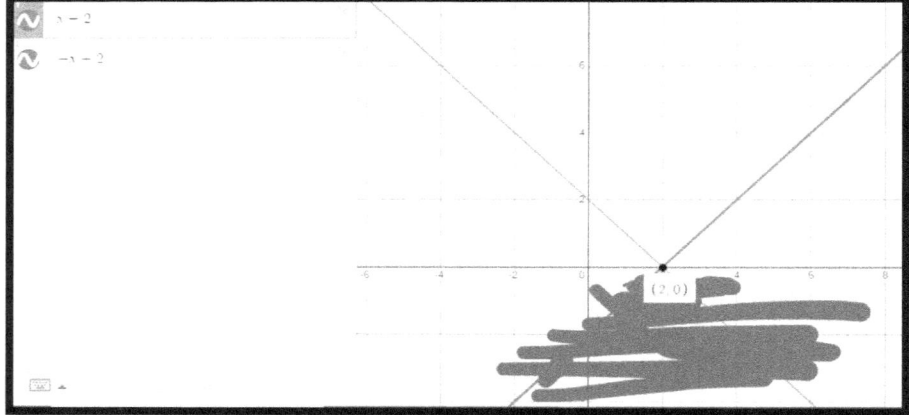

5.

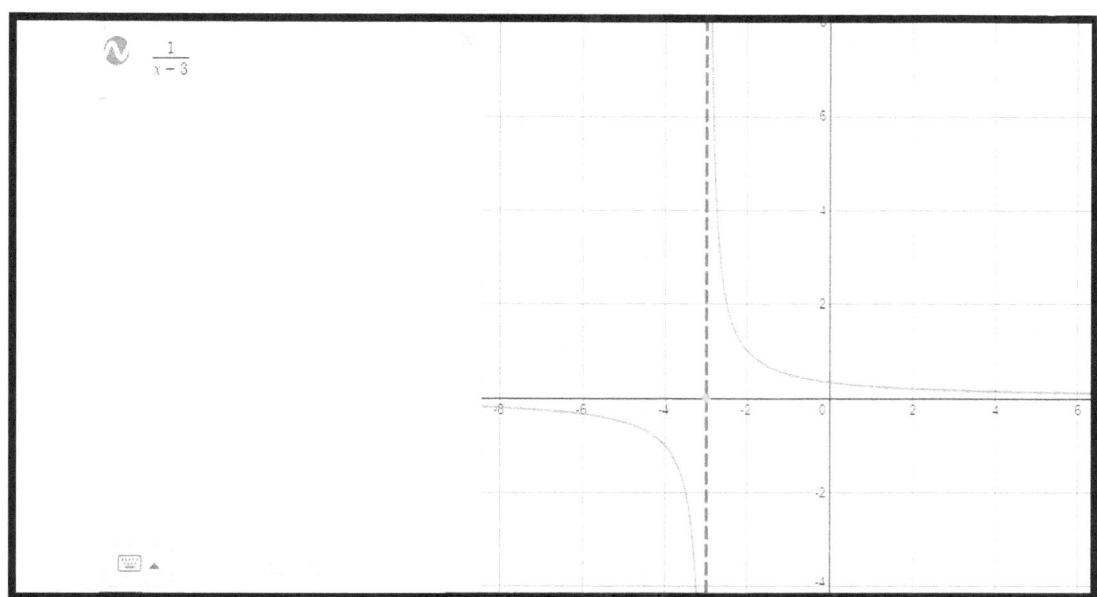

Vertical asymptote: $x = -3$
because $x + 3 \neq 0$
$\qquad x \neq -3$

Chapter 6: Linear and Quadratic Inequalities

1. A) $-x \geq 2$
   $-\dfrac{x}{-1} \leq \dfrac{2}{-1}$
   $x \leq -2$

   b) $x - 3 < 5$
   $x < 5 + 3$
   $x < 8$

   c) $4 - x < 2$
   $-x < 2 - 4$
   $-\dfrac{x}{-1} < -\dfrac{2}{-1}$
   $x > 2$

   d) $-3x \leq -6$
   $-\dfrac{3x}{-3} \geq -\dfrac{6}{-3}$
   $x \geq 2$

2. A) $y - 2 > x$

$y > x + 2$

b) $-2y \geq 4x + 6$

$-\dfrac{2y}{-2} \geq \dfrac{4x}{-2} + \dfrac{6}{-2}$

$y \leq -2x - 3$

3.

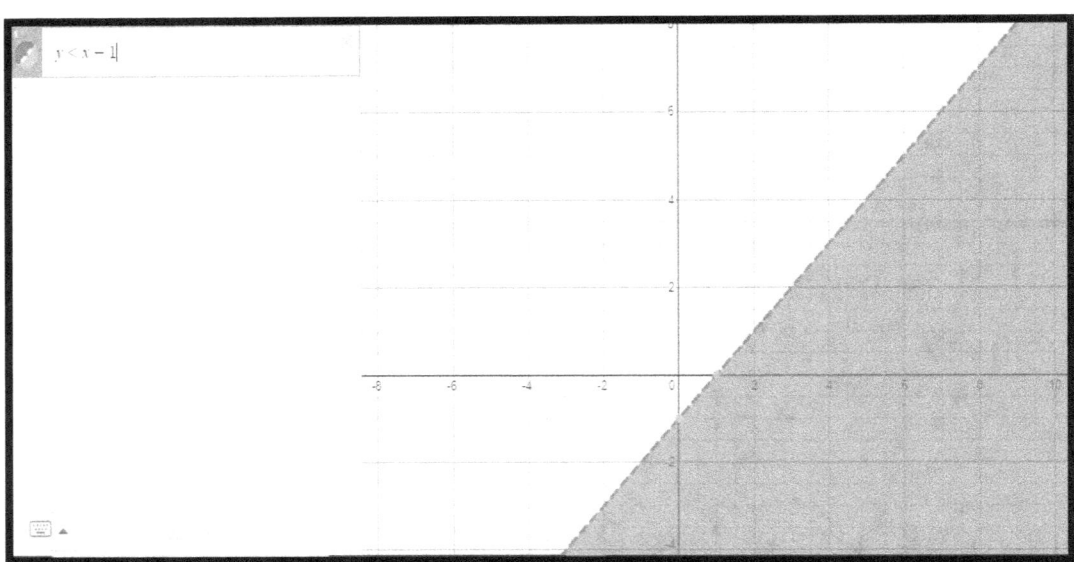

4.

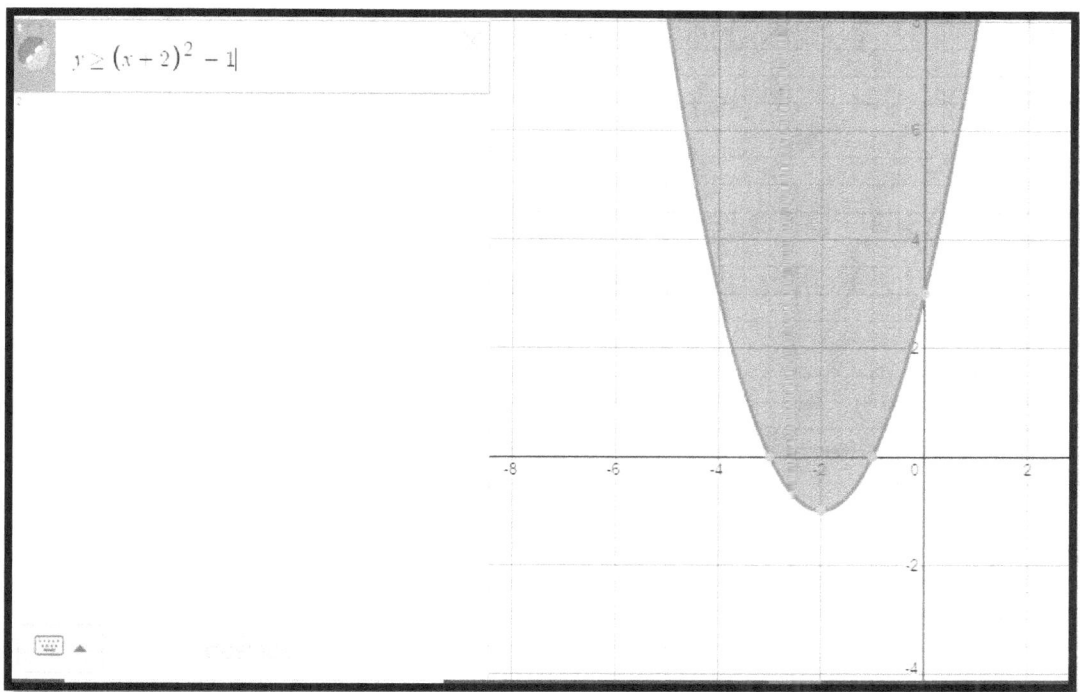

Part 2: Math 12

Chapter 7: Transformations Of Functions

1. A) Vertical expansion by a factor of 3, horizontal compression by a factor of 1/6, and 2 units right.

   b) Reflection in the x-axis and 3 units left

c) Reflection in the y-axis, Horizontal compression by a factor of 1/7, and a vertical translation of 1 unit down.

d) Vertical expansion by a factor of 3, horizontal expansion by a factor of 3, and a vertical translation of 2 units up

2.

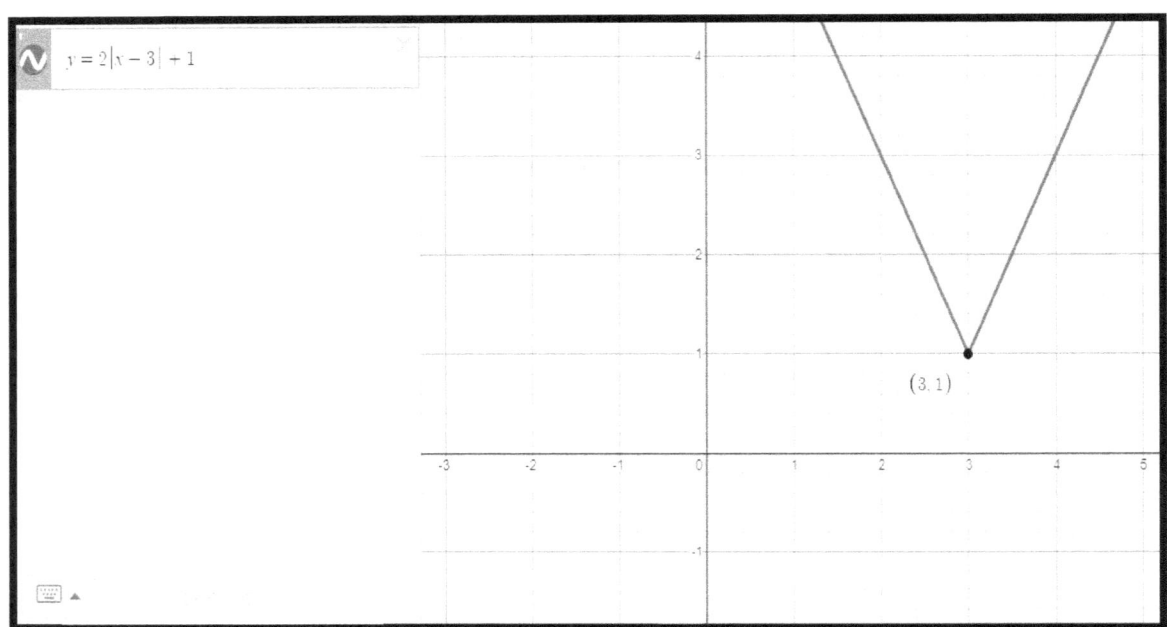

3.

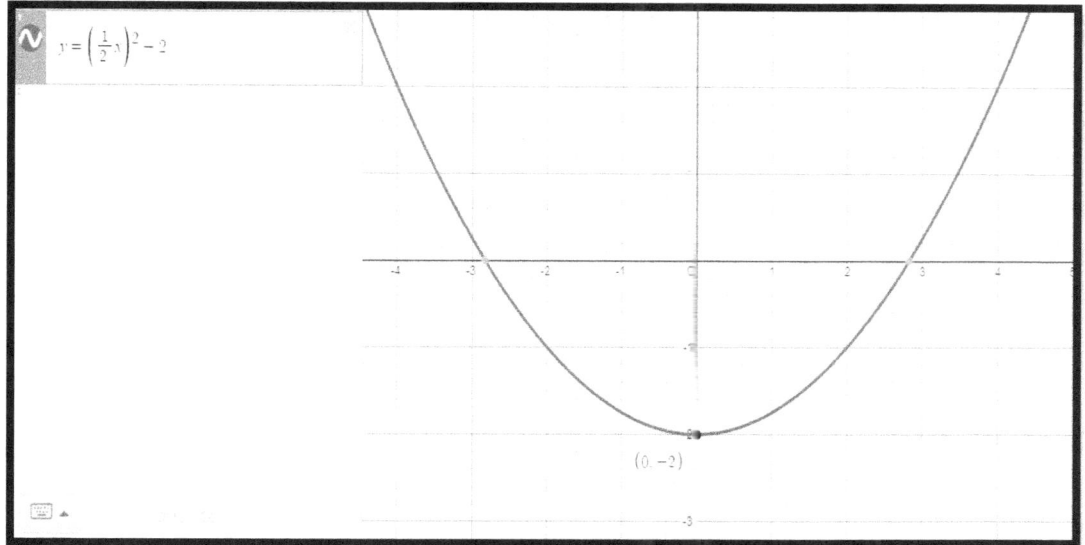

4. A)

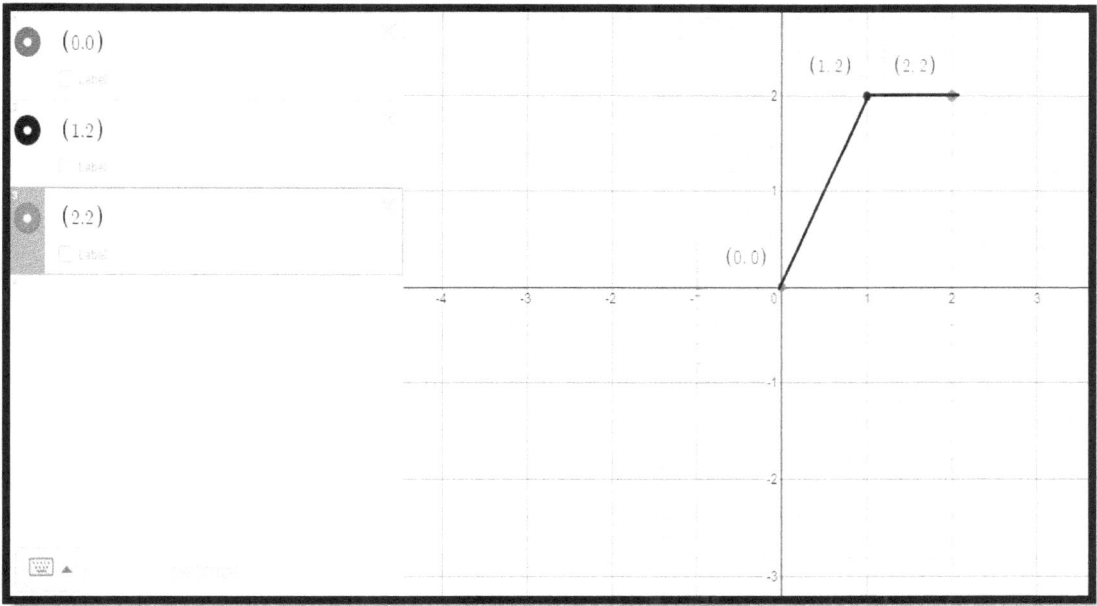

b)

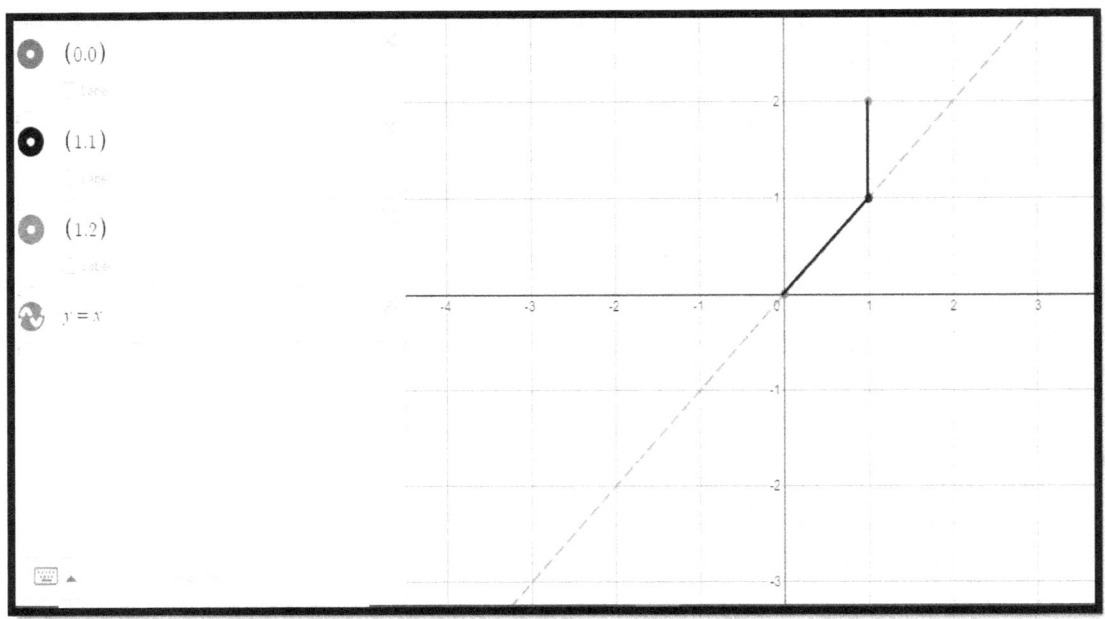

Chapter 8: Radical and Rational Functions

1. $\sqrt{x-4} - 4 = 0$
   $(\sqrt{x-4})^2 = (4)^2$
   $x - 4 = 16$
   $x = 16 + 4$
   $x = 20$

2. $x = \sqrt{x-4} + 6$
   $x - 6 = \sqrt{x-4}$
   $(x-6)^2 = x - 4$
   $x^2 - 12x + 36 = x - 4$
   $x^2 - 12x + 36 + 4 - x$
   $x^2 - 13x + 40 = 0$
   $(x-5)(x-8) = 0$
   Extraneous root: $x = 5$ Real solution: $x = 8$

3. $\sqrt{x+5} = x - 7$

$$x + 5 = (x - 7)^2$$
$$x + 5 = x^2 - 14x + 49$$
$$0 = x^2 - 14x + 49 - x - 5$$
$$0 = x^2 - 15x + 44$$
$$0 = (x - 11)(x - 4)$$

Extraneous root: $x = 4$

Real solution: $x = 11$

4. Vertical expansion by 2, reflection in y-axis, 2 units right, and 1 unit up.

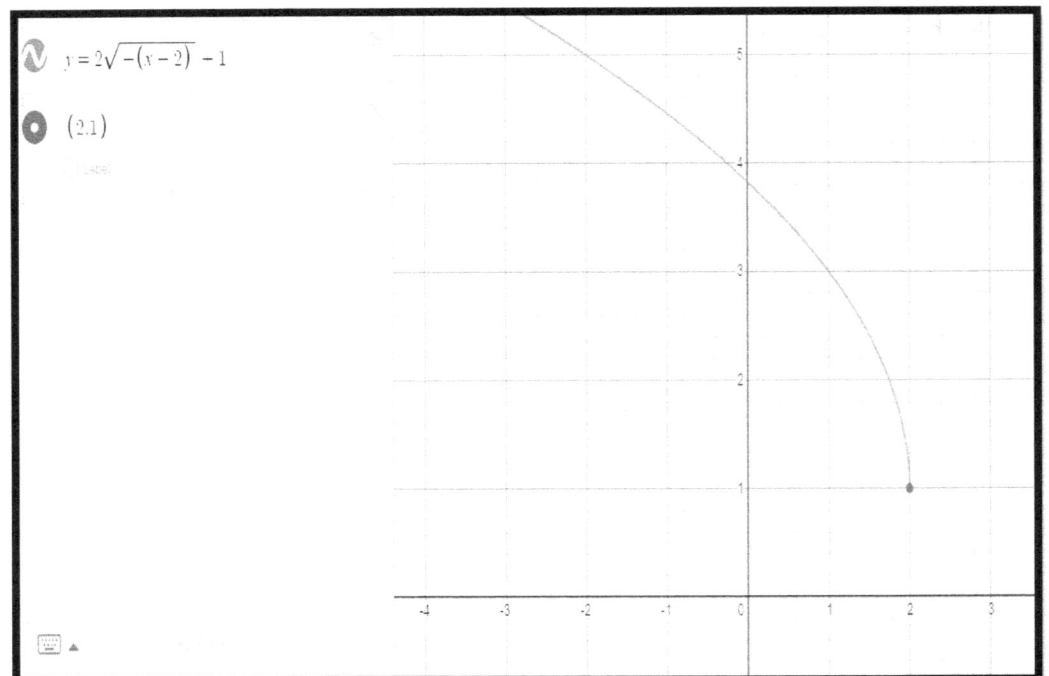

5. A) Horizontal asymptote: $\dfrac{2x^3}{5x^3} = \dfrac{2}{5}$

b) Horizontal asymptote: $\dfrac{x}{x} = 1$

6. A) $y = \dfrac{x^2 + x - 2}{x + 2} = \dfrac{(x+2)(x-1)}{x+2}$

$y = x - 1$

Restriction: $x \neq -2$

Point of discontinuity: $f(-2) = -2 - 1 = -3$

Point of discontinuity: $(-2, -3)$

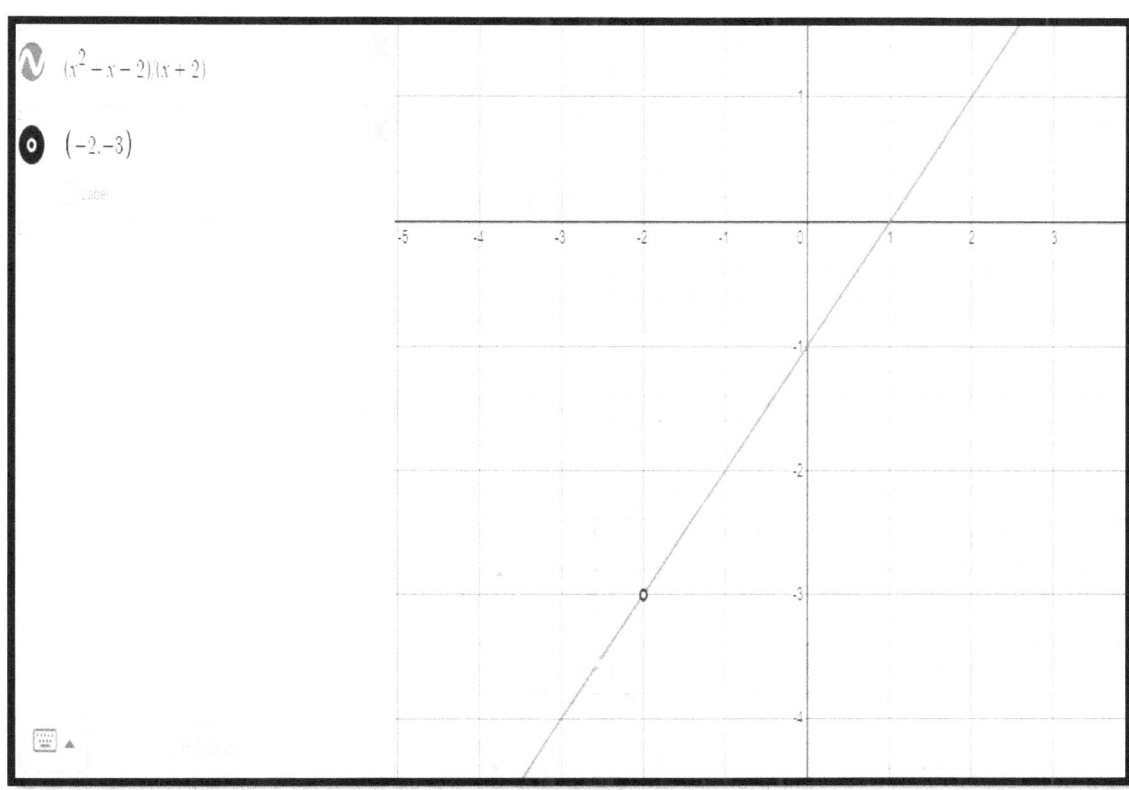

6.b) $y = \dfrac{2}{x-2} + 1$

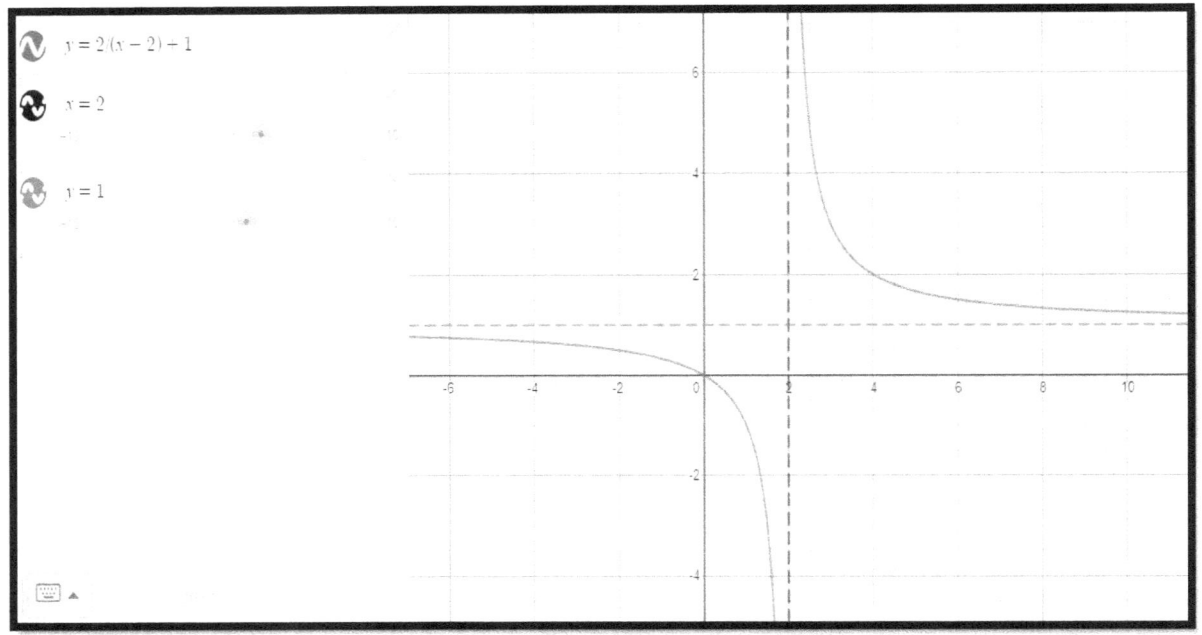

Chapter 9: Polynomials

1.

$$x^3 + 2x^2 - 7x - 9 \div (x-1)$$

$$\underline{1} \begin{array}{|cccc} 1 & 2 & -7 & -9 \\ & 1 & 3 & -4 \\ \hline 1 & 3 & -4 & \boxed{-13} \end{array}$$

Remainder

$$(x^3 + 2x^2 - 7x - 9) = (x^2 + 3x - 4)(x-1) - 13$$

2.

A) $(x^2 - 2x) \div (x - u)$

$P(u) = (u)^2 - 2(u)$
$P(u) = u^2 - 2u$

b) $(x^3 - x + 2) \div (x + 4)$
$P(-4) = (-4)^3 - (-4) + 2$
$P(-4) = -64 + 4 + 5$
$P(-4) = -58$

3. $(x^2 + cx - 7) \div (x - 2)$ has a remainder of 3
$P(2) = 3$
$P(2) = (2)^2 + c(2) - 7$
$P(2) = 4 + 2c - 7$
$3 = 4 + 2c - 7$
$3 = 2c - 3$
$3 + 3 = 2c$
$6 = 2c$
$\dfrac{6}{2} = \dfrac{2c}{2}$
$c = 3$

4. $x^3 + 3x^2 - 4x + 6$
Potential roots: $\pm 1, \pm 2, \pm 3, \pm 6$

5. $P(b) = (b)^3 - (b)^2 b - (b)b^2 + b^3$
$P(b) = b^3 - b^3 - b^3 + b^3$
$P(b) = 0$ Yes, $(x - b)$ is a factor because the remainder is zero.

6. $P(w) = w^2 - 5w + 6$
$P(3) = (3)^2 - 5(3) + 6$
$P(3) = 9 - 15 + 6$
$P(3) = 0$ Yes, $(w - 3)$ is a factor because the remainder is zero.

7.
a) $x^3 + 5x^2 + 6x$
$x(x^2 + 5x + 6) = 0$
$x(x + 2)(x + 3) = 0$

b) $x^3 - 8d^3$
$(x)^3 - (2d)^3$
$(x - 2d)((x)^2 + (x)(2d) + (2d)^2)$
$(x^3 - 8d^3) = (x - 2d)(x^2 + 2dx + 4d^2)$

c) $27v^3 + b^3$
$(3v)^3 + (b)^3 = (3v + b)((3v)^2 - (3v)(b) + (b)^2)$
$(27v^3 + b^3) = (3v + b)(9v^2 - 3bv + b^2)$

d) $x^3 - 2x^2 - 4x + 8$
$[x^3 - 2x^2] - [4x - 8]$
$x^2(x - 2) - 4(x - 2) = (x^2 - 4)(x - 2)$
$(x - 2)(x - 2)(x + 2)$ or $(x + 2)(x - 2)^2$

8. $x = -4$ $x = 3$ $x = 2$
    $(x + 4) = 0$ $(x - 3) = 0$ $(x - 2) = 0$

$(x - 2)$ has a multiplicity of two, therefore, it is written as $(x - 2)^2$

Factored form of equation: $(x + 4)(x - 3)(x - 2)^2$
Or $(x + 4)(x - 3)(x - 2)(x - 2)$

Chapter 10: Logarithms and Exponents PART A

1. A) $(3x^{-2}y^4)^2 = (9x^{-4}y^8)$

$$= \frac{9y^8}{x^4}$$

b) $\left(\dfrac{x^2}{b^{-1}}\right)^{-2} = (x^2 b)^{-2}$

$(x^{-4}b^{-2}) = \dfrac{1}{x^4 b^2}$

c) $v\left(\dfrac{a}{c}\right)\left(\dfrac{a^{-1}}{c^{-1}}\right)\left(\dfrac{b}{d}\right)\left(\dfrac{d}{b}\right)$

$= v\left(\dfrac{a}{c}\right)\left(\dfrac{c}{a}\right)\left(\dfrac{b}{d}\right)\left(\dfrac{d}{b}\right)$

$= v$

2. A) $2^{x-2} = 8^x$

$2^{x-2} = (2^3)^x \qquad 2^{x-2} = 2^{3x}$

$x - 2 = 3x$

$-2 = 3x - x$

$-2 = 2x$

$-\dfrac{2}{2} = \dfrac{2x}{2}$

$x = -1$

b) $3^{c-6} = 81$

$3^{c-6} = 3^4$

$c - 6 = 4 \qquad c = 4 + 6$

$c = 10$

3. $x^{\tfrac{3}{2}} = 64$

$\left(x^{\tfrac{3}{2}}\right)^{\tfrac{2}{3}} = (64)^{\tfrac{2}{3}}$

$x = (\sqrt[3]{64})^2$

$x = (4)^2$

$x = 16$

Chapter 10: Logarithms and Exponents PART B

1.

A) $\log 6 + \log 5$
$= \log(6 * 5)$
$= \log 30$

b) $\log 7 - \log 2$
$= \log\left(\dfrac{7}{2}\right)$

c) $2\log x - 3\log y$
$= \log x^2 - \log y^3$
$= \log\left(\dfrac{x^2}{y^3}\right)$

d) $3^{\log_3 7}$

Since $a^{\log_a x} = x$

$3^{\log_3 7} = 7$

2. A) $\log(x-2) = 1$
$x - 2 = 10^1$
$x - 2 = 10$
$x = 10 + 2$
$x = 12$

b) $\log 15 - \log 5 = \log x$
$\log\left(\dfrac{15}{5}\right) = \log x$
$\log 3 = \log x$
$3 = x \qquad x = 3$

c) $\log x + \log x = \log 36$
$\log(x * x) = \log 36$
$\log x^2 = \log 36$
$x^2 = 36$
$x = 6$ because $x \neq -6$

3. A) $3^x = 11$
$x \log 3 = \log 11$
$\dfrac{x \log 3}{\log 3} = \dfrac{\log 11}{\log 3}$
$x = \dfrac{\log 11}{\log 3}$
$x = 2.1827$

b) $7^{3x} + 3 = 21$

$7^{3x} = 21 - 3$

$7^{3x} = 18$

$3x \log 7 = \log 18$

$\dfrac{x * 3\log 7}{3\log 7} = \dfrac{\log 18}{3\log 7}$

$x = \dfrac{\log 18}{3\log 7}$

$x = 0.4951$

c) $3^x = 27^{x-1}$

$3^x = (3^3)^{x-1}$

$3^x = 3^{3x-3}$

$x = 3x - 3$

$x - 3x = -3$

$-2x = -3$

$x = \dfrac{2}{3}$

Chapter 10: Logarithms and Exponents PART C

1. A) $A = 1200\left(1 + \dfrac{0.05}{12}\right)^{12t}$

$A = 1200\left(1 + \dfrac{0.05}{12}\right)^{12*8}$

$A = 1200(1.004166...)^{96}$

$A = 1200(1.49058...)$

$A = \$1{,}788.70$

b)
$$4000 = 1200\left(1 + \frac{0.05}{12}\right)^{12t}$$

$$\frac{4000}{1200} = \frac{1200(1.004166\ldots)^{12t}}{1200}$$

$$\frac{10}{3} = 1.004166^{12t}$$

$$\log\frac{10}{3} = 12t\log 1.004166$$

$$\frac{\log\left(\frac{10}{3}\right)}{12\log 1.004166} = \frac{12t\log 1.004166}{\log 1.004166}$$

$$t = 24.13 \text{ years}$$

2.
$$1400 = 900\left(1 + \frac{r}{4}\right)^{4*12}$$

$$\frac{1400}{900} = \frac{900\left(1 + \frac{r}{4}\right)^{48}}{900}$$

$$\frac{14}{9} = \left(1 + \frac{r}{4}\right)^{48}$$

$$\left(\frac{14}{9}\right)^{\frac{1}{48}} = \left(\left(1 + \frac{r}{4}\right)^{48}\right)^{\frac{1}{48}}$$

$$\sqrt[48]{\frac{14}{9}} = 1 + \frac{r}{4}$$

$$\sqrt[48]{\frac{14}{9}} - 1 = \frac{r}{4}$$

$$4(0.00092473\ldots) = \frac{r}{4}(4)$$
$$r = 0.037 \quad 3.7\%$$

3. $30 = 100(1 - 0.04)^d$
   $30 = 100(0.96)^d$
   $$\frac{30}{100} = \frac{100(0.96)^d}{100}$$
   $0.3 = 0.96^d$
   $\log 0.3 = d\log 0.96$
   $$\frac{\log 0.3}{\log 0.96} = d\frac{\log 0.96}{\log 0.96}$$
   $d = 29.49 \; meters$

4.

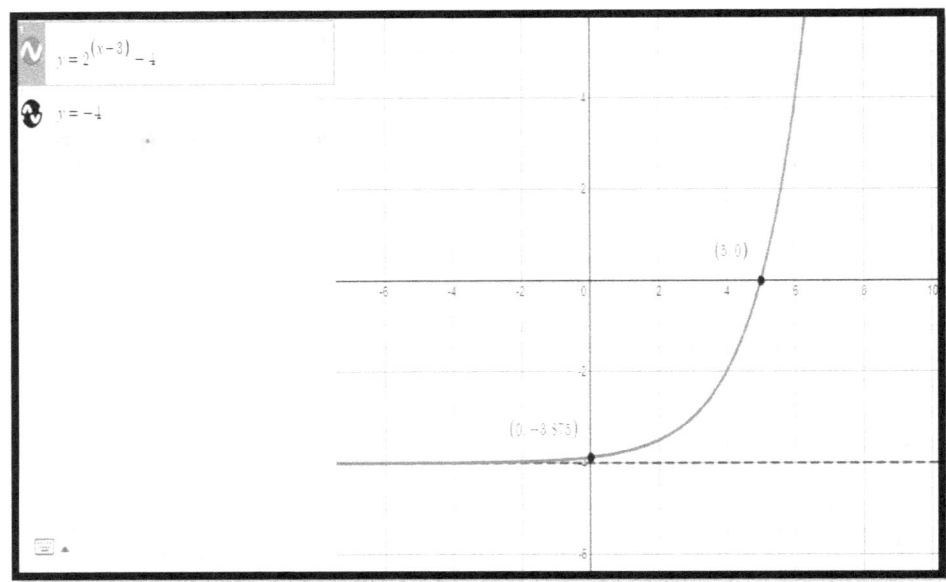

Domain: $\{x | X \in R\}$
Range: $\{y | y > -4, y \in R\}$
Asymptote: $y = -4$

5.

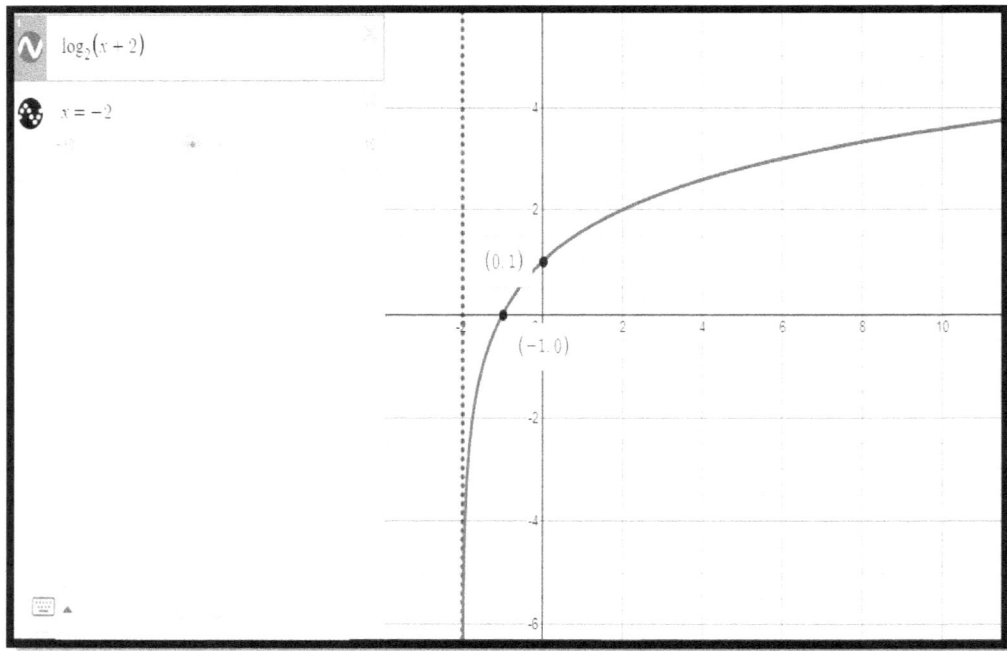

Question 5. Domain $\{x | x > -2, x \in R\}$

Range: $\{y|y \in R\}$
Asymptote: $x = -2$

6.

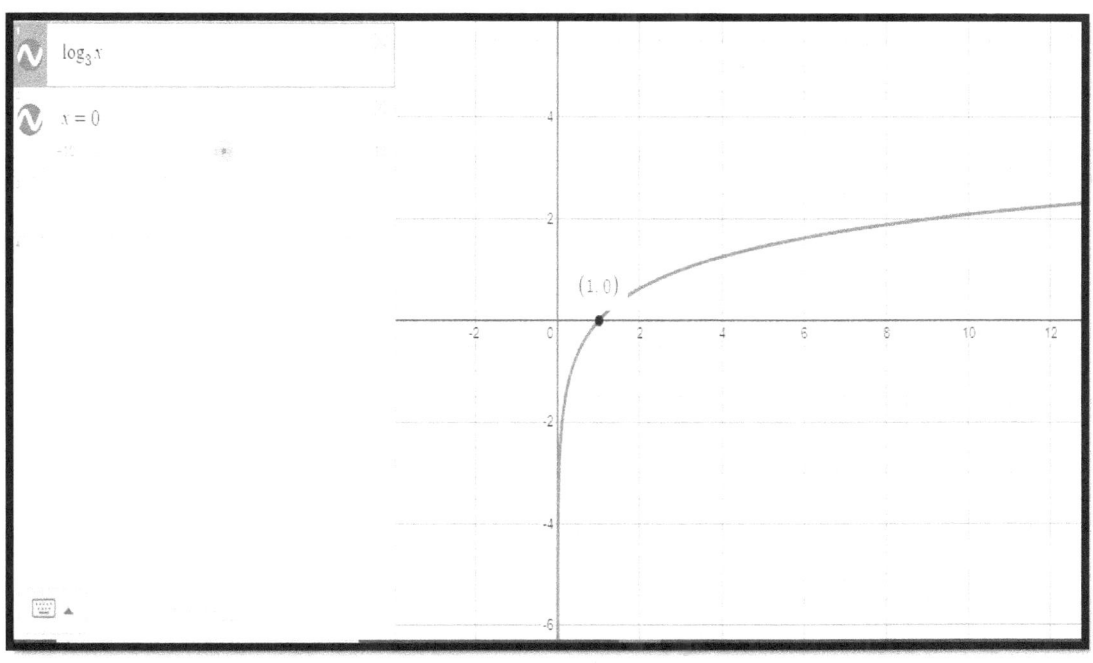

7. $y = 8^x$
$x = 8^y$
$\log x = \log 8^y$
$\dfrac{\log x}{\log 8} = \dfrac{y \log 8}{\log 8}$
$y = \dfrac{\log x}{\log 8}$
$y = \log_8 x$

8. $y = 4^{x-2}$
   $x = 4^{y-2}$
   $\log x = \log 4^{y-2}$
   $\log x = (y-2)\log 4$
   $\dfrac{\log x}{\log 4} = \dfrac{(y-2)\log 4}{\log 4}$
   $\dfrac{\log x}{\log 4} = y - 2$
   $y - 2 = \log_4 x$
   $y = \log_4 x + 2$

9. $y = \log_3 x$
   $x = \log_3 y$
   $y = 3^x$

Chapter 11: Circular Functions PART A

1. A) $1° = \dfrac{\pi}{180}$

   $30(1°) = \dfrac{\pi}{180}(30)$

   $30° = \dfrac{30\pi}{180}$

$$30° = \frac{\pi}{6}$$

1. B) $(240)1° = \frac{\pi}{180}(240)$

$$240° = \frac{240\pi}{180}$$

$$240° = \frac{4\pi}{3}$$

c) $-10(1°) = \frac{\pi}{180}(-10)$

$$-10° = -\frac{10\pi}{180}$$

$$-10° = -\frac{\pi}{18}$$

2. A) $\pi = 180°$

$$\frac{1}{12}\pi = 180°\left(\frac{1}{12}\right)$$

$$\frac{\pi}{12} = 15°$$

b) $\pi = 180°$

$$\left(\frac{1}{3}\right)\pi = 180°\left(\frac{1}{3}\right)$$
$$\frac{\pi}{3} = 60°$$

c) $\pi = 180°$
$$\left(\frac{1}{18}\right)\pi = 180°\left(\frac{1}{18}\right)$$
$$\frac{\pi}{18} = 10°$$

3. Radius = 5cm $\theta = 3\ radians$
$a = r\theta$ $a = (5)(3)$
$a = 15cm$

4. A) $\dfrac{3\pi}{4}$

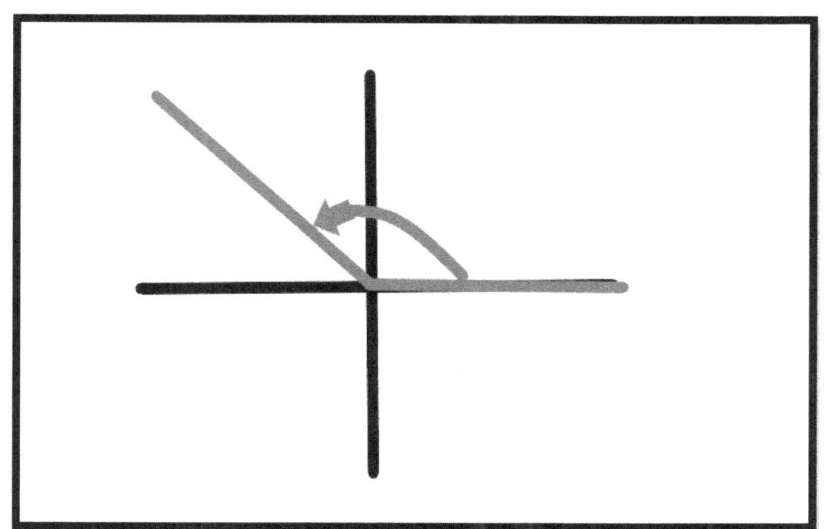

b) $-\dfrac{\pi}{4}$

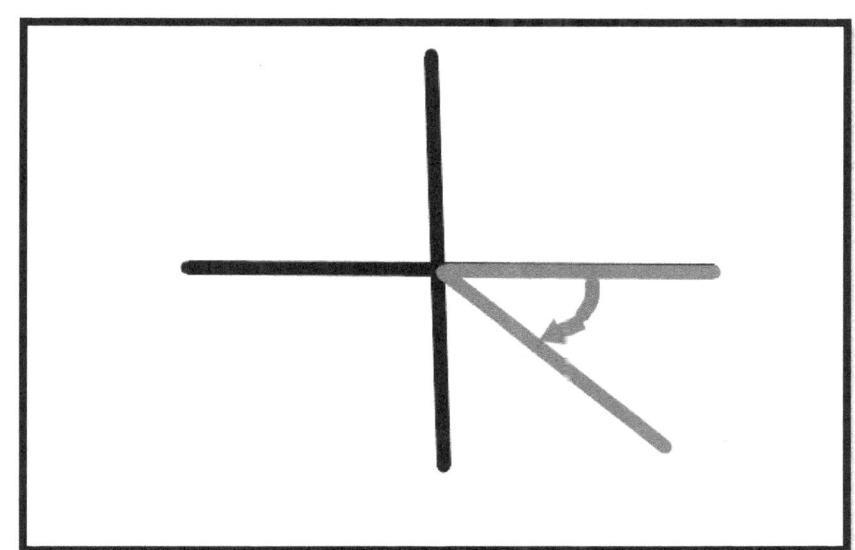

Chapter 11: Circular Functions PART B

1. $y = 3\sin 2x$ Amplitude: 3 (vertical expansion by 3), Horizontal compression by ½

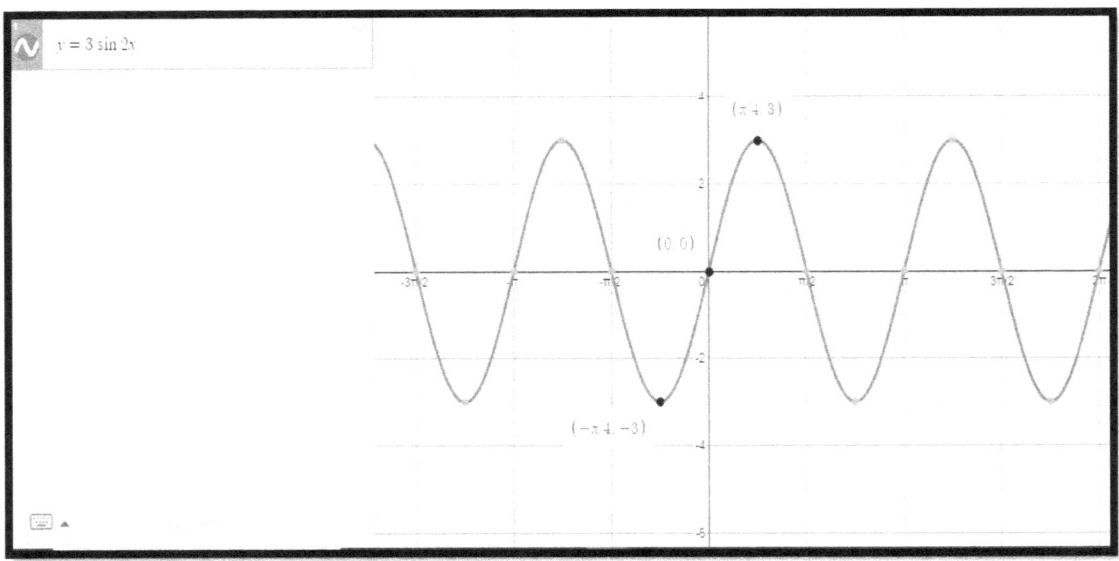

2. Amplitude: 2, phase shift of $\frac{\pi}{2}$ radians left.

$$y = 2\sin\left(x + \frac{\pi}{2}\right)$$

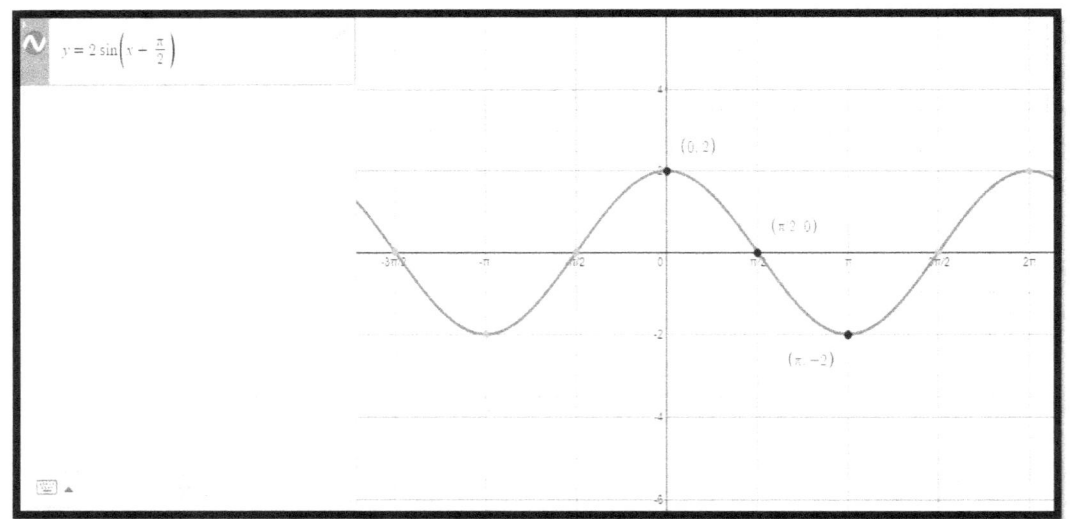

3. Amplitude: 2, horizontal compression by ½, phase shift of $\frac{\pi}{2}$ radians right.

$$y = 2\cos 2\left(x - \frac{\pi}{2}\right)$$

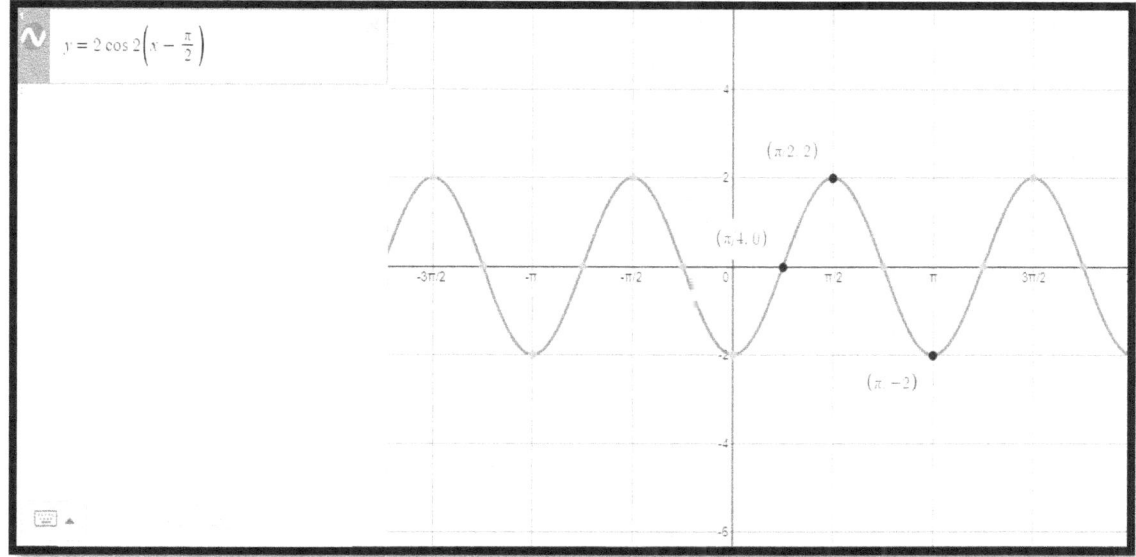

4. Amplitude 3, vertical displacement of 1 unit up,
   $y = 3\cos(x) + 1$

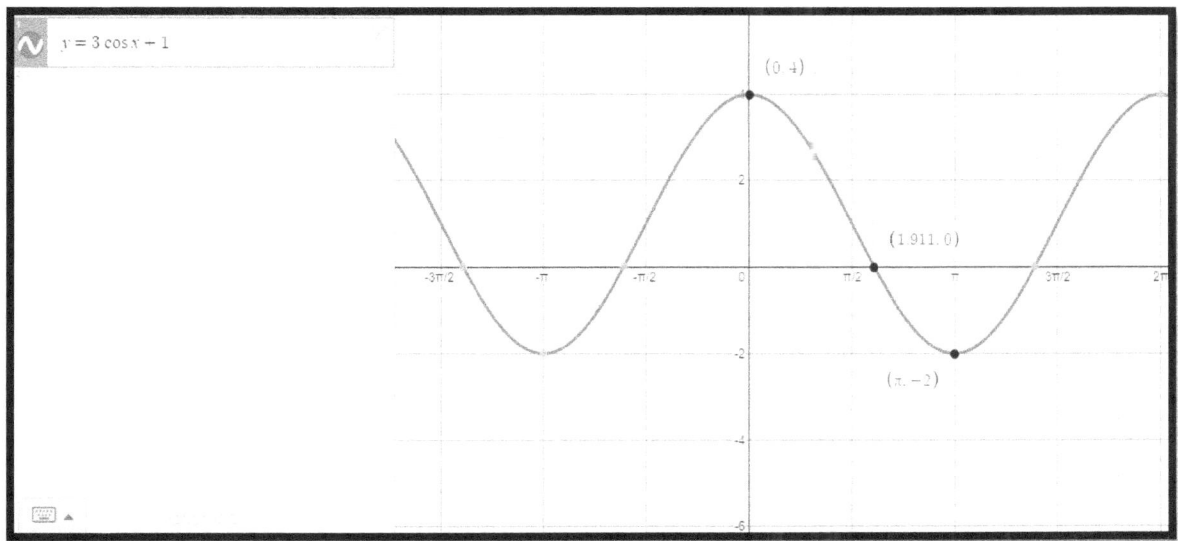

5.

a) $y = 5\sin 3\left(x - \dfrac{\pi}{6}\right) + 4$

Amplitude: 5 (Vertical expansion by 5), Horizontal compression by 1/3, a horizontal phase shift of $\dfrac{\pi}{6}$ radians right and a vertical displacement of 4 units up.

Period: $\dfrac{2\pi}{b} = \dfrac{2\pi}{3}$

b) $y = -2\cos 5\left(x - \dfrac{\pi}{7}\right) - 1$

Amplitude: 2 (Vertical expansion by 2), a reflection in the x-axis, a horizontal compression of 1/5, a horizontal phase shift of $\dfrac{\pi}{7}$ radians to the right, and a vertical displacement of 1 unit down.

6. Radius 20 meters,

$d = Radius + 3$
$d = 20 + 3$
$d = 23$

$a = \dfrac{43 - 3}{2} = \dfrac{40}{2} = 20$

$c = 0$

$p = 48$ one rotation every 48 seconds.

$h(t) = 20\sin 2\pi\dfrac{t}{48} + 23$

The graph can be found on the next page,

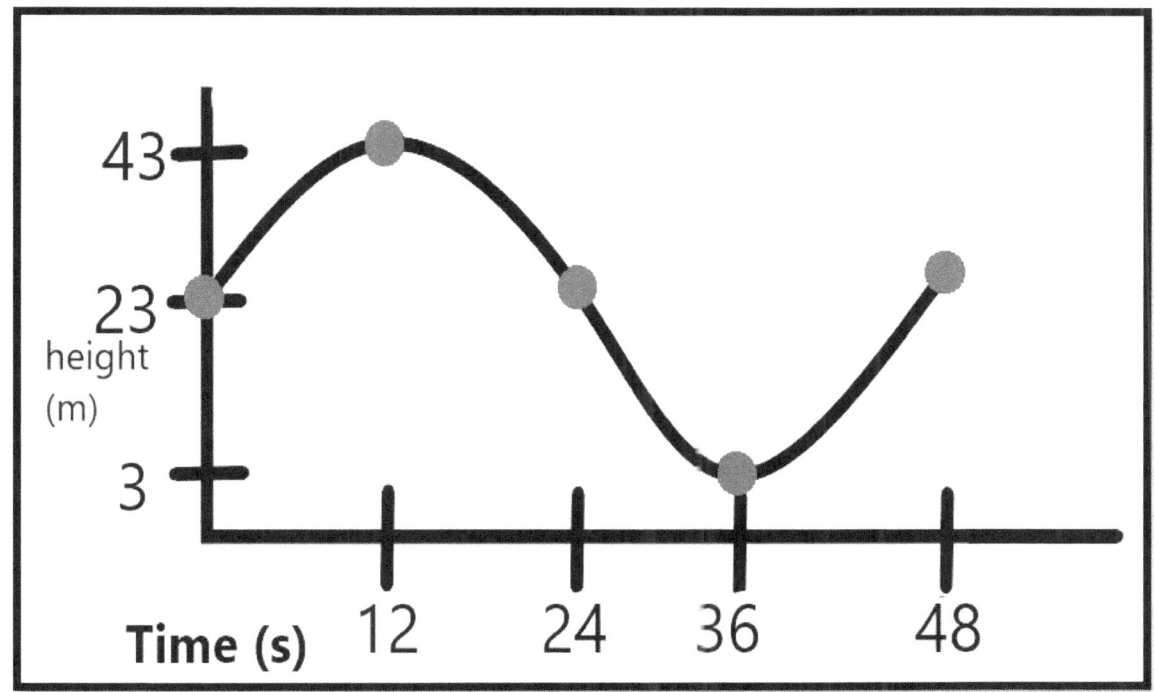

7. $\tan 2(x - \dfrac{\pi}{6})$

$b = 2, \quad c = \dfrac{\pi}{6}$

$\left[bx - bc \neq \dfrac{\pi}{2}(2n+1) \right] b$

$\left[(2)x - (2)\left(\dfrac{\pi}{6}\right) \neq \dfrac{\pi}{2}(2n+1) \right] 2$

$\left[2x - \dfrac{2\pi}{6} \neq \dfrac{\pi}{2}(2n+1) \right] 2$

$\left[2x - \dfrac{\pi}{3} \neq \dfrac{\pi}{2}(2n+1) \right] 2$

$$4x - \frac{2\pi}{3} \neq \frac{2\pi}{2}(2n+1)$$

Let's multiply this entire equation by 3 so we can simplify the expression,

$$3(4x - \frac{2\pi}{3} \neq \frac{2\pi}{2}(2n+1)$$

$$12x - \frac{6\pi}{3} \neq \frac{6\pi}{2}(2n+1)$$

$$12x - 2\pi \neq 3\pi(2n+1)$$
$$12x - 2\pi \neq 6n\pi + 3\pi$$
$$12x - 2\pi + 2\pi \neq 6n\pi + 3\pi + 2\pi$$
$$12x \neq 6n\pi + 5\pi$$
$$\frac{12x}{12} \neq \frac{6n\pi + 5\pi}{12}$$

$$x \neq \frac{6n\pi + 5\pi}{12}$$

Chapter 12: Trigonometric Equations and Identities

1. $\cos 3x$ has 6 solutions, $2b = 2(3) = 6$

2. A) $3\sec^2 x + 6\sec x$
   $= 3secx(secx + 2)$

   b) $4\cot^2 x - 9$ difference of squares,
   $= (2\cot x + 3)(2\cot x - 3)$

c) $\cos^2 x + 3\cos x + 2$ Two numbers that multiply to 2 and add up to 3,   2 & 1
$= (\cos x + 2)(\cos x + 1)$

d) $2\sin^2 x + 4\sin x - 6$ two numbers that multiply to -12 and add up to 4?   6 & -2
$2\sin^2 x - 2\sin x + 6\sin x - 6$
$2\sin x(\sin x - 1) + 6(\sin x - 1)$
$(2\sin x + 6)(\sin x - 1)$ we can remove "2"
$= 2(\sin x + 3)(\sin x - 1)$

e) $\csc^2 x - 16$ this is a difference of squares
$= (\csc x + 4)(\csc x - 4)$

f) $\cos^2(j)\sin(j) - \cos(j)\sin^2(j) = 0$
$0 = \cos(j)\sin(j)(\cos(j) - \sin(j))$

g) $4\cos^2(f) - 25 = 0$ (difference of squares)
$(2\cos(f) - 5)(2\cos(f) + 5) = 0$

h) $2\sin^2(x) + 7\sin(x) + 3 = 0$
We need to find two numbers that multiply to 6, and add up to 7. The numbers that satisfy these conditions are 6 & 1
$2\sin^2(x) + 6\sin(x) + 1\sin(x) = 3 = 0$
$2\sin(x)(\sin(x) + 3) + 1(\sin(x) + 3) = 0$
$(2\sin(x) + 1)(\sin(x) + 3) = 0$

i) $v^2\csc^2(b) + v\csc(b)\csc(g) = 0$
$v\csc(b)(v\csc(b) + \csc(g)) = 0$

j) $9\sec^2(x) - 4\cot^2(x) = 0$
$(3\sec(x) - 2\cot(x))(3\sec(x) + 2\cot(x)) = 0$

k) $\cot^2(x) + 6\cot(x) + 8 = 0$
We need to find two numbers that multiply to 8, and add up to 6. The two numbers that satisfy these conditions are 4 & 2.
$(\cot(x) + 4)(\cot(x) + 2) = 0$

l) $\tan^2(x) - \tan(x) - 12 = 0$
We need to find two numbers that multiply to -12 and add up to -1. These two numbers are -4 & 3.
$(\tan(x) - 4)(\tan(x) + 3) = 0$

m) $\sin^2(m) - 7 = 0$
$(\sin(m) + \sqrt{7})(\sin(m) - \sqrt{7})$

n) $4\csc^4(v) - 12\csc^3(v) = 0$
$4\csc^3(v)(\csc(v) - 3)$

o) $3\sin^4(x) - 3\sin^3(x) - 36\sin^2(x)$
$3\sin^2(x)(\sin^2(x) - \sin(x) - 12)$
We need to find two numbers that multiply to -12 and add up to -1. These two numbers are -4 & 3.
$3\sin^2(x)(\sin(x) - 4)(\sin(x) + 3) = 0$

3. A) $\sin x \cos x = \sin^2 x$ set the right side to 0
   $\sin x \cos x - \sin^2 x = 0$ factor "$\sin x$" from both terms,
   $\sin x (\cos x - \sin x) = 0$ deal with each factor separately,

   $\sin x = 0$ $\cos x - \sin x = 0$
   $x = 0, \pi, 2\pi$ $\cos x = \sin x$

                        $\dfrac{\cos x}{\cos x} = \dfrac{\sin x}{\cos x}$

   $x = n\pi$ $1 = \tan x$

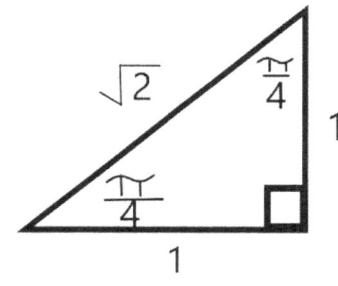

   $\tan \dfrac{\pi}{4} = 1$

   $x = \dfrac{\pi}{4} + n\pi$

   $x = n\pi$

4. A) $2\sin \theta = 1$

   $\sin \theta = \dfrac{1}{2}$ $\sin \dfrac{\pi}{6} = \dfrac{1}{2}$

   Which two angles in standard position have a reference angle of $\dfrac{\pi}{6}$ and have a positive sine ratio of $\dfrac{1}{2}$?

$\dfrac{\pi}{6}$ and $\dfrac{5\pi}{6}$

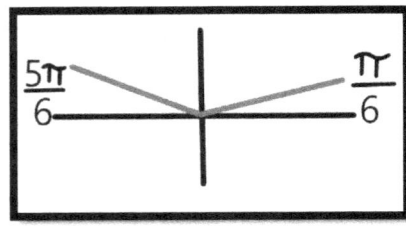

$\theta = \dfrac{\pi}{6}, \dfrac{5\pi}{6}$

b) $2\cos\theta - \sqrt{3} = 0$

$2\cos\theta = \sqrt{3}$

$\cos\theta = \dfrac{\sqrt{3}}{2}$

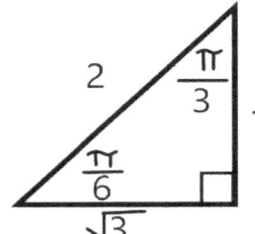

$\cos\dfrac{\pi}{6} = \dfrac{\sqrt{3}}{2}$

Cosine is positive in Quadrant 1 and 4,

$\theta = \dfrac{\pi}{6}, \dfrac{11\pi}{6}$ $2\pi - \dfrac{\pi}{6} = \dfrac{12\pi}{6} - \dfrac{\pi}{6} = \dfrac{11\pi}{6}$

c) $3\tan\theta + 3 = 0$
$3\tan\theta = -3$
$\tan\theta = -\dfrac{3}{3}$
$\tan\theta = -1$

$\tan\dfrac{\pi}{4} = 1$

In which quadrants is tangent negative?

Quadrant 2 and 4,

Quadrant 2 solution: $\pi - \dfrac{\pi}{4} = \dfrac{4\pi}{4} - \dfrac{\pi}{4} = \dfrac{3\pi}{4}$

Quadrant 4 Solution: $2\pi - \dfrac{\pi}{4} = \dfrac{8\pi}{4} - \dfrac{\pi}{4} = \dfrac{7\pi}{4}$

$\theta = \dfrac{3\pi}{4}, \dfrac{7\pi}{4}$

d) $4\cos^2 \theta - 1 = 0$
Difference of squares,
$(2\cos \theta - 1)(2\cos \theta + 1) = 0$

$2\cos \theta - 1 = 0 \qquad\qquad 2\cos \theta + 1 = 0$
$2\cos \theta = 1 \qquad\qquad\quad 2\cos \theta = -1$
$\cos \theta = \dfrac{1}{2} \qquad\qquad\quad \cos \theta = -\dfrac{1}{2}$

Solutions lie in all four quadrants, first find the reference angle,

$\cos \dfrac{\pi}{3} = \dfrac{1}{2}$

Quad 1 solution $= \dfrac{\pi}{3}$

Quad 2 solution $= \pi - \dfrac{\pi}{3} = \dfrac{3\pi}{3} - \dfrac{\pi}{3} = \dfrac{2\pi}{3}$

Quad 3 solution $= \pi + \dfrac{\pi}{3} = \dfrac{3\pi}{3} + \dfrac{\pi}{3} = \dfrac{4\pi}{3}$

Quad 4 solution $= 2\pi - \dfrac{\pi}{3} = \dfrac{6\pi}{3} - \dfrac{\pi}{3} = \dfrac{5\pi}{3}$

$\theta = \dfrac{\pi}{3}, \dfrac{2\pi}{3}, \dfrac{4\pi}{3}, \dfrac{5\pi}{3}$

5. Prove $\cot\theta + \tan\theta = \csc\theta\sec\theta$

Answers may vary, I will give you an example solution,

$$\cot\theta + \tan\theta = \csc\theta\sec\theta$$

$$\frac{\cos\theta}{\sin\theta} + \tan\theta \quad\Big|\quad \left(\frac{1}{\sin\theta}\right)\left(\frac{1}{\cos\theta}\right)$$

$$\frac{\cos\theta}{\sin\theta} + \frac{\sin\theta}{\cos\theta}$$

$$\left(\frac{\cos\theta}{\cos\theta}\right)\left(\frac{\cos\theta}{\sin\theta}\right) + \left(\frac{\sin\theta}{\cos\theta}\right)\left(\frac{\sin\theta}{\sin\theta}\right)$$

$$\frac{\cos^2\theta}{\sin\theta\cos\theta} + \frac{\sin^2\theta}{\sin\theta\cos\theta} \quad \text{Pythagorean}$$

Identity, numerator = 1

$$\frac{1}{\sin\theta\cos\theta} = \frac{1}{\sin\theta\cos\theta}$$

6. A) $\cos\left(\dfrac{\pi}{4}+\dfrac{\pi}{6}\right)=\cos\dfrac{\pi}{4}\cos\dfrac{\pi}{6}-\sin\dfrac{\pi}{4}\sin\dfrac{\pi}{6}$

$=\left(\dfrac{1}{\sqrt{2}}\right)\left(\dfrac{\sqrt{3}}{2}\right)-\left(\dfrac{1}{\sqrt{2}}\right)\left(\dfrac{1}{2}\right)$

$=\dfrac{\sqrt{3}}{2\sqrt{2}}-\dfrac{1}{2\sqrt{2}}$

$=\dfrac{\sqrt{3}-1}{2\sqrt{2}}$

b) $\sin(x-\pi)$
$=\sin x\cos\pi-\cos x\sin\pi$
$=(\sin x)(-1)-(\cos x)(0)$
$=-\sin x$

c) $\cos\left(\dfrac{\pi}{2}-x\right)$

$\cos\left(\dfrac{\pi}{2}-x\right)=\cos\dfrac{\pi}{2}\cos x+\sin\dfrac{\pi}{2}\sin x$

$=(0)\cos x+(1)\sin x$
$=\sin x$

d) $\sin\left(x-\dfrac{\pi}{4}\right)$

$=\sin x\cos\dfrac{\pi}{4}-\cos x\sin\dfrac{\pi}{4}$

$=\sin x\left(\dfrac{1}{\sqrt{2}}\right)-\cos x\left(\dfrac{1}{\sqrt{2}}\right)$

$=\dfrac{\sin x}{\sqrt{2}}-\dfrac{\cos x}{\sqrt{2}}$

$$= \frac{\sin x - \cos x}{\sqrt{2}}$$

e) $\dfrac{\tan 7z - \tan 2z}{1 + \tan 7z \tan 2z}$

$a = 7z \quad b = 2z$

$= \tan(a - b)$

$\tan((7z) - (2z))$

$= \tan 5z$

7.

a) $4\sin 5x \cos 5x$

$2(\sin 2(5x)) = 2(2\sin 5x \cos 5x)$

$= 2\sin 10x$

b) $2 - 4\sin^2 \frac{1}{2}x$

$2(\cos 2(\frac{1}{2}x)) = 2(1 - 2\sin^2 \frac{1}{2}x)$

$= 2\cos x$

c) $6\sin 6x \cos 6x$

$\sin 2\theta = 2\sin \theta \cos \theta$

$3(\sin 2\theta) = 3(2\sin \theta \cos \theta)$

$= 3(\sin 2(6x))$

$= 3\sin 12x$

d) $8\sin\theta\cos\theta$
$\sin 2\theta = 2\sin\theta\cos\theta$
$4(\sin 2\theta) = 4(2\sin\theta\cos\theta)$
$= 4\sin 2\theta$

e) $14\cos^2 7x - 7$
$\cos 2\theta = 2\cos^2\theta - 1$
$7(\cos 2(7x)) = 7(2\cos^2\theta - 1)$
$= 7\cos 14x$

8.

a)
$$\frac{\sin 2\theta}{2} = \cot\theta \sin^2\theta$$
$$\frac{2\sin\theta\cos\theta}{2} = \left(\frac{\cos\theta}{\sin\theta}\right)(\sin\theta)(\sin\theta)$$
$$\sin\theta\cos\theta = \sin\theta\cos\theta$$

Chapter 13: Intervals

1.
a) $-2 < x \leq 6$ $\quad(-2, 6]$
b) $3 \leq x < 9$ $\quad[3, 9)$

c) $x > \pi$ $\qquad$ (π, ∞)
d) $x < 7$ $\qquad$ $(-\infty, 7)$
e) $x > 6$ & $x \leq 18$ $\qquad$ $(6, 18]$ here's why,

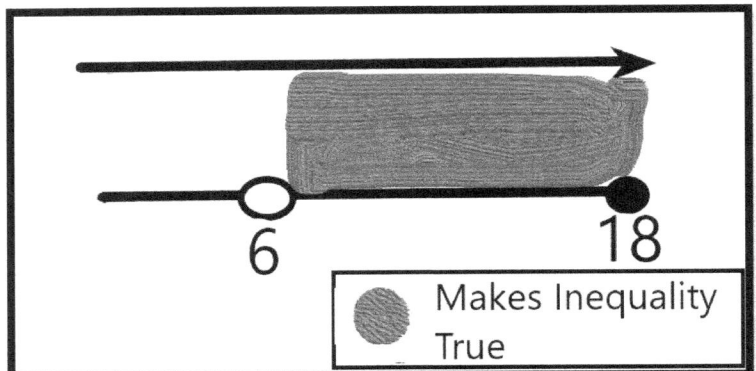

2.

a) $(3, \infty)$ $\qquad$ $x > 3$
b) $[\pi, 3\pi)$ $\qquad$ $\pi \leq x < 3\pi$
c) $(-\infty, \sqrt{3})$ $\qquad$ $x < \sqrt{3}$
d) $[2\sqrt{2}, 7]$ $\qquad$ $2\sqrt{2} \leq x \leq 7$

www.ingramcontent.com/pod-product-compliance
Lightning Source LLC
Chambersburg PA
CBHW080449220526
45465CB00006B/2212